MICROLITHOGRAPHY FUNDAMENTALS IN SEMICONDUCTOR DEVICES AND FABRICATION TECHNOLOGY

MICROLITHOGRAPHY FUNDAMENTALS IN SEMICONDUCTOR DEVICES AND FABRICATION TECHNOLOGY

Saburo Nonogaki
Protech, Ltd.
Tokyo, Japan

Takumi Ueno
Hitachi, Ltd.
Tokyo, Japan

Toshio Ito
Oki Electric Industry Co., Ltd.
Tokyo, Japan

MARCEL DEKKER, INC.　　　NEW YORK · BASEL · HONG KONG

Library of Congress Cataloging-in-Publication Data

Nonogaki, Saburo.
  Microlithography fundamentals in semiconductor devices and fabrication technology / Saburo Nonogaki, Takumi Ueno, Toshio Ito.
    p. cm.
  Includes bibliographical references and index.
  ISBN 0-8247-9951-8
  1. Semiconductors—Design and construction. 2. Microlithography—Industrial applications. 3. Manufacturing processes. 4. Photoresists. I. Ueno, Takumi. II. Ito, Toshio. III. Title.
TK7871.85.N66   1998
621.3815'2—dc21   98-4222
                                                                                  CIP

This book is printed on acid-free paper.

Headquarters
Marcel Dekker, Inc.
270 Madison Avenue, New York, NY 10016
tel: 212-696-9000; fax: 212-685-4540

Eastern Hemisphere Distribution
Marcel Dekker AG
Hutgasse 4, Postfach 812, CH-4001 Basel, Switzerland
tel: 44-61-261-8482; fax: 44-61-261-8896

World Wide Web
http://www.dekker.com

The publisher offers discounts on this book when ordered in bulk quantities. For more information, write to Special Sales/Professional Marketing at the headquarters address above.

Copyright © 1998 by Marcel Dekker, Inc. All Rights Reserved.

Neither this book nor any part may be reproduced or transmitted in any form or by any means, electronic or mechanical, including photocopying, microfilming, and recording, or by any information storage and retrieval system, without permission in writing from the publisher.

Current printing (last digit):
10 9 8 7 6 5 4 3 2 1

PRINTED IN THE UNITED STATES OF AMERICA

Preface

Microlithography is a fabrication technology which utilizes etching technique to produce very fine patterns of metals, insulators and semiconductors. It is one of the key technologies in the manufacture of semiconductor devices such as diodes, transistors and integrated circuits (ICs). The microlithography applied to the fabrication of an IC is a highly advanced photolithography utilizing ultraviolet or visible light to generate fine patterns. Photolithography itself had been developed as a technology to fabricate printing plates far before the IC was invented.

In order to improve the performance of an IC, it is necessary to increase the integration density (the number of components per unit area of an IC chip). To achieve high density of integration, it is necessary to reduce the minimum feature size of the components. Therefore, photolithography used in the fabrication of high-performance IC has been constantly required to have much higher resolution capability. To satisfy this requirement, many improvements in this technology have been made. As a result, the minimum feature size has been reduced to the submicrometer region and is now approaching the theoretical limit of resolution in photolithography.

This book reviews the recent state of the art in microlithography with an emphasis on the basic principles underlying the technology. The book will help the people engaged in the fabrication of semiconductor devices understand the theoretical basis of the processes they are dealing with. It is hoped that the book is also useful to beginners as a guide from basic principles to the forefront of microlithography. The readers with general interest in science will find here a good example of applied science where optics, photochemistry, electron physics and other fields of science are ingeniously combined to develop a new revolutionary technology.

Chapter 1 presents an overview of microlithography applied to semiconductor device fabrication together with its brief history. The ways of expressing the sensitivity characteristics of resist materials and a necessary condition for correct pattern transfer in photolithography are also presented in this chapter.

Chapter 2 describes optical exposure tools used in the past and at the present. The future trend is also presented in this chapter.

Chapter 3 presents fundamental principles of optical image transfer with an emphasis on the theoretical limit of resolution. A rigorous solution of two-dimensional light wave diffraction is used here to make the discussion clear and quantitative.

Chapter 4 discusses photoresists, including photochemical reactions taking place in the resist systems and dissolution mechanisms in developers. Chemical amplification resists for KrF excimer laser lithography as well as those for ArF excimer laser lithography are also described. To add to these, new approaches to improve the resolution using base-catalysed reactions during development are presented in this chapter.

Chapter 5 describes practical microlithographic processes including adhesion enhancement of substrate, resist coating, exposure and development of photoresists, and etching of substrate.

Chapter 6 covers fundamentals of x-ray lithography, including x-ray sources, masks and resists. Although this chapter mainly deals with 1:1 proximity printing, reduction projection systems are briefly described.

Chapter 7 presents electron beam lithography together with the trend of improvement in throughput. The resolution limit in this lithography is also discussed in this chapter.

Chapter 8 describes various types of microlithographic processes. It includes resolution enhancement approaches in resist materials and processes such as contrast enhancement materials, multilayer resist, the dry development process, and so on.

The Appendix provides a mathematical basis of the discussion given in Chapter 3. The equations used to calculate the light intensity distributions shown in Sections 3.1 and 3.2 are derived here by solving two-dimensional diffraction problems rigorously on the basis of the Huygens-Fresnel principle.

The authors are grateful to Dr. Takao Iwayanagi for providing information on historical developments in microlithography and Drs. Shinji Okazaki and Kiyotake Naraoka for giving information on practical processes in microfabrication of semiconductor devices.

Saburo Nonogaki
Takumi Ueno
Toshio Ito

Contents

Preface .... *iii*

1. Introduction 1
2. Exposure Systems in Photolithography 11
3. Optical Pattern Transfer 25
4. Chemistry of Photoresist Materials 65
5. Practical Processes in Microlithography 133
6. X-ray Lithography 159
7. Electron Beam Lithography 199
8. Variations in the Microlithographic Process 229

Appendix: Two-dimensional Diffraction *303*

Index *317*

1
Introduction

1.1 PHOTOLITHOGRAPHY TO FABRICATE SEMICONDUCTOR DEVICES

Before the transistor action was first confirmed by W. Shockley, J. Bardeen and W. H. Brattain at Bell Laboratories in 1948, photolithography had been well developed and widely used to fabricate printing plates. Therefore, it seems natural that this technology was readily applied to the fabrication of planar transistors which were designed about ten years after the invention of transistors. An example of photolithographic process applied to the fabrication of semiconductor devices is shown in Fig. 1.1, where the process to etch patternwise the oxide layer formed on a silicon wafer is illustrated.

The surface of oxide layer is coated with a solution of photosensitive polymeric material generally called "photoresist" or simply "resist" by spin-coating where the wafer is rotated in high speed, for example, 3000 revolutions per minute. Most of the solvent in the spread film of solution evaporates during the spin-coating process to leave a solid film of resist on the oxide layer. The wafer is then baked to ensure the perfect evaporation of solvent.

The resist-coated wafer is contacted with an optical mask bearing fine patterns and exposed to light. The light source used in this type of exposure (contact printing) is a high-pressure mercury lamp which emits a strong ultraviolet line called i-line (peaked at a wavelength of 0.365 µm).

After the exposure, the wafer is treated with a developing solution. There are two types of photoresist, one being called "negative" and the other "positive". In the case of negative photoresist, the resist on the exposed area is insolubilized by the exposure and remains undissolved by developing solution while the resist on the unexposed area is dissolved away from the substrate, as illustrated in Fig. 1.1D. On the contrary, positive photoresist on the exposed area is rendered soluble by the exposure and dissolved away from the substrate while the resist on the unexposed area remains on the substrate. The definition of negative or positive type of photoresist is in accordance with that of silver halide emulsion in photography if the resist is assumed to be opaque.

The resist pattern obtained by the development is used as a mask for the etching of the oxide layer. In the case of wet etching where a liquid etchant is used, the etching is isotropic, or, in other words, it proceeds laterally as well

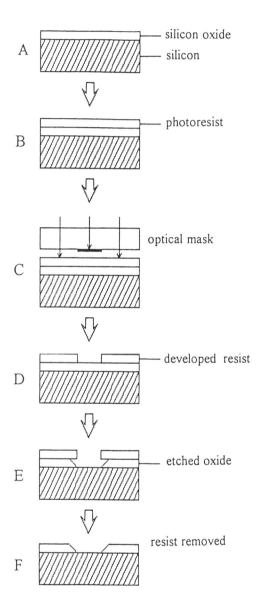

Fig. 1.1. Photolithographic process to etch patternwise the oxide layer formed on the surface of a silicon wafer.

Introduction 3

as vertically resulting in sloped edges of the oxide layer as shown in Fig. 1.1E.
Finally, the resist pattern is removed away from the wafer surface by using a stripping solution or oxygen plasma.

Another important microlithographic process in semiconductor device fabrication is the patterning of conducting layer to form fine patterns of integrated circuits. The process is very similar to that described above except for the material to be etched and its etchant.

1.2 HISTORICAL DEVELOPMENT OF MICROLITHOGRAPHY

In the beginning of semiconductor device microfabrication at Bell Laboratories, a negative type of photoresist produced by Kodak under the trade name Kodak Photo Resist was used. The frequently used term "photoresist" comes from this trade name. The resist was afterward replaced with Kodak Thin Film Resist because of its unsatisfactory adhesion to silicon and silicon oxide [1]. Kodak Thin Film Resist was also a negative type of photoresist. Because of its good adhesion and stiffness, the same type of negative photoresist is still widely used as a most suitable photoresist in contact printing.

Since miniaturization is the essential way to improve the performance of semiconductor devices, the microfabrication technology in the semiconductor industry has been constantly required to have higher resolution capability together with higher alignment accuracy. In early 1970s, the minimum feature size of semiconductor devices had been reduced to several micrometers by the use of improved exposure systems. At that time, the practical limit of resolution in microlithography was considered to be about 2 micrometers. This estimation was based mainly on the resolution capability of photoresists such as Kodak Thin Film Resist.

The use of electron beams to delineate fine patterns of photoresist was first reported in 1965 [2,3]. The minimum feature size of resist patterns obtained by this method was about 1 micrometer, which was probably limited by the resolution capability of the resist used there. A high resolution capability of electron-beam exposure was demonstrated by M. Hatzakis [4] in 1969. He obtained 0.5-µm-wide line patterns by using a scanning electron microscope as a delineating tool and poly(methyl methacrylate) as a positive electron-beam resist.

The first practical electron-beam exposure system was developed by Bell Laboratories in 1975 [5]. The system was designed not for the direct fabrication of integrated circuits but for the fabrication of optical masks used in the photolithography to fabricate integrated circuits. The use of this system significantly simplified the process of mask fabrication, eliminating the use of optical pattern generators and reducing printing process. At present, optical masks for the fabrication of highly integrated circuits are routinely fabricated by electron-beam exposure systems.

The use of soft-x ray in the pattern transfer from mask to resist layer was first reported by D. L. Spears and H. I. Smith [6] in 1972. As the wavelength of soft x-ray used in this technique is around 1 nm, a very high resolution can be expected. As a result of many improvements in x-ray sources, masks and resist materials, the lithography using this technique (x-ray lithography) is now approaching the stage of practical application.

Contact printing (illustrated in Fig. 1.1C) had been the unique practical method of pattern transfer from mask to resist layer until 1973 when a mirror-projection type of printing system was first introduced by Perkin-Elmer [7]. This type of printing system was followed by a lens-projection type developed by GCA in 1977. In the contact printing process, the optical mask is pressed firmly against the resist-coated wafer. This often results in mechanical defects in the mask. The use of a projection printing system totally eliminates this difficulty. Furthermore, improvements in optical system and alignment mechanism have led to the development of projection printing systems with high resolution and high alignment accuracy.

Positive type of photoresist had been scarcely used in the microfabrication of semiconductor devices before the development of projection printing systems because the resist is mechanically fragile and apt to be damaged by the hard contact with the mask. As the fragility of positive photoresist is not a problem in projection printing, the resist has been used there almost exclusively because of its excellent resolution capability.

Wave optics teaches us that the resolution of an optical system increases with shortening the wavelength of light used. Therefore, the resolution in photolithography is expected to be improved by shortening the wavelength of light. The result of improvement was first demonstrated by B. J. Lin [8] in 1975. Since then, printing systems and resist materials for this technique have been developed. A very sophisticated projection printing system using a KrF excimer laser oscillating at a wavelength of 0.248 µm was developed at Bell Laboratories in 1986 [9]. The lithography based on this type of projection printing system is expected to be successfully applied to the fabrication of high-density integrated circuits.

The use of a phase-shifting mask to improve the resolution in photolithography was reported by Levenson and others [10] in 1982. The principle of improvement in resolution is illustrated in Fig. 1.2, where the imaging of a double slit pattern of a conventional mask is compared with that of a phase-shifting mask with the same pattern dimensions. As the space between slits narrows, the images of slits overlap with each other. In the case of

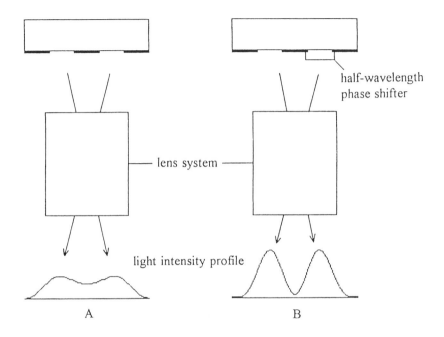

Fig. 1.2. Imaging of double-slit pattern of a conventional mask (A) and a phase-shifting mask (B).

the conventional mask, the electric vectors of the images are in phase with each other on the image plane. Therefore, the increase in light intensity by overlapping is maximized in this case. On the contrary, in the case of a phase-shifting mask, the wave of light passing through the slit with phase shifter is phase-shifted by one-half wavelength. Consequently, the electric vectors of slit images are in an inverse phase with each other on the image plane, resulting in the maximum decrease in light intensity and the minimum loss in resolution by overlapping.

Vapor-phase etching was first introduced in the microfabrication of semiconductor devices in the middle of 1970s [11]. Silicon nitride, which is chemically stable and resistant to liquid-phase etching, is easily etched by vapor-phase etching. In the early stage of development, vapor-phase etching was isotropic, giving the same sloped edges of an etched substrate as those obtained by liquid-phase etching (illustrated in Fig. 1.1E). At present, several types of anisotropic vapor-phase etching have been developed. Among them, the most widely used is reactive ion etching (abbreviated to RIE) . By the use of these anisotropic etching techniques, an ideal etching (as illustrated in Fig. 1.3), has become possible.

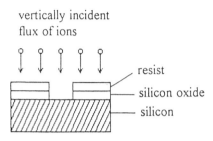

Fig. 1.3. Anisotropic vapor-phase etching.

Chemically amplified resists for semiconductor device fabrication were first reported by H. Ito and C. G. Willson [12] in 1983. Since then, many kinds of chemically amplified resists (based on the same concept) have been developed, and some of them are now commercially available. The resists are considered to be most suited to deep-uv, electron-beam and x-ray lithography because of their extraordinarily high sensitivities.

1.3 SENSITIVITY CHARACTERISTICS OF THE RESIST LAYER

The sensitivity characteristics of the resist layer are usually expressed by the relationship between developed resist film thickness and exposure dose, as illustrated in Fig. 1.4. The unit used to express the exposure dose is usually mJ/cm^2 for photo- and x-ray resists and C/cm^2 for electron-beam resist. The threshold dose, denoted by D_T, is defined as the dose at the intersection of the characteristic curve and the abscissa. The sensitivity of the resist is usually expressed by D_T, although the sensitivity is inversely proportional to D_T and should be expressed by $1/D_T$ in the strict sense of notation. In both cases of positive and negative resists, D_T is generally dependent on the initial resist film thickness and the development conditions.

The contrast of the resist, usually denoted by γ, is defined as the tangent of the characteristic curve at the point of threshold dose, normalized by the maximum film thickness R_0, that is,

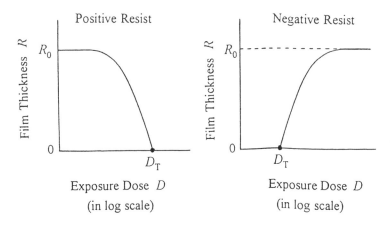

Fig. 1.4. Sensitivity characteristics of positive and negative photoresists.

$$\gamma = \frac{1}{R_0} \cdot \frac{dR}{d(\log D)} \quad \text{at} \quad D = D_T, \qquad (1.1)$$

where R is the resist film thickness, D the exposure dose, and the base of logarithm is 10. The contrast is also dependent on the resist film thickness and the conditions of development, especially in the case of the positive resist.

1.4 CORRECT PATTERN TRANSFER FROM MASK TO RESIST

A simple example of pattern transfer from the mask to resist layer is illustrated in Fig. 1.5, where the process to transfer a half-plane pattern correctly by contact printing is shown.

A theoretical calculation described in the Appendix shows that the light intensity at the edge of the geometrical shadow is very close to one fourth of that at an illuminated position far apart from the edge, as indicated by I_0 and $I_0/4$ in Fig. 1.5. Therefore, if the resist layer is exposed to light so that the dose at the fully illuminated position is four times the threshold dose D_T, the dose at the edge of geometrical shadow is very close to the threshold dose, and, consequently, the edge of resist pattern obtained by the development after exposure will be also very close to the edge of geometrical shadow cast on the resist layer, as illustrated in Fig. 1.5.

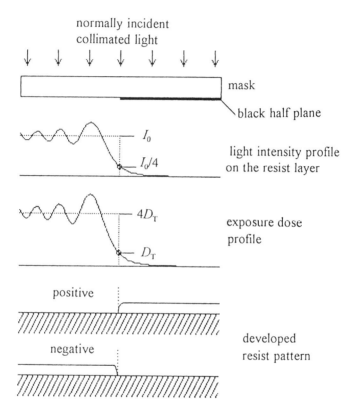

Fig. 1.5. Correct transfer of half-plane pattern from mask to resist layer by contact printing.

REFERENCES

1. Arnost Reiser, *Photoreactive Polymers, The Science and Technology of Resists*, p. 18, John Wiley & Sons, 1989.
2. I. M. Mackintosh, Application of the scanning electron microscope to solid-state devices, *Proc. IEEE*, **53**, 370-377 (1965).

3. R. F. M. Thornley and T. Sun, Electron beam exposure of photoresists, *J. Electrochem. Soc.*, **112**, 1151-1153 (1965).
4. M. Hatzakis, Electron resists for microcircuit and mask production, *J. Electrochem. Soc.*, **116**, 1033-1037 (1969).
5. Donald R. Herriot, R. J. Collier, David S. Alles, and J. W. Stafford, EBES: A practical electron lithographic system, *IEEE Trans. Electron Devices*, **ED-22**, 385-392 (1975).
6. D. L. Spears and Henry I. Smith, High-resolution pattern replication using soft x-rays, *Electronics Letters*, **8**, 102-104 (1972).
7. Peter Moller, All-reflective 1:1 projection printing system, *Technical Papers: Regional Technical Conference "Photopolymers: Principles, Processes, and Materials"* sponsored by Soc. Plastic Engineers, Inc. Mid-Hudson Section, October 24-26, 1973, pp. 56-62 (1973).
8. B. J. Lin, Deep uv lithography, *J. Vac. Sci. Technol.*, **12**, 1317-1320 (1975).
9. Victor Pol, James H. Bennewitz, Gary C. Escher, Martin Feldman, Victor A. Firton, Tanya E. Jewell, Bruce E. Wilcomb, and James T. Clemens, Excimer laser-based lithography: a deep ultraviolet wafer stepper, *Proc. SPIE*, **633**, 6-16 (1986).
10. M. D. Levenson, N. S. Viswanathan, and R. A. Simpson, Improving resolution in photolithography with a phase-shifting mask, *IEEE Trans. Electron Devices*, **ED-29**, 1828-1836 (1982).
11. R. G. Poulsen, Plasma etching in integrated circuit manufacture, *J. Vac. Sci. Technol.*, **14**, 266-274 (1977).
12. H. Ito and C. G. Willson, Chemical amplification in the design of dry developing resist materials, *Polymer Eng. Sci.*, **23**, 1012-1018 (1983).

2

Exposure Systems in Photolithography

The demand to shrink the minimum feature size for large integration semiconductor circuits has driven the improvement in performance of optical exposure tools. Exposure systems consist of a light source, a filter, illumination optics, a mask, projection optics, an alignment system and a wafer stage. Since fundamental optics will be described in Chapter 3, typical exposure tools are described here.

2.1 LIGHT SOURCES

2.1.1 Mercury Lamps

The source of radiation for photolithography has traditionally been a Hg or Hg-rare gas discharge lamp. The emission spectra from a high pressure mercury lamp is shown in Fig.2.1 [1]. The equilibrium pressure of Hg and the rare gas determine the spectral distribution of the light output. This output is high in the near-UV region (350-450 nm) and in the mid-UV region (300-350 nm), and lower in the DUV region (200-300 nm).

The conventional discharge lamp consists of a quartz envelop that encloses two refractory metal electrodes (usually W) with a small separation that are charged with a carefully metered amount of elemental Hg and a rare gas (usually Xe). Electrons produced by high voltage pulse (ignition) gain their energy from an electric field induced by an external power supply and cause ionization and excitation of mercury and/or rare gas incorporated in the lamp. Excited mercury atoms can emit light based on the energy diagram as shown in Fig.2.2 [2]. Secondary electrons generated by the ionization process also gain their energy from the electric field and can ionize the gas atoms. Electron collisions including elastic and inelastic scattering lead to the increase in temperature of the lamp, which increases the vapor pressure of mercury. These continuous processes reach the local equilibrium of the discharge at the center of the electrodes. Such discharge is classified as arc discharge in the range of several

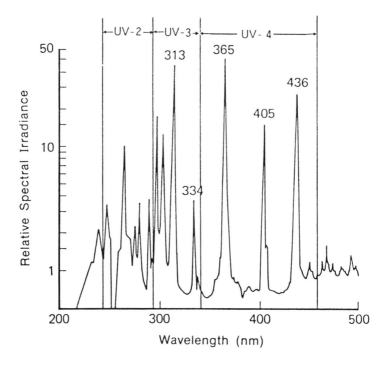

FIG.2.1 Spectral irradiance of an Xe-Hg lamp. (Courtesy of SVG).

tens amperes current and several tens of voltage. These lamps are maintained by external power supplies in the range of 0.5-2.0 kW.

2.1.2 Excimer Lasers

Excimer lasers have been attracted as strong light sources for short wavelength lithography because of low light output from high pressure Hg lamp in deep-UV region. This relatively new class of very efficient and extremely powerful pulsed lasers have become commercially available. They operate at several characteristic wavelengths ranging from vacuum ultraviolet light to greater than 400 nm (as shown in Table 2.1), depending on the gas mixtures used. The output is typically 10-20 ns wide pulses with repetition rates from ten to several hundred Hertz.

An excimer is defined as a bound-state excited molecule formed between an excited atom with a ground-state atom, while its ground state is repulsive or very weakly bounded state as shown in Fig.2.3 [3]. Although the term excimer comes from "excited dimer", an excited molecule of an excited atom and its

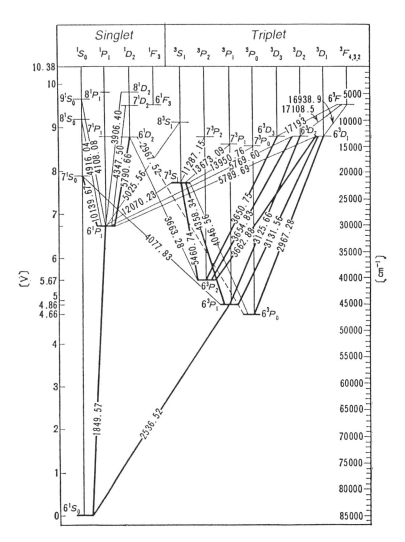

FIG.2.2 Energy diagram of a mercury atom.

ground state, the excimer also includes exciplex (an excited complex of an excited atom and a different ground-state atom). An excimer can emit light to yield separated atoms in the ground state. This nature meets the requirement of reverse population (higher concentration in excited state than lower state) for laser media.

An important characteristic of excimer lasers that sets them apart from traditional UV lasers is their lack of spatial coherence. The interference phenomena that result from the high spatial coherence of traditional single-mode continuous wave lasers produces a random intensity variation in projected patterns called speckle. This speckle phenomenon has historically made use of lasers in high-resolution lithography very difficult. The beam of excimer lasers is so highly multimode that speckles are, for all practical purposes, nonexistent in projected patterns.

Table 2.1 Wavelengths for excimer lasers

Excimer	Wavelength
XeF	351 nm
XeCl	308 nm
KrF	248 nm
KrCl	222 nm
ArF	193 nm
F_2	157 nm

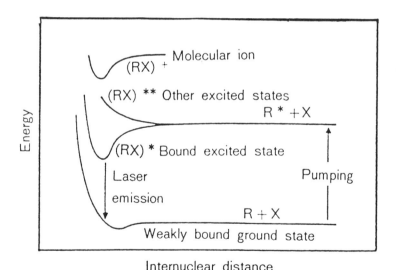

FIG.2.3 Potential energy curves of rare gas and halogen atom.

The application of excimer laser sources to DUV lithography has been demonstrated for contact printing [3] and mirror projection printing [4]. The promising application of excimer lasers, especially the KrF excimer laser, is a light source for a reduction projection printing in step-and-repeat fashion by using specially constructed refracting lenses. However, the number of optical materials used to correct chromatic aberration is limited to quartz as optical materials. Therefore, spectrally narrowed laser less than 3 pm is needed for a stepper to avoid chromatic aberration.

The use of excimer lasers in production has been hampered by the high cost and inconvenience of operating these devices in a production environment. The lasers in early stage suffered from pulse-to-pulse power reproducibility and limited operational stability on a single gas fill. However, significant progress has been made in all of these areas. Improvements in electrode materials, gas handling systems, and pulse-control electronics have demonstrated major extensions in power stability. Now that most of the engineering difficulties have been overcome, excimer lasers have become an important part of the lithographic process.

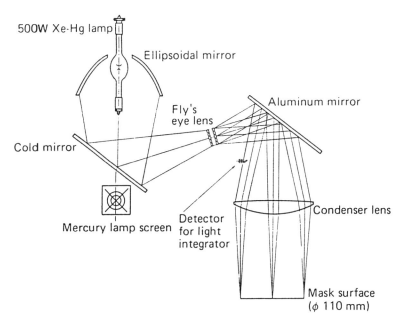

FIG.2.4 Optical layout of the illuminator in the Canon PLA520FA contact/proximity printer. [Reprint with permission from *Proceedings of 16th Symposium on Semiconductor and Integration Circuit*, p.84 (1979).]

2.2 EXPOSURE SYSTEMS

2.2.1 Contact/Proximity Printers

In contact printing, a photomask and a resist-coated wafer are brought into tight contact. The wafer is then exposed to UV or DUV radiation of controlled collimation through the mask. In proximity printers, some mechanism is provided that allows exposure with a small, controlled gap between the mask and the wafer surface. The optical system of this printer for DUV is shown in Fig.2.4 [5]. Due to the problems of the defects caused by the mask-wafer contact and the difficulty in alignment, this type of exposure system is not used in advanced Large Scale Integrated circuits (LSI) factories.

2.2.2 1:1 Projection Printers

The all-reflecting 1:1 projection system was first reported by Moller [6] in 1973. The optical and scanning configuration of such a system is illustrated in Fig.2.5, which uses reflective spherical mirror surfaces to project images of the mask onto the wafer substrate. The illuminating radiation generated from a high-pressure mercury lamp passes through a condenser and a slit of a few milli-

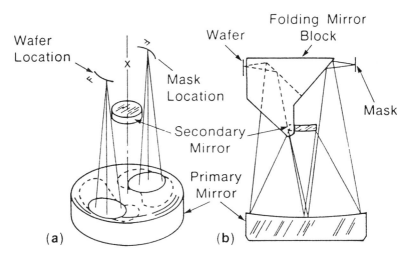

FIG.2.5 All-reflecting 1:1 projection optical system. [Reprint with permission from *Proc. Reg. Tech. Conf. Photopolymers*, SPE, p.56 (1973).]

meters in width. A set of spherical mirrors is used to generate a narrow, ring-shaped aberration-free arc of light. The slit or arc of radiation is imaged by primary and secondary mirrors to cover the entire width of the mask. Wafers are scanned with this arc to produce an image as the mask is moved synchronously in the object plane. This scanning approach is designed to minimize distortion and aberration of the optical system by keeping the imaging illumination always within the zone of good optical correction.

One of the outstanding features of this system is the flexibility it offers. In particular, the mirror lens system in this configuration has essentially no chromatic aberration. Hence, one can vary the exposure wavelength simply by imposing the appropriate transmission filter between the condenser system and the mask. The wavelength region proposed for this system is shown in Fig.2.1. Many improvements have been made in the design of these tools since their first commercial introduction by Perkin-Elmer (now SVG Lithography). The projection optics of the Perkin-Elmer Micralign™ 500 are shown in Fig.2.6. The NA (Numerical Aperture) of this system is 0.16. The refracting elements are all single elements and fabricated of high quality quartz.

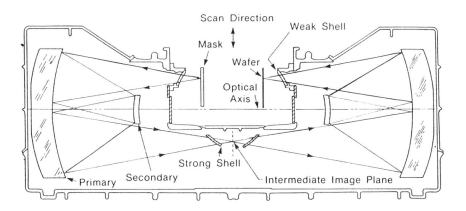

FIG.2.6 Micralign™ 500 projection optics system. (Courtesy of SVG.)

2.2.3 Reduction Projection Aligner with Step and Repeat System for g-line and i-line

A schematic representation of a reduction projection system is shown in

Fig.2.7. The system projects a mask image onto a limited area of wafer, consisting of one or two chips of DRAM devices. In order to expose the entire area of a wafer, the wafer is stepped with precise mechanical movement to a new position before each exposure. A series of exposures involves repeating of stepping, alignment and exposure, which is called as a step-and-repeat exposure system, or a stepper for short.

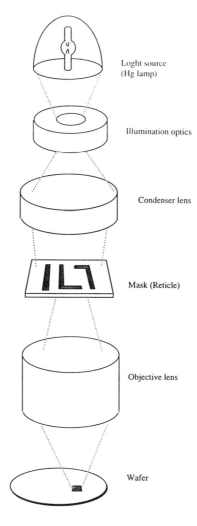

FIG.2.7 Reduction projection aligner with step-and-repeat system (or stepper).

Exposure Systems in Photolithography 19

A high pressure mercury lamp is used as a light source for g-line (436 nm) and i-line (365 nm) steppers. One of the characteristic emission lines from the mercury lamp is selected by a combination of a mirror and a filter, since the chromatic aberration prevents the construction of projection optics for all the emission lines. Illumination optics is designed for uniform exposure intensity at the mask and the wafer plane. A series of masks of chromium patterns on quartz substrates are usually fabricated by electron beam lithography.

As will be described in Chapter 3, the higher numerical aperture of projection optics offers a higher resolution capability. The design of the projection optics determines the performance of the stepper (such as resolution capability, exposure field size, depth of focus, etc.). Although only one element for an objective lens is shown in Fig.2.7, the lens for the projection optics consists of more than ten elements to optimize chromatic aberration and other aberrations such as spherical aberration, coma, astigmatism, curvature of field and distortion. Alignment accuracy becomes important for shrinking design rule of LSI because large process latitude should be incorporated in LSI design for poor overlay accuracy.

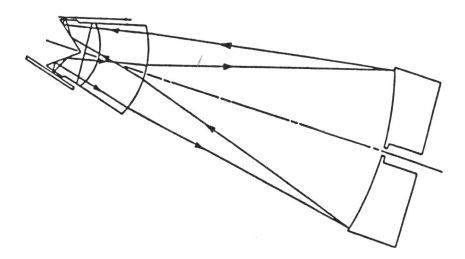

FIG.2.8 Basic Wynne-Dyson system.

2.2.4 1:1 Wyne-Dyson Exposure System

Wyne-Dyson exposure system consists of a mirror and three-component lens as shown in Fig.2.8, while Dyson's original concept consisted of a single thick lens and a mirror. This simple arrangement is free from Seidal aberrations, although Dyson showed that it suffers from high order astigmatism growing as the fourth power of the field radius. Wyne-Dyson optics offers a small separation between the object and image planes and the nearest glass surface. Although this optics can be utilized at high NA and at deep-UV region, 1:1 mask fabrication prevents this system from use at the resolution required for leading edge circuits. Therefore, this system may be applied to non-critical processes with high throughput, taking advantage of scanning a large field.

2.2.5 Exposure Systems for Deep-UV Lithography

For most of the technologies for DUV, production-worthy solutions are available. Deep-UV exposure systems are classified into two types, a KrF reduction projection system of step-and repeat mode and a reduction projection of step-and-scan mode. The arrangement for the KrF step-and-repeat exposure system is essentially the same as i-line and g-line steppers except that the KrF excimer laser is installed instead of a high pressure Hg lamp as a light source. A typical example of a KrF excimer laser stepper is shown in Fig.2.9. High quality, wide-field quartz lenses with high NA and the required low distortion and aberration are beginning to appear. A stable and accurate DUV alignment system has been developed using TTL (through-the-lens) alignment at the helium-neon laser wavelength.

The early concerns about excimer lasers as light sources for the stepper described in 2.1 are gradually disappearing. The special demands on dose control and wavelength stability that are inherent to these pulsed narrow band lasers have been solved adequately. Cavity lifetimes exceeding 2×10^9 pulses and wavelength stability of 0.25 pm are available from principal laser suppliers.

In contrast to the step-and-repeat system, Perkin-Elmer (now SVGL) has designed catadioptic (reflective and refractive) optics with a step-and-scan reticle/wafer stage system named Micrascan™ [7]. The system was designed for 4:1 reduction optics with reflective and refractive optical elements as shown in Fig.2.10. Unlike steppers, step-and-scan systems do not project the entire reticle field image to the wafer level. Instead, they project only a slit-defined portion of the field. To generate a complete image at each location on the wafer, the reticle and the wafer stage are concurrently translated until the mask is fully scanned by a slit.

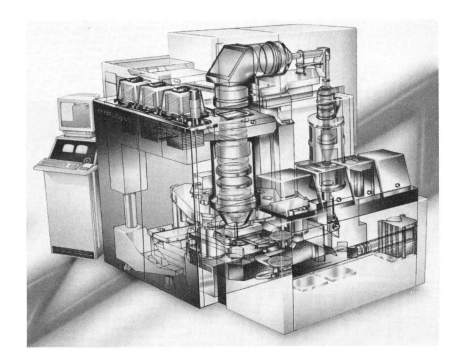

FIG.2.9 KrF excimer laser stepper (Courtesy of ASM).

Scanning offers a major advantage in that it generates a large field size for advanced or multiple chip imaging (while field size is the issue associated with lens resolution and distortion).

SVGL has announced Micrascan™ II for a deep UV lithography step-and-scan exposure system as shown in Fig.2.11 [8]. The basic concept of the system can be traced to the former version that employed catadioptic (reflective and refractive) optics with a step-and-scan reticle/wafer stage system. Micrascan™ II's current Hg-Xe source system for 0.35-μm lithography provides a usable band width of 8 nm, centered at 244 nm. For 0.25-μm generation, they announced Micrascan™III with automated NA control for process optimization;

22   Chapter 2

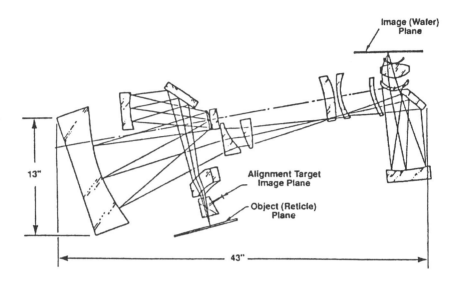

FIG.2.10 Catadioptic optical system, Micrascan™ I. [Reprint with permission from *J. Vac. Sci. Technol.*, **B7**, 1607(1989).]

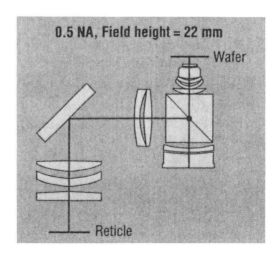

FIG.2.11 Basic catadioptic optical system of the SVGL Micrascan™ II. [Reprint with permission from *Solid State Technology*, **38**(9), 96(1995).]

an increase of maximum NA; off-axis illumination and /or pupil filtering for DOF (depth-of-focus) enhancement. This system uses a 248 nm KrF laser instead of a lamp, and is a "broadband" laser compared to a KrF excimer laser reduction projection aligner.

2.2.6 Exposure Systems for Future Lithography

It becomes more and more difficult to design the higher NA exposure system with a large field size, which is driven by the combined forces of shrinking geometries and growing chip sizes. Nikon announced its DUV step-and-repeat scanning technology as shown in Fig.2.12 [9]. The image field is reduced 4x the projection of reticle, and the image is built up by a simultaneous scan of reticle and the wafer through the illumination and projection optics. The reticle stage and the wafer stage are synchronously scanned and the reticle scans four times faster than wafer. Moreover, dynamic control of wafer leveling and lens focus is also carried out during the scan. The dose delivered to the wafer by excimer laser pulses is also controlled.

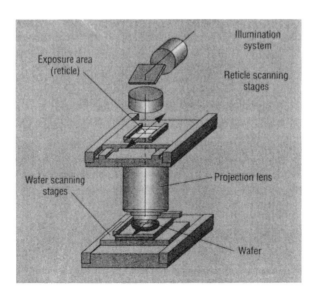

FIG.2.12 Nikon NSR-S201A scan and control metrology. [Reprint with permission from *Solid State Technology*, **38**(9), 92(1995).]

Recently, interest in 193 nm DUV lithography has increased sharply. Both refractive and catadioptic systems have been proposed which are essentially the same optics as described above. SVGL delivered to MIT Lincoln Laboratory a 193 nm tool which is actively used for evaluation of 193 nm resists. At present, 193 nm technology is limited by serious material issues of absorption both for resist and optical materials. Absorption of exposure light by optical materials results in temperature rise. The temperature rise in optics induces the change in refractive index and curvature of lens and leads to distortion in optics. In addition, a severe spectrally narrowed ArF laser is required. In the catadioptic system, spectral narrowing is not so severe as that for the refractive lens system.

REFERENCES

1. C. G. Willson, Organic resist materials- theory and chemistry, *ACS Symp. Ser.*, **219**, ch.3 (1983).
2. S. Sato and S. Shida, *Hikarikagaku to Hosyasenkagaku (Photochemistry and Radiation chemistry)*, Tokyo Douzin (196).
3. K. Jain, C. G. Willson, and B. J. Lin, Ultrafast high resolution contact lithography with excimer lasers, *IBM J. Res. Dev.*, **26**, 151 (1982).
4. R. T. Kerth, K. Jain and M. R. Latta, Excimer laser projection lithography on a full-field scanning projection system, IEEE Electron Device Lett., **EDL-7**, 299(1986).
5. I. Kano and K. Momose, Deep-UV illumination system for proximity aligner, *Proceeding of 16th Symposium on Semiconductor and Integration Circuit*, p.84 (1979).
6. P. Moller, All-reflective 1:1 projection printing system, *Proc. Reg. Tech. Conf. Photopolymers*, SPE, Ellenville, New York, p.56 (1973).
7. J. D. Buckley, D. N. Galburt, and C. Karatzas, Step-and-scan lithography using reduction optics, *J. Vac. Sci. Technol.*, **B7**, 1607(1989).
8. J. J. Shamaly, SVGL: Scanning the lithography roadmap for DUV and beyond, *Solid State Technology*, **38**(9), 96(1995).
9. A. Dickinson, Nikon: DUV lithography into the mainstream, *Solid State Technology*, **38**(9), 92(1995).

3
Optical Pattern Transfer

In the microlithography for semiconductor device fabrication, the minimum feature size of the pattern printed optically on a layer of photoresist is often in the same order of magnitude as the wavelength of light used to print the pattern. Therefore, the knowledge of wave optics, where light is regarded as an electromagnetic wave, is necessary to find the theoretical limit of resolution in this technology.

It has been well established that all kinds of photochemical reactions take place by the stimulation of electric oscillation involved in the electromagnetic wave. Therefore, in this book, the magnitude of electric oscillation is regarded as a measure of light intensity relevant to photochemical reactions.

3.1 CONTACT PRINTING

The method of contact printing (illustrated in Fig. 1.1C) is very simple. A layer of photoresist formed on a substrate is contacted with an optical mask bearing opaque-and-transparent pattern. The layer is then exposed to ultraviolet or visible light through the mask. After that, the layer is developed with developer to form the pattern in accordance with the mask pattern.

Although the method is called contact printing, the contact is usually imperfect, and most areas of the resist surface remain uncontacted with the mask surface. Therefore, in the strict sense of the words, the method should be called "close proximity printing". If the resist surface is perfectly contacted with the mask surface, the transferred pattern will be little affected by the diffraction of light. However, in the usual case where the two surfaces are separated by a thin film of air, the effect of diffraction on the pattern transfer becomes significant. The rigorous solution of the two-dimensional diffraction problem (described in the Appendix), is useful for the estimation of the effect of diffraction on the pattern transfer by contact printing.

3.1.1 Straight Line Edges of Large Pattern

The diffraction by a rectilinear edge of a large pattern is well approximated by the diffraction by the edge of a half plane. In the case where a collimated ray of light is normally incident on a perfectly black half plane, the intensity distribution of diffracted light is obtained from eq. (A.40) in the Appendix. An example of the calculated result is shown in Fig. 3.1.

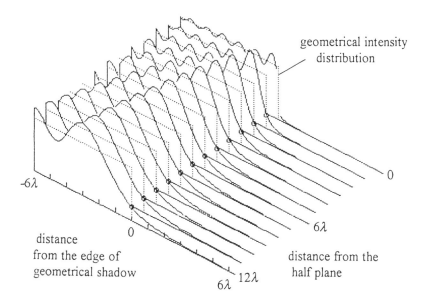

Fig. 3.1. Calculated intensity distribution of light diffracted by a black half plane. The wavelength of light is denoted by λ. Small circles indicate the positions where the intensity is equal to one-fourth of that of incident light.

As indicated by small circles in Fig. 3.1, the spatial positions where the intensity of diffracted light is equal to one fourth of that of incident light are very close to the edge of the geometrical shadow. If the light is linearly polarized so that the electric vector is parallel to the edge of the half plane (the case of transverse electric wave expressed by eq. (A.23) in the Appendix), the intensity at the edge of the geometrical shadow is exactly equal to one fourth of the intensity of incident light regardless of the distance from the half plane. In contact printing, where the mask pattern contains a rectilinear edge of relatively large opaque area, the layer of photoresist is considered to be exposed to the diffracted light with the intensity distribution as shown in Fig. 3.1. As explained in Section 1.4, if the exposure dose measured on the fully illuminated resist surface is equal to four times the threshold dose, the edge of the photoresist pattern obtained by development after the exposure is very close to or (in the case of the transverse electric wave) coincident with the edge of the geometrical shadow.

Generalization of the consideration described above leads to an important conclusion that the exposure dose in the contact printing should be four times the threshold dose if a precise pattern transfer is required. In the following sections, it will be shown that this rule is also applicable to many cases including projection printing.

3.1.2 Isolated Bright Line

The mask pattern of isolated bright line is regarded as a slit constructed by a pair of black half planes whose edges are parallel to each other. In the case where a collimated ray of light is incident on the mask plane, the intensity distribution of diffracted light is obtained from eq. (A.59) in the Appendix. An example of the calculated result is shown in Fig. 3.2, where the intensity of light diffracted by a four-wavelength-wide slit is plotted against a two-dimensional spatial position.

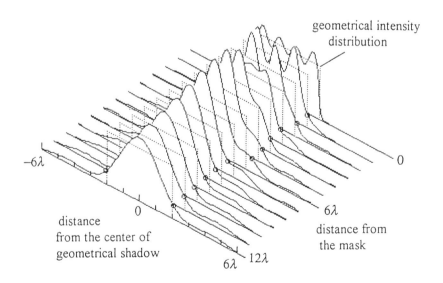

Fig. 3.2. Calculated intensity distribution of light diffracted by a four-wavelength-wide slit. The wavelength of light is denoted by λ. Small circles indicate the positions where the intensity is equal to one-fourth of that of incident light.

As seen in Fig. 3.2, the positions indicated by small circles where the intensity of diffracted light is equal to one-fourth of that of incident light deviate from the edges of geometrical shadow to some extent. Therefore, even if the exposure dose on the (supposed) fully illuminated area is controlled to be four times the threshold dose, the width of the resist pattern obtained by the development is not always the same as that of the original mask pattern. The calculated result regarding the change in line width of resist pattern with the change in distance between the mask and resist surface is shown in Fig. 3.3.

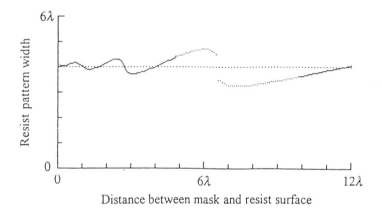

Fig. 3.3. Calculated line width of resist pattern transferred from a four-wavelength-wide slit mask pattern as a function of distance between the mask and resist surface. The wavelength of light is denoted by λ. The exposure dose on the (supposed) fully illuminated area is four times the threshold dose. The solid curves indicate that the deviation of resist pattern width from that of mask pattern is within the range of ±10%.

As seen in Fig. 3.3, the width of resist pattern swings around the width of the original mask pattern as the distance between the mask and resist surface changes, indicating that the four times the threshold exposure dose is also appropriate in the present case. When the exposure dose is fixed to this appropriate value, the deviation of resist pattern width from the mask pattern width is dependent on the mask pattern width and the distance between mask and resist surface. An example of expression for such dependency is shown in Fig. 3.4, where the width of mask pattern which is transferred to resist pattern within the deviation range of ±10% is plotted against the distance between the mask and resist surface.

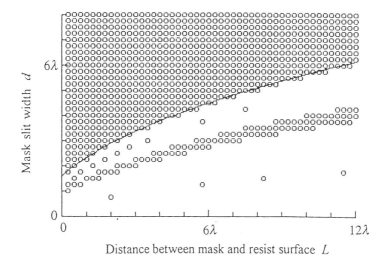

Fig. 3.4. The width of mask slit pattern which is transferred to resist pattern within the deviation range of ±10% by four times the threshold exposure dose as a function of the distance between the mask and resist surface.

The parabolic curve shown in Fig. 3.4 indicates the correlation between the mask slit width (d) and the distance between the mask and resist surface (L), which is expressed by

$$d = 1.72\sqrt{\lambda(L+0.84\lambda)}, \qquad (3.1)$$

where λ is the wavelength of light.

As seen in Fig. 3.4, a mask slit pattern with a width larger than that given by eq. (3.1) is transferred to the resist pattern within the deviation range of ±10% by four times the threshold exposure dose.

3.1.3 Isolated Black Line

Under the condition that the mask is illuminated by normally incident collimated light, the intensity distribution of the light diffracted by an isolated black straight line on the mask is calculated by using eq. (A.60) in the Appendix. An example of the calculated result is shown in Fig. 3.5, where the intensity of light diffracted by a four-wavelength-wide black straight line is plotted against a two-dimensional spatial position.

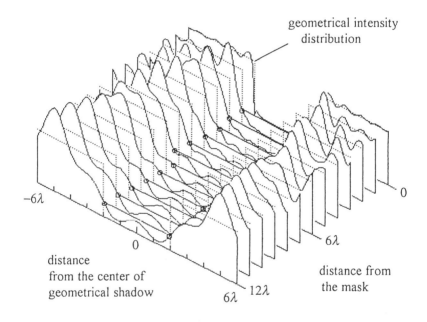

Fig. 3.5. Calculated intensity distribution of light diffracted by a four-wavelength-wide black straight line. Small circles indicate the positions where the intensity is equal to one-fourth of that of incident light.

As seen in Fig. 3.5, the positions indicated by the small circles, where the intensity is equal to one-fourth of that of incident light, deviate from the edges of the geometrical shadow to some extent. Therefore, even if the exposure dose (on the fully illuminated area of resist surface) is controlled to be four times the threshold dose, the line width of resist pattern obtained by the development is not always the same as that of the original mask pattern. An example of the calculated result regarding the change in line width of resist pattern with the change in distance between the mask and resist surface is shown in Fig. 3.6.

Optical Pattern Transfer

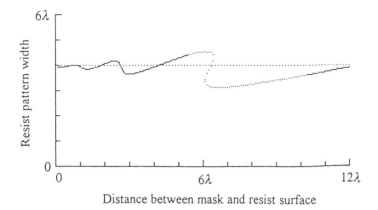

Fig. 3.6. Calculated line width of resist pattern transferred from a four-wavelength-wide black line mask pattern as a function of the distance between mask and resist surface. The exposure dose is supposed to be four times the threshold dose. The solid curves indicate that the deviation of resist pattern width from that of the mask pattern is within the range of ±10%.

In the same manner as in the case of the slit mask pattern, the resist pattern width swings around that of the original mask pattern as seen in Fig. 3.6. Therefore, the four times the threshold exposure dose is also considered to be appropriate in the present case. The curve shown in the Fig. 3.6 has a Z-shaped portion, which indicates that a false resist line pattern with a side line on each side will be formed if the resist layer is exposed at the distance corresponding to that portion.

The width of the mask line pattern which is transferred to the resist pattern within the deviation range of ±10% by the four times the threshold exposure dose is plotted against the distance between the mask and resist surface in Fig. 3.7. The parabolic curve shown in the figure indicates the correlation between the mask line width (d) and the distance between the mask and resist surface (L). The correlation is expressed by

$$d = 2.15\sqrt{\lambda(L+0.84\lambda)} \ . \tag{3.2}$$

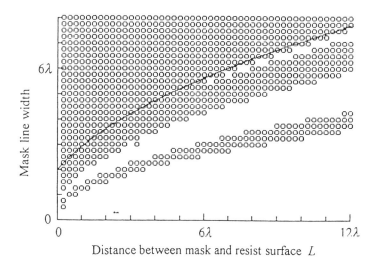

Fig. 3.7. The width of the mask black line pattern which is transferred to the resist pattern (within the deviation of ±10% by four times the threshold exposure dose) as a function of the distance between the mask and resist surface.

As seen in Fig. 3.7, a mask line pattern with a width larger than that given by eq. (3.2) is transferred to the resist pattern with the deviation range of ±10% by the four times the threshold exposure dose.

3.1.4 Line and Space

In this subsection, we consider the case where an optical mask bearing an equally wide line-and-space pattern is illuminated with a normally incident beam of collimated light. The pattern is optically equivalent to an array of parallel slits. The intensity of light diffracted by the pattern is obtained by three steps of calculation as follows. First, the electric vector of light diffracted by each slit is calculated by using the solution of a two-dimensional diffraction problem described in the Appendix. Secondly, the electric vectors are summed to give the electric vector of light diffracted by the whole pattern. Finally, the squared amplitude of electric oscillation is calculated as a relative value of light intensity.

Optical Pattern Transfer 33

As an example of the calculated result, a center part of intensity distribution of light diffracted by an array of 10 three-wavelength-wide slits is shown in Fig. 3.8.

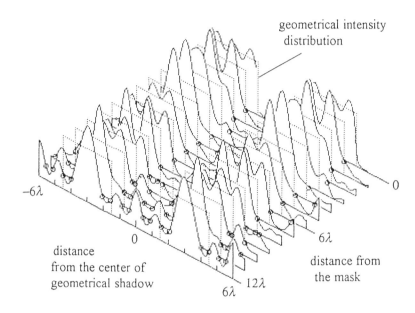

Fig. 3.8. Calculated intensity distribution of light diffracted by an array of 10 three-wavelength-wide slits. Small circles indicate the positions where the intensity is equal to one-fourth of that of incident light.

As seen in Fig. 3.8, the intensity distribution profiles are complex, indicating that a precise pattern transfer over a wide range of slit width and distance between the mask and resist surface can not be expected. This supposition is ascertained by the calculated result shown in Fig. 3.9, where the width of slit (which is transferred to the resist pattern within the deviation range of ±10% by four times the threshold exposure dose) is plotted against the distance between the mask and resist surface. The blank areas in Fig. 3.9 correspond to the conditions under which no precise pattern transfer is successful. These areas are scattered over a wide range of slit width and distance from the mask to resist surface, indicating that the contact printing is not suited for the transfer of line-and-space pattern.

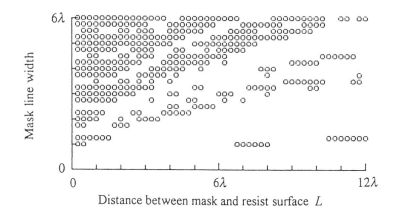

Fig. 3.9. The slit width in an array of 11 slits which is transferred to the resist pattern within the deviation range of ±10% by four times the threshold exposure dose as a function of the distance between the mask and resist surface.

3.1.5 Phase Shifter Grating

A special type of phase-shifting mask is useful for the contact printing of very fine line-and-space patterns [1]. The mask pattern is composed of a large number of phase shifter stripes. The cross-section of the mask is illustrated in Fig. 3.10. The phase shifter shifts the phase of light by one-half wavelength. The stripes are parallel to each other, and the space between them is equal to the width of the stripe.

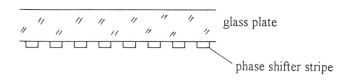

Fig. 3.10. Cross-sectional view of phase-shifting mask.

Optical Pattern Transfer 35

In the case where a beam of collimated light is normally incident on the mask, the intensity distribution of diffracted light is calculated in a similar way to that employed in the case of line-and-space patterns. An example of the calculated result is shown in Fig. 3.11, where the intensity of light diffracted by an array of one-wavelength-wide phase shifter stripes is plotted against a two-dimensional spatial position.

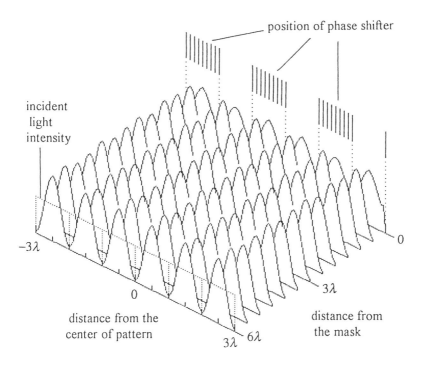

Fig. 3.11. Calculated intensity distribution of light diffracted by an array of one-wavelength-wide phase shifter stripes. The number of stripes taken into calculation is 20.

The profile of intensity distribution shown in Fig. 3.11 is nearly independent of the distance from the mask, indicating that the mask is suited for contact printing. An example of a scanning electron micrograph of the line-and-space resist pattern obtained by using this type of mask is shown in Fig. 3.12. A very high resolution capability (about 0.1μm) of a positive photoresist is demonstrated in this micrograph. As this method is feasible without using any expensive instruments such as a reduction projection printing system, it is practically useful in evaluating the resolution capability of the photoresist material. As shown in Fig. 3.11, the spatial frequency of the pattern is doubled by the transfer from the mask to resist. This is another unique feature of this method.

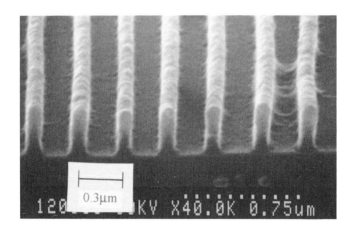

Fig. 3.12. Scanning electron micrograph of a resist pattern obtained by i-line exposure through a mask containing a 0.3-μm line-and-space pattern phase shifter. The resist is NPR-Λ18SH 3 (positive photoresist) obtained from Nagase Electronic Chemicals.

3.2 PROXIMITY PRINTING

Proximity printing is very similar to contact printing except for that the former keeps the resist surface uncontacted with the mask while the latter allows contact. Since practical contact printing should be regarded as a close proximity printing, the case where the resist surface is apart from the mask by a distance up to 12 wavelengths has been dealt with in the foregoing section. As the increase in distance between the mask and resist surface results in the loss in resolution, no further discussion on proximity printing using visible or ultraviolet light will be necessary. However, proximity printing in x-ray lithography, where the distance between the mask and resist surface is three to four orders of magnitude larger than the wavelength, must be dealt with in this section.

Optical Pattern Transfer

Although the potential resolution capability of x-ray lithography is very high, the practical resolution is significantly lowered by increasing the distance between the mask and resist surface. The intensity distribution of x-ray on the resist surface is obtained by using the rigorous solution of a two-dimensional diffraction problem as described in the Appendix.

3.2.1 Straight Line Edges of Large Pattern Transferred by X-Ray

In the same manner as described in 3.1.1, the intensity distribution of x-ray diffracted by a straight line edge of large pattern is calculated. In the present case, eq. (A.25) in the Appendix is used instead of eq. (A.40) because the influence of polarization is negligible. An example of calculated result is shown in Fig. 3.13, where the intensity of x-ray diffracted by a straight line edge of a large pattern is plotted against a two-dimensional spatial position.

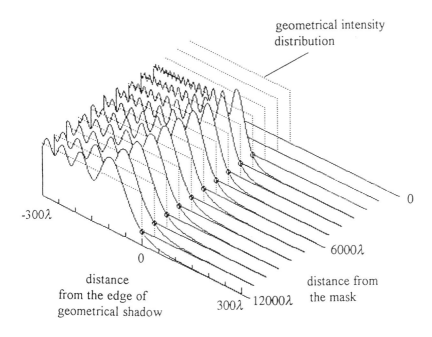

Fig. 3.13. Calculated intensity distribution of-x ray diffracted by a straight line edge of large pattern. The wave length of x-ray is denoted by λ. Small circles indicate the positions where the intensity is equal to one-fourth of that of incident x-ray.

As indicated by small circles in Fig. 3.13, the positions where the intensity of x-ray is equal to one fourth of that of incident x-ray coincide with the edge of the geometrical shadow. Therefore, the edge of the resist pattern obtained by the exposure of four times the threshold dose coincides with the edge of the geometrical shadow.

3.2.2 Isolated Bright Line Transferred by X-Ray

The intensity distribution of x-ray diffracted by a slit, which represents an isolated bright line mask pattern, is calculated by using eq. (A.49) in the Appendix. An example of the calculated result is shown in Fig. 3.14, where the intensity of x-ray diffracted by a 200-wavelength-wide slit is plotted against a two-dimensional spatial position.

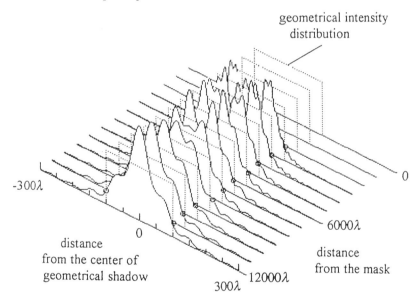

Fig. 3.14. Calculated intensity distribution of x-ray diffracted by a 200-wavelength-wide slit. Small circles indicate the positions where the intensity is equal to one-fourth of that of the incident x-ray.

As seen in Fig. 3.14, the positions indicated by small circles (where the intensity is equal to one-fourth of that of incident x-ray) deviate from the edge of the geometrical shadow. Therefore, even if the exposure dose on a (supposed) fully irradiated area is controlled to be four times the threshold dose, the width

Optical Pattern Transfer 39

of the resist pattern developed after the exposure is not always the same as that of the original mask pattern. The width of mask pattern which is transferred to the resist pattern within the deviation range of ±10% by four times the threshold exposure dose is plotted against the distance between the mask and resist surface in Fig. 3.15.

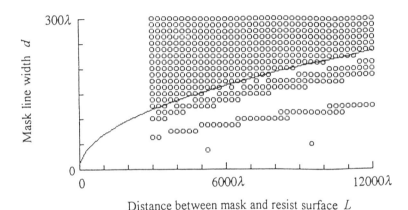

Fig. 3.15. The width of isolated bright line mask pattern transferred to resist pattern within the deviation range of ±10% by four times the threshold exposure dose as a function of distance between the mask and resist surface.

The parabolic curve shown in Fig. 3.15 indicates the correlation between the mask pattern width (d) and the distance between the mask and resist surface (L), and is expressed by

$$d = 2.15\sqrt{\lambda L}. \tag{3.3}$$

Equation (3.3) is practically identical to eq. (3.2) because λ is three to four orders of magnitude smaller than L in the present case, and hence, the term 0.84λ in eq. (3.2) can be neglected.

As seen in Fig. 3.15, a mask line pattern with a width larger than that given by eq. (3.3) is transferred to a resist pattern within the deviation range of ±10% by four times the threshold exposure dose.

3.2.3 Isolated Black Line Transferred by X-Ray

The intensity distribution of x-ray diffracted by an opaque stripe, which represents an isolated black line mask pattern, is calculated by using eq. (A.61) in the Appendix. An example of the calculated result is shown in Fig. 3.16, where the intensity of x-ray diffracted by a 200-wavelength-wide stripe is plotted against a two-dimensional spatial position.

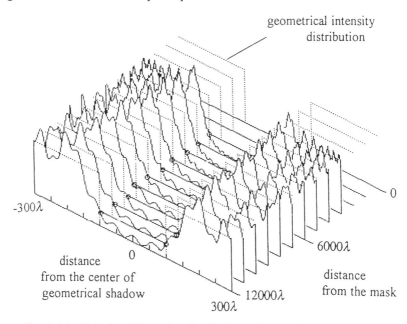

Fig. 3.16. Calculated intensity distribution of x-ray diffracted by a 200-wavelength-wide opaque stripe. Small circles indicate the positions where the intensity is equal to one-fourth of that of the incident x-ray.

As indicated by small circles in Fig. 3.16, the positions where the intensity is equal to one-fourth of that of the incident x-ray deviate from the edge of the geometrical shadow. Therefore, even if the exposure dose on a fully illuminated area is controlled to be four times the threshold dose, the width of the resist pattern obtained by the development is not always the same as that of the original mask pattern. The width of mask pattern which is transferred to the resist pattern within the deviation range of ±10% by four times the threshold exposure dose is plotted against the distance between the mask and resist surface in Fig. 3.17.

Optical Pattern Transfer 41

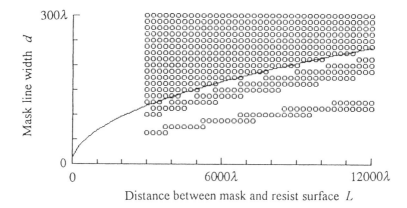

Fig. 3.17. The width of isolated opaque line mask pattern transferred to resist pattern within the deviation range of ±10% by four times the threshold exposure dose as a function of distance between the mask and resist surface.

The parabolic curve shown in Fig. 3.17, indicating the minimum line width of mask pattern which is transferred to the resist pattern within the deviation range of ±10% by four times the threshold exposure dose, is identical to that shown in Fig. 3.15 and expressed by eq. (3.3).

3.2.4 Line-and-Space Pattern Transferred by X-Ray

We consider the case where an equally wide line-and-space pattern on an x-ray mask is transferred to the resist layer with a perpendicularly incident x-ray. The intensity of x-ray diffracted by the mask pattern is calculated in the same manner as described in 3.1.4. An example of the calculated result is shown in Fig. 3.18, where the intensity of x-ray diffracted by an array of four slits, which are 150-wavelength-wide and spaced by the same width, is plotted against a two-dimensional spatial position.

As seen in Fig. 3.18, the profiles of intensity distribution are complex, indicating that a precise pattern transfer over a wide range of mask line width and the distance between the mask and resist surface can not be expected. The line width of mask pattern which is transferred to the resist pattern within the deviation range of ±10% by four times the threshold exposure dose is plotted against the distance between the mask and resist surface in Fig. 3.19. The figure shows a considerable loss in resolution compared with the case of an isolated line.

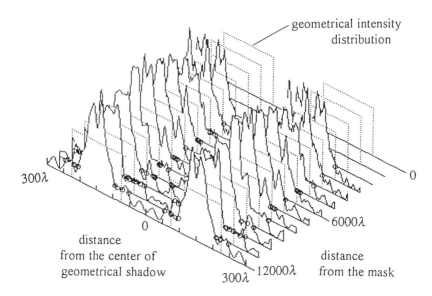

Fig. 3.18. Calculated intensity distribution of x-ray diffracted by an array of four slits which are 150-wavelength-wide and spaced by the same width. Small circles indicate the positions where the intensity is equal to one-fourth of that of the incident x-ray.

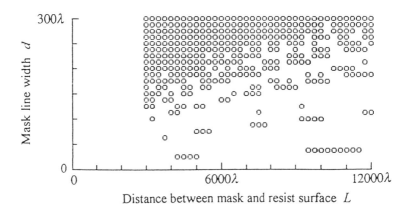

Fig. 3.19. Slit width in an array of five slits which is transferred to the resist pattern within the deviation range of ±10% by four times the threshold exposure dose as a function of the distance between the mask and resist surface.

3.3 PROJECTION PRINTING

In the projection printing process, the image of the optical mask pattern is projected by a lens or mirror system. The same types of patterns as considered in the foregoing sections are also considered in this section. The projection of these patterns can be dealt with as a two-dimensional diffraction and interference problem by assuming that the longitudinal dimension of each pattern is several orders of magnitude larger than the wavelength of light used to project the image.

3.3.1 Infinitely Narrow Slit

The projection of an infinitely narrow slit is schematically illustrated in Fig. 3.20.

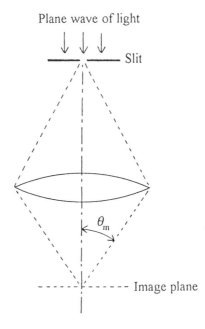

Fig. 3.20. Projection of an infinitely narrow slit.

A plane wave of illuminating light is perpendicularly incident on the screen where the slit is opened. The wave is diffracted by the slit to form a cylindrical wave. A portion of the cylindrical wave entering the lens is converged on the image plane to form a line image. The numerical aperture denoted by NA is defined by using the convergence angle θ_m shown in Fig. 3.20 as

$$NA = \sin \theta_m. \qquad (3.4)$$

Since the optical path from lens to image plane is six orders of magnitude larger than the wavelength of light, the converging wave is well approximated by the superposition of an infinite number of plane wavelets with different propagation directions.

For quantitative discussion, we set an orthogonal coordinate system and define some variables as shown in Fig. 3.21. The image plane is expressed by $z = 0$, and the geometrical image position on the x-z plane is indicated by $P(\xi, 0, 0)$. The angle between propagating direction of wavelet and z-axis is expressed by θ measured counterclockwise from the direction of the z-axis.

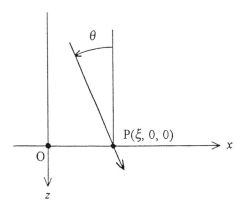

Fig. 3.21. Setting of coordinate system and definition of ξ and θ.

The plane wave of ordinary, unpolarized light is considered to be the superposition of two linearly polarized waves whose electric vectors are perpendicular to each other. In this section, we decompose the electric wave of ordinary light into p- and s-waves, where the electric vector of the p-wave is parallel to the y-axis while that of the s-wave is perpendicular to it.

First, we consider the case of p-wave. The electric vector components d^2E_x, d^2E_y, d^2E_z of the wavelet with the propagating direction θ [within the interval (θ, $\theta +d\theta$) and the convergence point ξ on the x-axis within the interval (ξ, $\xi+d\xi$)] is expressed by

$$d^2 E_x = 0, \quad (3.5)$$

$$d^2 E_y = K E_0 \exp(i\varphi)\, d\theta\, d\xi, \quad (3.6)$$

$$d^2 E_z = 0, \quad (3.7)$$

where K is a constant, E_0 the amplitude of electric oscillation on the image plane in the case of flood exposure where the mask is totally transparent, and φ is defined as

$$\varphi = (k \sin \theta)(x - \xi) + (k \cos \theta) z - \omega t, \quad (3.8)$$

where k is 2π times the wavenumber, ω the angular velocity, and t the time. In this book, we use complex representation for an electromagnetic wave as shown in eq. (3.6), where the component of the electric or magnetic vector is regarded as the real part of an associated complex wave. Although the amplitude KE_0 is slightly dependent on the convergence angle θ, the dependency is neglected in eq. (3.6) for simplicity by assuming that K is a constant.

Secondly, we consider the case of the s-wave. Since the wave of ordinary light is decomposed into p- and s-waves with equal intensity, the s-wavelet associated with the p-wavelet expressed by eqs. (3.5) - (3.7) is given by

$$d^2 E'_x = \cos \theta \, d^2 E_y, \quad (3.9)$$

$$d^2 E'_y = 0, \quad (3.10)$$

$$d^2 E'_z = -\sin \theta \, d^2 E_y. \quad (3.11)$$

To find the value of K, we suppose the case where the slit is infinitely widened to make the mask totally transparent for flood exposure. When the width of the geometrical image of slit is increased to $2L$ while the center position of the image is fixed at the origin, the electric vector of the p-wave E_y is expressed as

$$E_y = \int_{-L}^{L} \int_{-\theta_m}^{\theta_m} K E_0 \exp(i\varphi)\, d\theta\, d\xi. \quad (3.12)$$

Calculation of integrals and limits gives the result,

$$\lim_{L\to\infty} E_y = \lambda KE_0 \exp[i(kz-\omega t)], \tag{3.13}$$

where λ is the wavelength of light. As E_0 is defined to be the amplitude of electric oscillation on the image plane in the case of flood exposure where the mask is totally transparent, λK must be equal to unity, and hence, K is given by

$$K = \frac{1}{\lambda}. \tag{3.14}$$

By using eq. (3.14), eqs. (3.6), (3.9) and (3.11) are rewritten as follows.

$$d^2E_y = \frac{1}{\lambda}E_0 \exp(i\varphi)\,d\theta\,d\xi, \tag{3.15}$$

$$d^2E'_x = \cos\theta\,d^2E_y, \tag{3.16}$$

$$d^2E'_z = -\sin\theta\,d^2E_y. \tag{3.17}$$

3.3.2 Straight Line Edges of Large Pattern

The imaging of a rectilinear edge of a large pattern is well approximated by the imaging of the edge of a half-plane. In the case where the mask is illuminated by a perpendicularly incident beam of light and the bright area of the image is defined by $x \leq 0$ and $z = 0$, the electric vector components are given by integrating d^2E_y, $d^2E'_x$ and $d^2E'_z$ as expressed by eqs. (3.15)-(3.17) with regard to θ and ξ. As the intensity of light relevant to photochemistry is proportional to the sum of squared amplitudes of electric oscillation, a relative value of it, denoted by $I_1(x, z)$, is expressed as

$$\begin{aligned}I_1(x,z) = &\frac{1}{\lambda^2}\left|\int_{-\infty}^{0}\int_{-\theta_m}^{\theta_m} \exp(i\varphi)d\theta\,d\xi\right|^2 \\ &+ \frac{1}{\lambda^2}\left|\int_{-\infty}^{0}\int_{-\theta_m}^{\theta_m} \cos\theta\,\exp(i\varphi)d\theta\,d\xi\right|^2 \\ &+ \frac{1}{\lambda^2}\left|\int_{-\infty}^{0}\int_{-\theta_m}^{\theta_m} \sin\theta\,\exp(i\varphi)d\theta\,d\xi\right|^2,\end{aligned} \tag{3.18}$$

where φ is defined by eq. (3.8).

An example of the calculated result is shown in Fig. 3.22, where the light intensity in the vicinity of the edge of the geometrical image of the half-plane is plotted against a two-dimensional spatial position.

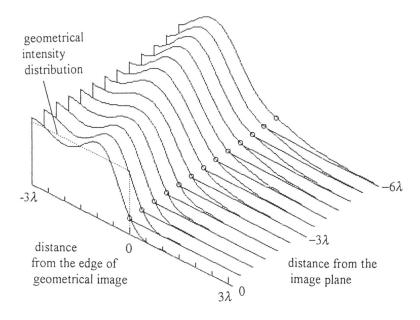

Fig. 3.22. Calculated light intensity distribution in the vicinity of the edge of the geometrical image of a half-plane. The small circles indicate the positions where the intensity is equal to one-fourth of that in the case of flood exposure where the mask is totally transparent.

As indicated by small circles in Fig. 3.22, the position where the light intensity is equal to one-fourth of that in the flood exposure where the mask is totally transparent are on a straight line passing the edge of the geometrical image and parallel to the z-axis. Therefore, if the exposure dose in the fully illuminated area is four times the threshold dose, the edge of the resist pattern formed by the development is on the edge of the geometrical shadow, and its position is not changed by defocusing.

3.3.3 Isolated Bright Line

In the case where a bright straight line mask pattern is illuminated with a perpendicularly incident beam of light, the intensity distribution in the vicinity of its image is calculated in the same manner as described above. When the edges of the geometrical image are parallel to the y-axis and defined by $x = -d/2$ and $x = d/2$, the relative light intensity, denoted by $I_2(x, z)$, is given by

$$I_2(x,z) = \frac{1}{\lambda^2}\left|\int_{-d/2}^{d/2}\int_{-\theta_m}^{\theta_m}\exp(i\varphi)\,d\theta\,d\xi\right|^2$$

$$+ \frac{1}{\lambda^2}\left|\int_{-d/2}^{d/2}\int_{-\theta_m}^{\theta_m}\cos\theta\,\exp(i\varphi)\,d\theta\,d\xi\right|^2$$

$$+ \frac{1}{\lambda^2}\left|\int_{-d/2}^{d/2}\int_{-\theta_m}^{\theta_m}\sin\theta\,\exp(i\varphi)\,d\theta\,d\xi\right|^2, \qquad (3.19)$$

where d is the width of the geometrical line image.

An example of the calculated result is shown in Fig. 3.23, where the light intensity in the vicinity of a two-wavelength-wide geometrical line image is plotted against a two-dimensional spatial position.

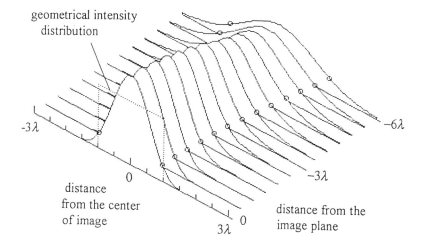

Fig. 3.23. Calculated light intensity distribution in the vicinity of the image of bright line projected by a lens with NA of 0.5. The small circles indicate the positions where the intensity is equal to one-fourth of that in the case of flood exposure.

Optical Pattern Transfer

The small circles in Fig. 3.23 indicate the positions where the light intensity is equal to one-fourth of that in the case of flood exposure. Therefore, when the resist on a (supposed) fully illuminated area is exposed with four times the threshold dose, the edges of resist line pattern will be formed on these positions by the development after exposure. As indicated in Fig. 3.23, the width of resist pattern thus obtained is very close to that of the geometrical image within some range of defocusing. Such range is called depth of focus (DOF).

The resolution and depth of focus are strongly dependent on the NA of the projection system. The dependency is exemplified in Fig. 3.24, where the minimum width of the geometrical line image (which is transferred to resist pattern within the deviation range of ±10% by the four times the threshold exposure dose) is calculated for various values of NA and plotted against the absolute distance from the image plane.

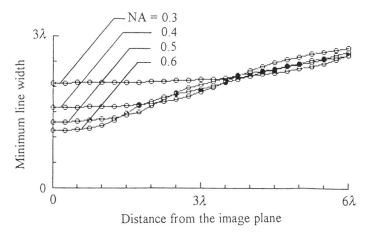

Fig. 3.24. Minimum width of geometrical bright line image transferred to resist pattern within the deviation range of ±10% by four times the threshold exposure dose as a function of the absolute distance from the image plane.

The flat part of each curve in Fig. 3.24 indicates both resolution and depth of focus by its vertical position and length, respectively. The resolution expressed by the minimum line width d_m at the image plane is approximately given, in the present case, by

$$d_m = 0.65 \frac{\lambda}{NA}. \qquad (3.20)$$

This type of equation is generally called Rayleigh's criterion for resolution. The depth of focus, denoted by DOF, is roughly expressed in the present case, as

$$\text{DOF} = 0.6 \frac{\lambda}{\text{NA}^2}. \qquad (3.21)$$

As the light intensity distribution is mirror-symmetrical with regard to the image plane, DOF is defined to be twice the length of the flat part of each curve in Fig. 3.24.

3.3.4 Isolated Black Line

When the image of an isolated black straight line is projected in the same situation as described above on the image plane $z = 0$ with the edges of the geometrical image defined by $x = -d/2$ and $x = d/2$, the relative light intensity, denoted by $I_3(x, z)$, is given by

$$I_3(x,z) = \left| \exp[i(kz-\omega t)] - \frac{1}{\lambda} \int_{-d/2}^{d/2} \int_{-\theta_m}^{\theta_m} \exp(i\varphi) d\theta\, d\xi \right|^2$$

$$+ \left| \exp[i(kz-\omega t)] - \frac{1}{\lambda} \int_{-d/2}^{d/2} \int_{-\theta_m}^{\theta_m} \cos\theta \exp(i\varphi) d\theta\, d\xi \right|^2$$

$$+ \left| \frac{1}{\lambda} \int_{d/2}^{d/2} \int_{-\theta_m}^{\theta_m} \sin\theta \exp(i\varphi) d\theta\, d\xi \right|^2 \qquad (3.22)$$

Here, the term $\exp[i(kz-\omega t)]$ in the equation is the result of double integration of $\exp(i\varphi)$ and $\cos\theta \exp(i\varphi)$ with regard to θ from $-\theta_m$ to θ_m and ξ from $-\infty$ to ∞.

An example of the calculated result is shown in Fig. 3.25, where the light intensity in the vicinity of a two-wavelength-wide geometrical line image is plotted against a two-dimensional spatial position.

As indicated by small circles in Fig. 3.25, the resist pattern obtained by development after the exposure will be very close to the geometrical image within the depth of focus if the resist on a fully illuminated area is exposed with four times the threshold dose.

The resolution and depth of focus characteristics are shown in Fig. 3.26, where the minimum width of the geometrical black line image (which is transferred to resist pattern within the deviation range of ±10% by four times the threshold exposure dose) is plotted against the distance from the image plane. In Fig. 3.25, there are some profile curves, each of which has more than two small circles. This means that false patterns will be formed in some cases. Such cases have been regarded as the case of incorrect pattern transfer in the plot in Fig. 3.26.

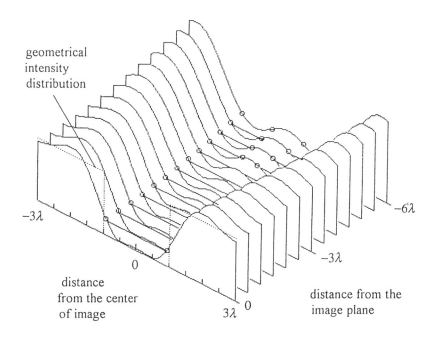

Fig. 3.25. Calculated light intensity distribution in the vicinity of the image of black line projected by a lens with NA of 0.5. The small circles indicate the positions where the intensity is equal to one-fourth of that in the case of flood exposure.

The resolution (d_m) and the depth of focus (DOF) in this case are approximately expressed as

$$d_m = 0.59 \frac{\lambda}{NA} \qquad (3.23)$$

and

$$DOF = 0.8 \frac{\lambda}{NA} . \qquad (3.24)$$

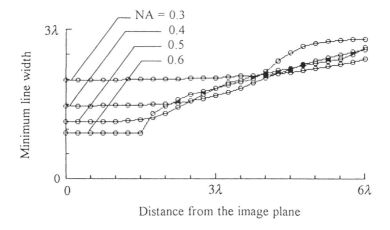

Fig. 3.26. Minimum width of geometrical black line image transferred to resist pattern within the deviation range of ±10% by four times the threshold exposure dose as a function of the absolute distance from the image plane.

3.3.5 Line and Space

We consider the projection of line-and-space pattern on the image plane $z = 0$. The projected geometrical image is illustrated in Fig. 3.27.

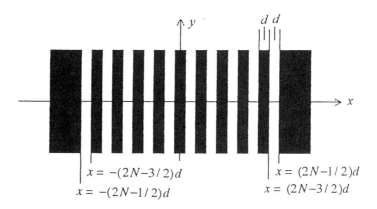

Fig. 3.27. Geometrical image of line-and-space pattern. The number of bright lines is $2N$, and the line width is d.

Optical Pattern Transfer

First, we consider the case of p-wave projection where the electric vector is parallel to the y-axis. The electric field E_y is given, by using eq. (3.15), as

$$E_y = \sum_{n=0}^{2N-1} \frac{E_0}{\lambda} \int_{x_n}^{x_n+d} \int_{-\theta_m}^{\theta_m} \exp(i\varphi) \, d\theta \, d\xi, \qquad (3.25)$$

where x_n is defined as

$$x_n = -(2N-1/2)d + 2nd. \qquad (3.26)$$

With increasing N, E_y approaches its limit as follows.

$$\lim_{N\to\infty} E_y = E_y^{(0)} + E_y^{(+1)} + E_y^{(-1)}, \qquad (3.27)$$

where

$$E_y^{(0)} = \frac{1}{2} E_0 \exp(ikz - \omega t), \qquad (3.28)$$

$$E_y^{(+1)} = -\frac{1}{\pi} E_0 \exp\left[i(k\sin\theta_d x + k\cos\theta_d z - \omega t)\right], \qquad (3.29)$$

$$E_y^{(-1)} = -\frac{1}{\pi} E_0 \exp\left[i(-k\sin\theta_d x + k\cos\theta_d z - \omega t)\right], \qquad (3.30)$$

and θ_d is defined as

$$d \sin\theta_d = \frac{\lambda}{2}. \qquad (3.31)$$

Equations (3.28)-(3.30) indicate that the image-forming waves converge into three plane waves as illustrated in Fig. 3.26, where the wave(+1), wave(0) and wave(-1) correspond to $E^{(+1)}{}_y$, $E^{(0)}{}_y$ and $E^{(-1)}{}_y$, respectively. These waves originate from the mask pattern as a result of diffraction by the pattern which acts as a grating. If θ_d exceeds θ_m, the wave(+1) and wave(-1) do not pass the lens system, and consequently, no line patterns are formed.

Secondly, we consider the case of the s-wave. The same consideration as described above leads to the conclusion that the image-forming waves converge into three plane waves expressed by

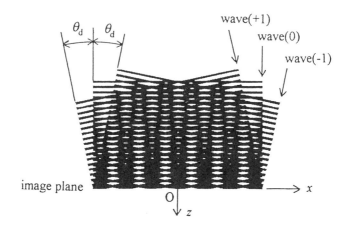

Fig. 3.28. Waves to form the image of line-and-space pattern.

$$E_x'^{(0)} = E_y^{(0)}, \tag{3.32}$$

$$E_x'^{(+1)} = \cos\theta_d E_y^{(+1)}, \tag{3.33}$$

$$E_z'^{(+1)} = -\sin\theta_d E_y^{(+1)}, \tag{3.34}$$

$$E_x'^{(-1)} = \cos\theta_d E_y^{(-1)}, \tag{3.35}$$

$$E_z'^{(-1)} = \sin\theta_d E_y^{(-1)}. \tag{3.36}$$

Here, all the other vector components are equal to zero.

Under the condition that the number of lines is very large, a relative light intensity distribution of line-and-space image, denoted by $I_4(x, z)$, is given by

$$I_4(x,z) = \frac{1}{E_0^2}\left|E_y^{(0)}+E_y^{(+1)}+E_y^{(-1)}\right|^2 + \frac{1}{E_0^2}\left|E_x'^{(0)}+E_x'^{(+1)}+E_x'^{(-1)}\right|^2$$

$$+ \frac{1}{E_0^2}\left|E_z'^{(+1)}+E_z'^{(-1)}\right|^2. \tag{3.37}$$

Optical Pattern Transfer

An example of the calculated result is shown in Fig. 3.29, where the light intensity in the vicinity of a two-wavelength-wide line-and-space image is plotted against a two-dimensional spatial position.

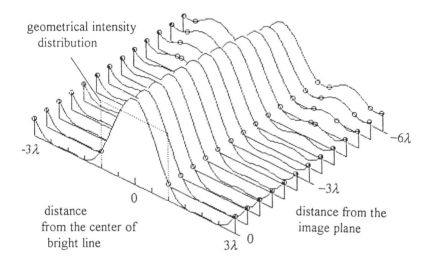

Fig. 3.29. Calculated light intensity distribution in the vicinity of the image of a two-wavelength-wide line-and-space pattern. The small circles indicate the positions where the intensity is equal to one-fourth of that in the case of flood exposure.

Figure 3.29 shows that if the exposure dose on a (supposed) fully illuminated area is kept to be four times the threshold dose, the resist pattern obtained by the development will be very close or identical to the geometrical image within some defocusing range. However, when the defocus exceeds a limit, false resist patterns will be formed on the dark line areas, as indicated in Fig. 3.29 by the small circles on the profile curves at the positions 4.5 to 6 wavelengths apart from the image plane. Therefore, in the present case, the depth of focus is confined not by dimensional deviation but by false pattern formation.

The resolution and depth of focus characteristics are shown in Fig. 3.30, where the minimum line width of the geometrical pattern, which is transferred to the resist pattern within the deviation range of ±10% without false pattern formation, is plotted against the absolute distance from the image plane.

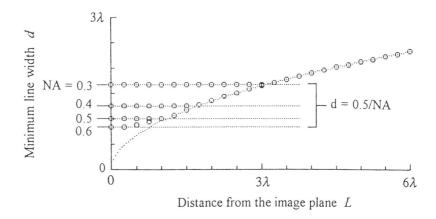

Fig. 3.30. Minimum line width of a geometrical two-wavelength-wide line-and-space image transferred to the resist pattern within the deviation range of ±10% by four times the threshold exposure dose as a function of the absolute distance from the image plane.

In Fig. 3.30, the horizontal dotted lines indicate the minimum line width d given by

$$d = 0.5\frac{\lambda}{NA}, \qquad (3.38)$$

and the parabolic dotted curve indicates the correlation between d and the absolute distance from the image plane L expressed as

$$d = 0.95\sqrt{\lambda L}. \qquad (3.39)$$

As the flat part of d-L correlation in Fig. 3.30 is well represented by the horizontal dotted line, eq. (3.38) is regarded as the Rayleigh's criterion for resolution in the present case. This relationship is also derived from the condition that the wave(+1) and wave(-1) pass the fringe of the lens system, that is, $\theta_d = \theta_m$.

The value of L at the crossing point of each of the horizontal lines with the parabolic line is equal to one-half of the depth of focus except for the case where NA = 0.6. This value of L is obtained by eliminating d from eqs. (3.38) and

(3.39). By using this value, the depth of focus (DOF) is expressed as

$$\text{DOF} = 0.55 \frac{\lambda}{\text{NA}^2}. \tag{3.40}$$

3.3.6 Spatial Coherency Factor

In the foregoing discussion, we have considered imaging under the condition that the mask is illuminated with a normally incident beam of light. This type of illumination is called spatially coherent illumination. Although this type of illumination is ideal for a precise pattern transfer, the loss of light is inevitable. Therefore, in a practical application, the mask is illuminated with uncollimated or spatially incoherent light as illustrated in Fig. 3.31.

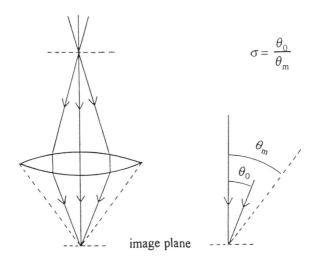

Fig. 3.31. Spatially incoherent illumination and the definition of spatial coherency factor σ.

Each point on the mask is illuminated with the rays of light confined within a cone. The spatial coherency factor σ is defined here as

$$\sigma = \frac{\theta_0}{\theta_m}, \tag{3.41}$$

where θ_0 is the maximum incidence angle of a geometrically traced illuminating ray shown in Fig. 3.31, and θ_m the angle of the cone formed by the lens aperture and the image point on the optical axis as shown in the figure, as well as in Fig. 3.20. The case of coherent illumination is expressed as $\sigma = 0$.

As illuminating rays of different directions are incoherent with each other, the light intensity in the vicinity of image is obtained by summing up the light intensity resulted from each ray. If the direction of the illuminating rays is expressed by direction cosines l, m, n, the range of direction is expressed by

$$\sqrt{l^2 + m^2} \leq \sin \theta_0, \qquad (3.42)$$

where θ_0 is the cone angle of illumination shown in Fig. 3.31. In the case where the direction is perpendicular to y-axis, expressed by $m = 0$, the light intensity in the vicinity of the image can be calculated in the same manner as described above. However, the calculation can not be applied to the cases where m is not equal to zero. Therefore, to take these cases into consideration approximately, we introduce a coefficient for the light intensity calculated by assuming that m is equal to zero. The coefficient is defined here to be proportional to the range of allowable m, that is $2(\sin^2 \theta_0 - l^2)^{1/2}$.

The effect of the spatial coherency factor σ on the resolution and the depth of focus is exemplified in Fig. 3.32, where the minimum line width of line-and-space pattern (which is transferred to resist pattern within the deviation range of ±10% by four times the threshold exposure dose), is calculated and plotted against the absolute distance from the image plane for various NA and σ values.

As seen in Fig. 3.32, the resolution decreases with increasing σ, while the depth of focus, indicated by the flat part of each curve, is little affected by the change of σ. By summarizing the calculated results shown in Fig. 3.32, the resolution represented by the minimum line width d is expressed as

$$d = 0.5 \frac{\lambda}{\text{NA}} + 0.57 \sigma \lambda. \qquad (3.43)$$

3.3.7 Phase Shifter Mask

The use of a phase shifter mask is very effective in improving the resolution of projection printing as already explained in Chapter 1. This method, reported by Levenson and others [2], applies a phase-inverting layer to one of the two adjacent bright areas on the mask to form a dark line between their projected images.

To calculate the light intensity in the vicinity of projected image, we consider

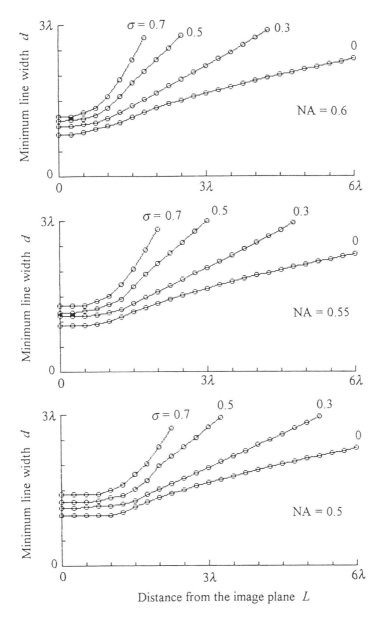

Fig. 3.32. Calculated minimum line width of line-and-space pattern transferred to the resist pattern within the deviation range of ±10% by four times the threshold exposure dose as a function of the absolute distance from the image plane.

the case of spatially coherent illumination where the mask is illuminated with a normally incident collimated beam of light. The same calculation procedure as described in 3.3.3 can be applied to the present case with a change in the phase of waves of light that forms the image of the area with the phase shifter.

An example of the calculated result is shown in Fig. 3.33, where the light intensity distribution in the vicinity of the image projected from a double slit mask pattern without a phase shifter is compared with that from the same pattern with a phase-shifter layer.

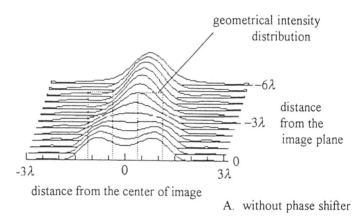

A. without phase shifter

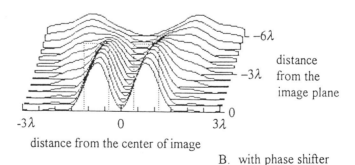

B. with phase shifter

Fig. 3.33. Calculated light intensity distribution in the vicinity of the image of double slit projected from the mask without a phase shifter (A) and with a phase shifter (B). The slits are 0.75-wavelength-wide and spaced by the same width.

As seen in Fig. 3.33, both the resolution and the focus latitude are significantly improved by the use of a phase shifter mask.

3.3.8 Focus-Latitude Enhancing Optical Filter

An ingenious method to enhance the focus latitude in projection printing has been proposed by Fukuda and others [3,4]. The method utilizes a specially designed optical filter on the pupil position of the lens system to modify the amplitude and phase of light waves. The principle of this method (in the case of spatially coherent illumination) is illustrated in Fig. 3.34.

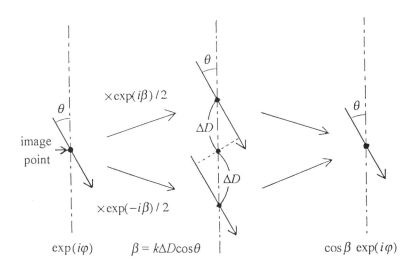

Fig. 3.34. Focus latitude enhancement by phase and amplitude modulation. The imaginary displacement is denoted by ΔD, and 2π times the wavenumber by k.

The image-forming plane wavelet is represented by an arrow with the incidence angle θ and oscillation $\exp(i\varphi)$ in Fig. 3.34. We suppose that the wave is divided into halves, and then their phases are shifted by β and $-\beta$ so that the image plane is displaced by ΔD and $-\Delta D$ to widen the depth of focus. As shown in Fig. 3.34, β is given by $\beta = k\Delta D\cos\theta$, where k is 2π times the wavenumber of light. The two phase-shifted waves are then superimposed into one wave. The resulting wave is expressed by $\cos\beta\exp(i\varphi)$. The amplitude of this obtained wave is $\cos\beta$. The phase of the wave is not changed by a positive value of $\cos\beta$, but inverted by a negative value of $\cos\beta$. If the filter causes the same changes in amplitude and phase corresponding to θ, an increase in the depth of focus, presumably as large as $2\Delta D$, can be expected. Such an increase in depth of focus is shown in Fig. 3.35, where the calculated light intensity distribution of the image of a bright line pattern projected by the conventional method is compared with that projected by using a filter with an optical character described above.

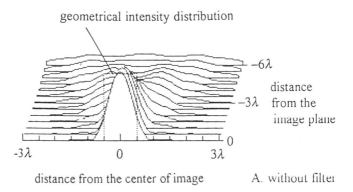

A. without filter

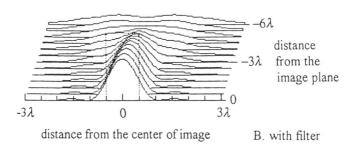

B. with filter

Fig. 3.35. Calculated light intensity in the vicinity of the image of a bright line pattern projected by the conventional method (A) and by using a focus-latitude enhancing filter (B). The geometrical image is one-wavelength wide. The NA of the lens is 0.6 and the image displacement (ΔD) is 2.1 times the wavelength of light.

As seen in Fig. 3.35, the use of the filter widens the focus latitude as expected, although the light intensity is decreased to some extent because of the absorption by the filter.

REFERENCES

1. S. Nonogaki and A. Imai, High resolution proximity exposure through a phase shifter mask, *Jpn. J. Appl. Phys.*, **32**, 4845-4849 (1993).
2. M. D. Levenson, N. S. Viswanathan, and R. A. Simpson, Improving resolution in photolithography with a phase-shifting mask, *IEEE Trans. Electron Devices*, **ED-29**, 1828-1836 (1982).
3. H. Fukuda, T. Terasawa, and S. Okazaki, Spatial filtering for depth of focus and resolution enhancement in optical lithography, *J. Vac. Sci. Technol.*, **B9**, 3113-3116 (1991).
4. H. Fukuda, Y. Kobayashi, K. Hama, T. Tawa, and S. Okazaki, Evaluation of pupil-filtering in high-numerical aperture i-line lens, *Jpn. J. Appl. Phys.* **32**, 5845-5849 (1993).

4

Chemistry of Photoresist Materials

4.1. INTRODUCTION

The aim of this chapter is to discuss the progress of photoresists in the past, present and future of lithography. It is worthwhile describing the history and the trend of lithography and resists. It can be recognized from Fig.4.1 [1] that the second turning point of resist materials will come soon. The first turning point was the replacement of a negative resist composed of cyclized rubber and a bisazide by a positive photoresist composed of a diazonaphthoquinone (DNQ) and a novolak resin. This was induced by changing the of exposure system from a contact printer to a g-line (436 nm) reduction projection step-and-repeat system, or the so-called stepper. The cyclized rubber system has poor resolution due to swelling during the development and has no sensitivity due to the lack of absorption at the g-line of the bisazide. On the other hand, the DNQ-based positive photoresist shows sensitivity at the g-line and high resolution capability. DNQ-novolak resist also renders resistance to dry-etching (plasma etching) which was begun in late 1970s.

Performance of the g-line stepper was improved by increasing NA (numerical aperture). Then a stepper using a shorter wavelength i-line (365 nm) was introduced. DNQ-novolak resist can still be used for the present workhorse of i-line lithography. Therefore, much effort has been made to improve the resolution capability of the DNQ-novolak resist as well as the depth-of-focus latitude. The effect of novolak resin and DNQ chemical structure on dissolution inhibition capability has been investigated mainly to get high dissolution contrast. The detail description will be given later in this chapter. The progress of this resist type and an i-line stepper is remarkable, achieving resolution below the exposure wavelength of the i-line ($0.365\mu m$). However, i-line lithography has difficulty in accomplishing 0.3 µm processes (64MDRAM), even using a high NA i-line stepper in conjunction with a DNQ-novolak resist.

Several competing lithographic technologies have been proposed for the future [2]: wavefront engineering of i-line, deep-UV lithography, and electron beam lithography. The wavefront engineering includes off-axis illumination (OAI), pupil filtering and phase-shifting lithography as discussed in Chapter 3. DNQ-novolak resists can be used for OAI and pupil filtering, although resists with

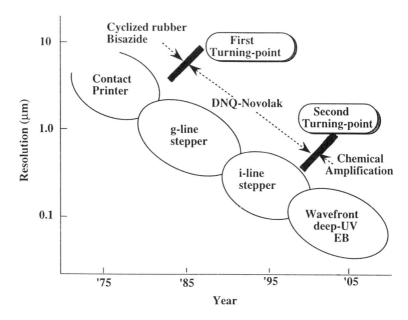

FIG. 4.1 Development trend of lithography and resists.

higher sensitivity are necessary. For phase-shifting lithography, the bridging of patterns at the end of the line-and-space patterns occurs when a positive resist is used. Therefore, negative resists with high sensitivity and high resolution are required. In deep-UV lithography positive and negative resists with high sensitivity, high resolution and high transmittance at the exposure wavelength are needed. Here, negative photoresists using azido compounds, DNQ-novolak positive photoresists, chemical amplification resists, and ArF excimer laser resists will be discussed.

4.2. PHOTOCHEMISTRY OF RESISTS

A molecule becomes energized to an excited state by the absorption of a photon of light and undergoes a chemical reaction [3]. Photo-induced chemical reactions which lead to a change in solubility in a developer are used in photoresists. The difference between photoreaction and ordinary thermal reactions lies in the fact that in photochemistry, individual molecules are promoted to excited states without immediate effect on surrounding molecules. In a thermally activated reaction, energy entering the system in the form of heat is distributed among all the molecules in the system according to statistical principles. It is a unique characteristic that selective excitation occurs on individual molecules in photochemistry. Therefore, the exposure to light through a mask can induce a photochemical reaction in a selected area. The subsequent development can form organic layer patterns which are used for underlying etching.

Optical exposure systems usually utilize a high-pressure mercury lamp and KrF excimer laser. As described in Chapter 2, a high-pressure mercury lamp emits a series of characteristic lines corresponding to the transition between excited states and lower states of mercury. A contact printer utilizes several emission lines from a mercury lamp to expose a resist film. On the other hand, several distinguished lines such as the g-line (436 nm) and the i-line (365 nm) are used for optical steppers. Therefore, photoresist systems should be designed for these wavelengths (photon energy). Since the late 1970s, g-line and i-line steppers have been used. Diazonaphthoquinone compounds are well suited for these wavelengths and widely used as photoactive compounds for these steppers.

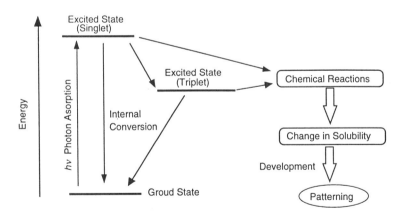

FIG. 4.2 Photochemistry for resists.

The initial step of photochemistry is absorption of photons. Absorption of light of these wavelengths excites electrons in the molecules to higher-energy orbitals producing electronically excited states (Fig.4.2). The degree of absorption is characterized by the absorption coefficient or extinction coefficient. The transmitted light intensity (I) passing through an absorbing media is described in terms of the Lambert-Beer absorption law as shown in Fig.4.3

$$I = I_0 10^{-\varepsilon c l} \quad (4.1)$$

where I_0 is the intensity of incident light, ε is molar extinction coefficient, c is the concentration of absorbing species and l is the sample thickness. The film thickness of a photoresist for lithographic process is about 1μm.

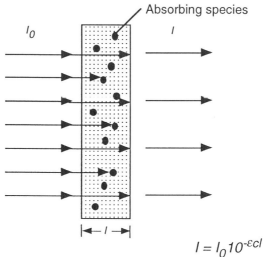

FIG. 4.3 Lambert-Beer law for light absorption.

The second step is a series of induced reactions from an excited state. It is necessary that the excitation energy is higher than a certain chemical bond energy of a molecule to induce bond scission, as this scission initiates subsequent chemical reactions. The yield of the photochemical reaction product, which changes the solubility in a developer, determines the sensitivity of a photoresist. The yield per one photon absorption is usually less than one. It implies that all excited states do not lead to the product desired. "Quantum yield", ϕ: the yield per one photon absorption, is defined in the following equation

$$\phi = \frac{k_p}{k_r + k_{nr} + k_p + k_{others}} \quad (4.2)$$

where k_p is the rate for the product desired, k_r is the rate of radiative transition, k_{nr} is the rate of nonradiative transition, and k_{others} is the rate for the other reactions. Since the absorption probability and yield for the product are proportional to ε and ϕ, it is generally accepted that the higher value of $\varepsilon\phi$ gives a higher sensitivity of photoresists as far as similar photochemical reactions are concerned.

Although high $\varepsilon\phi$ gives high sensitivity, too high absorbance (ε) in a resist film causes the high gradient of photoproduct along the resists film thickness. This gradient leads to the gradient of solubility for a developer, which sometimes deteriorates the resist profile. Bleaching character of photoactive compounds (increase in transmittance during exposure) makes less gradient of photoproduct. Optimization of molar extinction coefficient, high quantum yield and bleaching character are recommended for photoresist materials.

4.3 AZIDE-CYCLIZED RUBBER (POLYISOPRENE) PHOTORESISTS

Photosensitive systems composed of photoreactive aromatic azide compounds and a variety of host polymers have been well known since the 1930s [4]. Negative photoresists composing an azide and a cyclized cis-1,4-polyisoprene as a host polymer, such as Kodak's KTFR, had been widely used in the microelectronics industry. The most commonly used azide sensitizer for conventional near-UV lithography is 2,6-bis(4'-azidobenzal)-4-methylcyclohexanone (FIG.4.4).

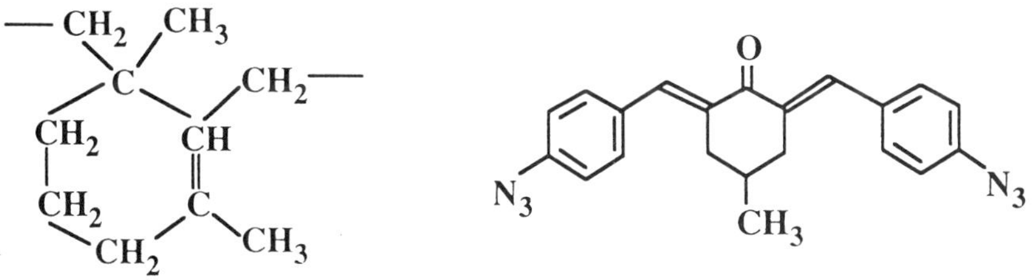

FIG. 4.4 Cyclized rubber and an azide for a contact printer.

Cyclized polyisoprene sensitized with an aromatic bisazide generates an insoluble, three-dimentional network via crosslinking upon irradiation. The photoinduced reactions associated with the network formation are shown in Fig.4.5. The primary event is the decomposition of the arylazide in the excited state into a reactive nitrene intermediate that can undergo a variety of reactions. The nitrene reactions include nitrene-nitrene coupling to form azo dyes, insertion into carbon-hydrogen bonds to form secondary amines, abstraction of hydrogen from the rubber backbone to form imino radical and a carbon radical that can subsequently undergo coupling reactions, and insertion into the double bond of polyisoprene to form a three-membered azirizine linkage.

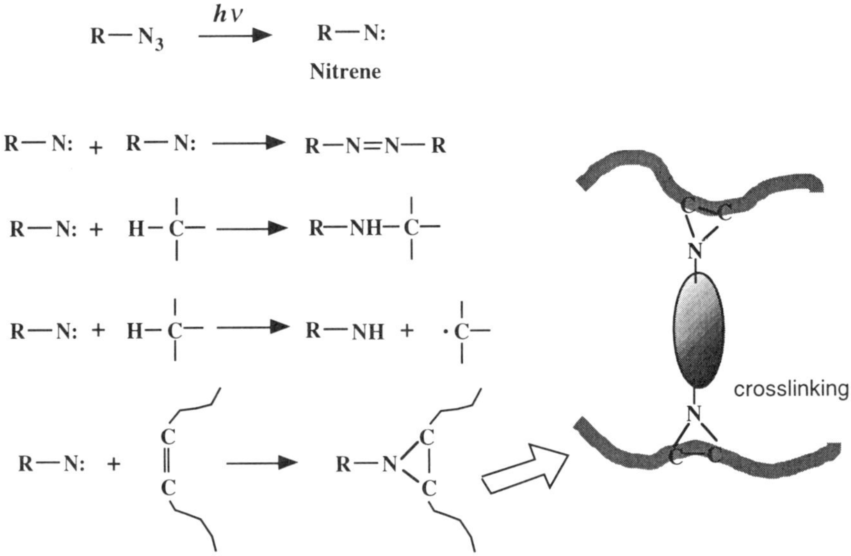

FIG. 4.5 Photochemistry of azide compounds.

As described above the photo-generated nitrene causes the reaction to insolubilize the exposed area. In the presence of oxygen, however, the nitrene predominantly reacts with oxygen. The reaction product does not lead to crosslinking of cyclized polyisoprene. When oxygen is exhausted by reaction with excess nitrene, a crosslinking reaction occurs. Therefore, the insolubilization is induced only in a high intensity area, where excess nitrenes are generated. In a low intensity area, diffused oxygen supplied from air deactivates

Chemistry of Photoresist Materials

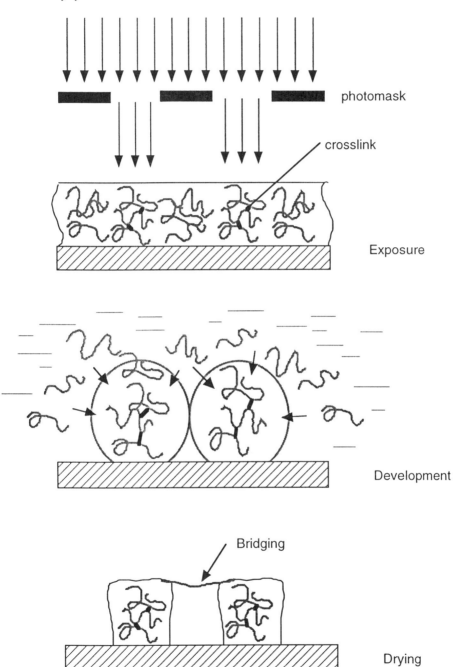

FIG. 4.6 Swelling behavior during development in an organic solvent.

the nitrene. This behavior is the so-called reciprocity law failure, which enhances resist contrast.

The cyclized rubber-bisazide formulations offer high sensitivity, ease of handling, and wide process latitude. However, the resolution of these systems is limited by relatively low contrast and swelling-induced deformation of resist patterns during development. Although the crosslinking reaction renders the polymer insoluble in the developer, the crosslinked polymer still has an affinity for the developer solvent. The crosslinked region absorbs solvent during development. The resulting increase in volume, or swelling, causes distortions of fine patterns in the form of "bridging" or "snaking" as shown in Fig.4.6. Therefore, it is difficult to use this type of resist for patterning below 1μm. This difficulty was one of the main reasons to change from a cyclized rubber resist to a DNQ-novolak resist.

4.4 AZIDE-PHENOLIC RESIN SYSTEMS

When cyclized-rubber resist was replaced by a DNQ-novolak reesist, it was said that the positive resist showed high resolution and the negative resist showed poor resolution. To get high resolution of negative resists, Iwayanagi et al. [5] proposed a nonswelling DUV resist. This resist, known as MRS (microresist for shorter wavelengths), consists of poly(p-vinylphenol), an alkali-soluble phenolic resin, and an azide as shown in Fig.4.7.

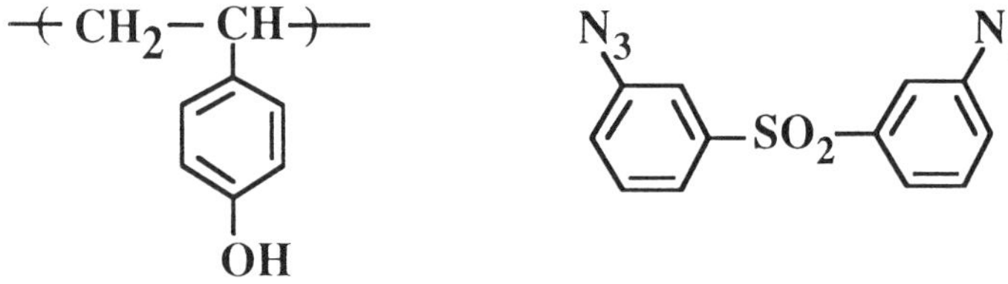

FIG. 4.7 Micro resist for shorter wavelength (MRS) composed of a phenolic resin and an azide for a deep-UV resist.

Dissolution inhibition exhibits in the exposed area, and the unexposed portion of the resist film is dissolved in an aqueous alkali solution in an etching-type dissolution that is devoid of swelling. Since the combined absorption of the azide and the resin is quite intense in the DUV region, the photochemical reaction leading to the decrease in solubility occurs mainly in the upper regions of the film. The insolubilized surface layer of the exposed area acts as a mask while the alkaline developer removes the unexposed area in an etching-type development process. Consequently, resist profiles change from overcut with increasing development time (Fig.4.8).

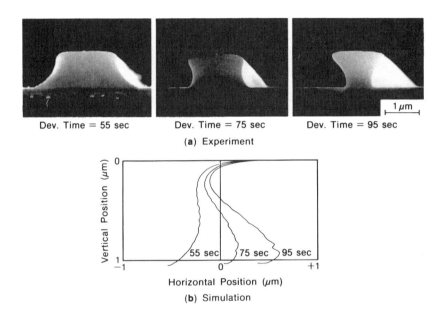

FIG. 4.8 Crosssectional view of MRS resist as a function of development time. [Reprint with permission from *IEEE Electron Device Lett.*, **EDL-3**, 58(1982).]

The decrease in dissolution upon exposure in this type of resist was first ascribed to the formation of a secondary amine generated from nitrene insertion into C-H bonds of the polymer. However, gel permeation chromatograhic analyses revealed that the molecular weight of poly(p-vinylphenol) increased upon irradiation in the presence of the azide [6]. Hydrogen abstraction from the polymer by nitrene and the subsequent polymer radical recombination resulted in an increase in the molecular weight of the polymer, rendering the exposed area less soluble in an aqueous base as shown in Fig. 4.9.

N3
N3
SO2
hν
N:
NH
CH2
CH
C
OH
OH
OH
OH
OH
increase in Mw

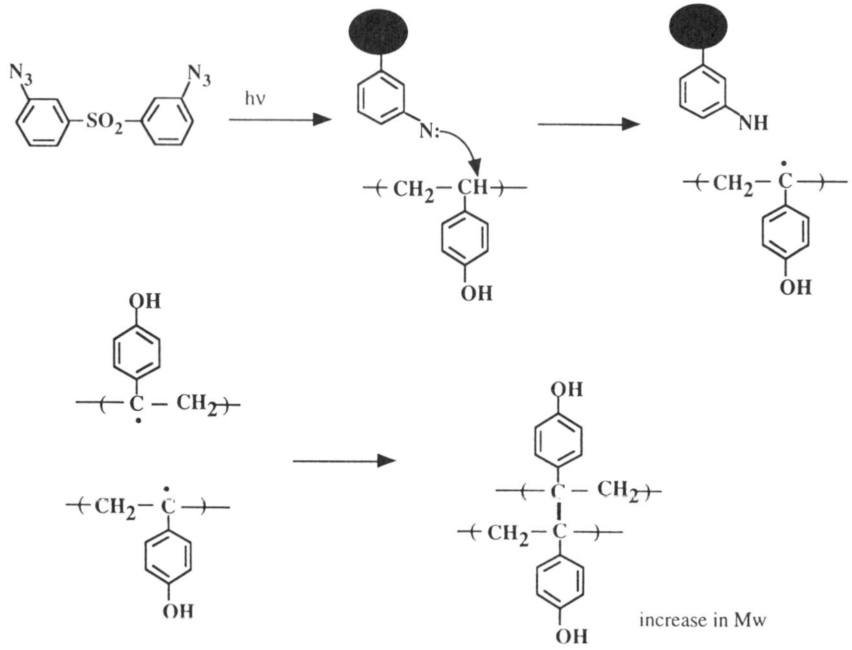

FIG.4.9 Reaction mechanism of MRS composed of a phenolic resin and an azide under exposure to deep-UV. Increase in molecular weight results in decrease in dissolution in aqueous base.

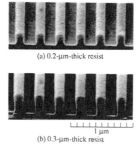

(a) 0.2-μm-thick resist
1 μm
(b) 0.3-μm-thick resist

FIG. 4.10 Line and space patterns of 0.16 μm obtained by KrF phase-shifting lithography. Exposure dose was 81 mJ/cm². [Reprint with permission from *J. Photopolym. Sci. Technol.*, **7**, 23(1994).]

As described in the introduction, a positive photoresist composed of DNQ and a novolak resin cannot be used for phase-shifting lithography, which requires negative resists with high resolution. Therefore, the resist composed of a phenolic resin and an azide has been evaluated for a resist for i-line phase-shifiting lithography. Among the azide compounds evaluated, 3,3'-dimethoxy-

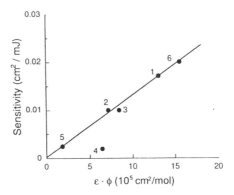

FIG. 4.11 Relationship between $\varepsilon\phi$ and resist sensitivity for various azide compounds, where ε is molar extinction coefficient and ϕ is quantum yield of azide photodecomposition. The resist sensitivity was defined as the reciprocal dose of the minimum dose, which starts to remain the resist film in resist sensitivity curves. [Reprint with permission from *Jpn. J. Appl. Phys.*, **31**, 4307(1992).]

Table 4.1 Azide Compounds

Structure	Exposure
bis(3-azidophenyl) sulfone; 3,3'-dimethoxy-4,4'-diazidobiphenyl	deep-UV
3,3'-dimethoxy-4,4'-diazidobiphenyl; 4-azidochalcone; 4-azido-4'-methoxychalcone; 2-(4-azidobenzylidene)-4,4-dimethylcyclohexanone	i-line
2-(4-azidocinnamylidene)-4,4-dimethylcyclohexanone	g-line

Chemistry of Photoresist Materials

4,4'-diazidobiphenyl gives high transmittance at i-line and acceptable transmittance at 248 nm (KrF laser) [7]. The resist showed a 0.16 μm line-and-space resolution combined with KrF phase-shifting lithography as shown in Fig.4.10. Systematic evaluation of bisazides of biphenyl derivatives for i-line showed a correlation between sensitivity and $\varepsilon\phi$, where ε is the molar extinction coefficient and ϕ is the quantum yield of a bisazide photodecomposition as shown in Fig.4.11. It is a good example of the relation between sensitivity and $\varepsilon\phi$. Azides developed for each optical lithography are summarized in Table 4.1.

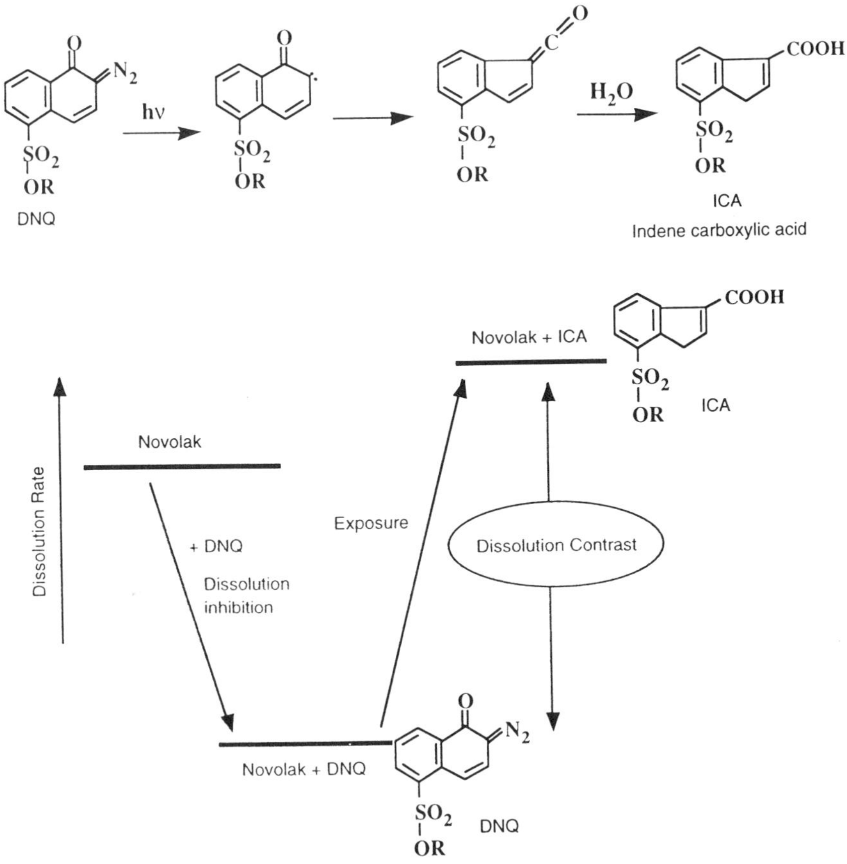

FIG. 4.12 Photochemistry of diazonaphthoquinone (DNQ).

4.5 DIAZONAPHTHOQUINONE (DNQ)-NOVOLAK POSITIVE PHOTORESISTS

The positive photoresist composed of DNQ and a novolak resin is the workhorse for semiconductor fabrication [8]. It is surprising that the advanced i-line positive photoresist of this type shows the resolution less than the exposure wavelength of i-line(0.365 μm) using i-line stepper with high NA. In this section the photochemical reaction of DNQ and improvement of the resist performance by newly designed novolak resins and DNQ inhibitors are discussed.

4.5.1 Photochemistry of DNQ

The Wolff rearrangement reaction mechanism of DNQ was proposed by Süs [9] in 1944. Upon irradiation to light, nitrogen is released to form an intermediate ketocarben species. The ketocarbene leads to ketene via the Wölff rearrangement.

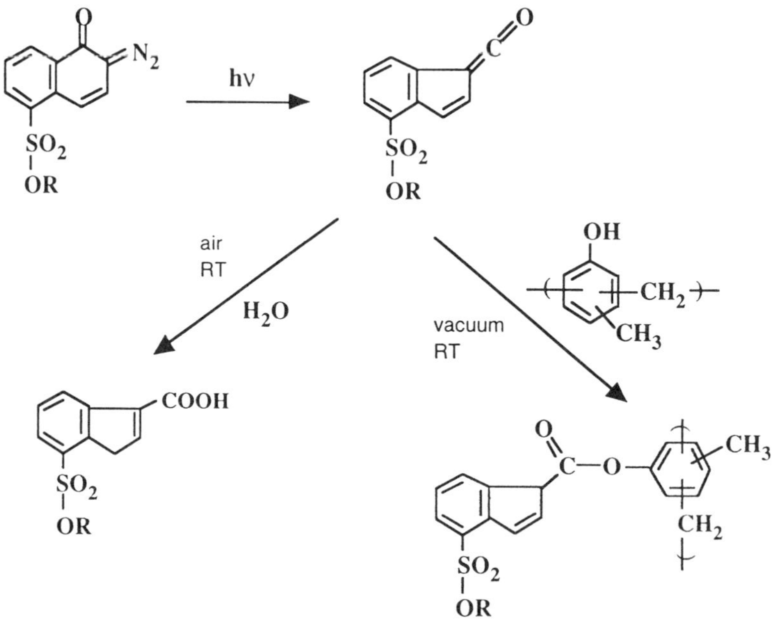

FIG. 4.13 UV induced reaction pathways for a DNQ compound in a novolak resin [10].

The ketene reacts with water to produce indenecarboxylic acid. The basic reaction was already established a half century ago, although Süs [9] suggested the chemical structure of the final product 1-indenecarboxylic acid, which was corrected to 3-indenecarboxylic acid [10] (Fig. 4.12). Packansky has reported that the reactivity of ketene depends on the conditions as shown in Fig. 4.13. Under ambient conditions, ketene reacts with water trapped in the novolak resin to yield 3-indenecarboxylic acid as described above. However, UV exposure in vacuum results in ester formation via ketene-phenolic OH reaction [10].

Many attempts to detect the intermediates in photochemistry of DNQ by using time-resolved spectroscopy have been reported [11-15]. It is still controversial whether ketocarbene is a reaction intermediate [16]. Since most of the attempts to detect the intermediates were performed in solution, further studies are needed to confirm the reaction intermediates in the resist film. Sheats described reciprocal failure, intensity dependence on sensitivity in DNQ-novolak resists with 364 nm exposure [17], which is postulated to involve the time-dependent absorbance of the intermediate ketene.

4.5.2 Improvement in Photoresist Performance

(1) Novolak resins

Even if the same DNQ compound is used, the sensitivity of the DNQ-novolak resist depends on the kind of novolak resin. Since the yield indene carboxylic acid is the same for the same amount of DNQ addition and the same exposure dose, the difference in sensitivity implies a difference in dissolution behavior in the developer depending on the matrix polymer.

A group at Sumitomo Chemical has made a systematic study on novolak resins to improve the lithographic performance of the positive resists [18]. It is difficult to satisfy all the requirements for high sensitivity, high resolution, high film thickness remaining after development and high heat resistance. Novolak resins having a molecular structure and a molecular weight different from existing materials have been designed and synthesized. They investigated the relation between lithographic performance and the characteristics of novolak resins such as the isomeric structure of cresol, the position of the methylene bond, the molecular weight and molecular weight distribution (Fig. 4.14). To clarify the lithographic performance, the dissolution rates of the unexposed and exposed resist were measured. It should be noted that dissolution rate can be defined in this type of resist, while the dissolution rate cannot be defined in

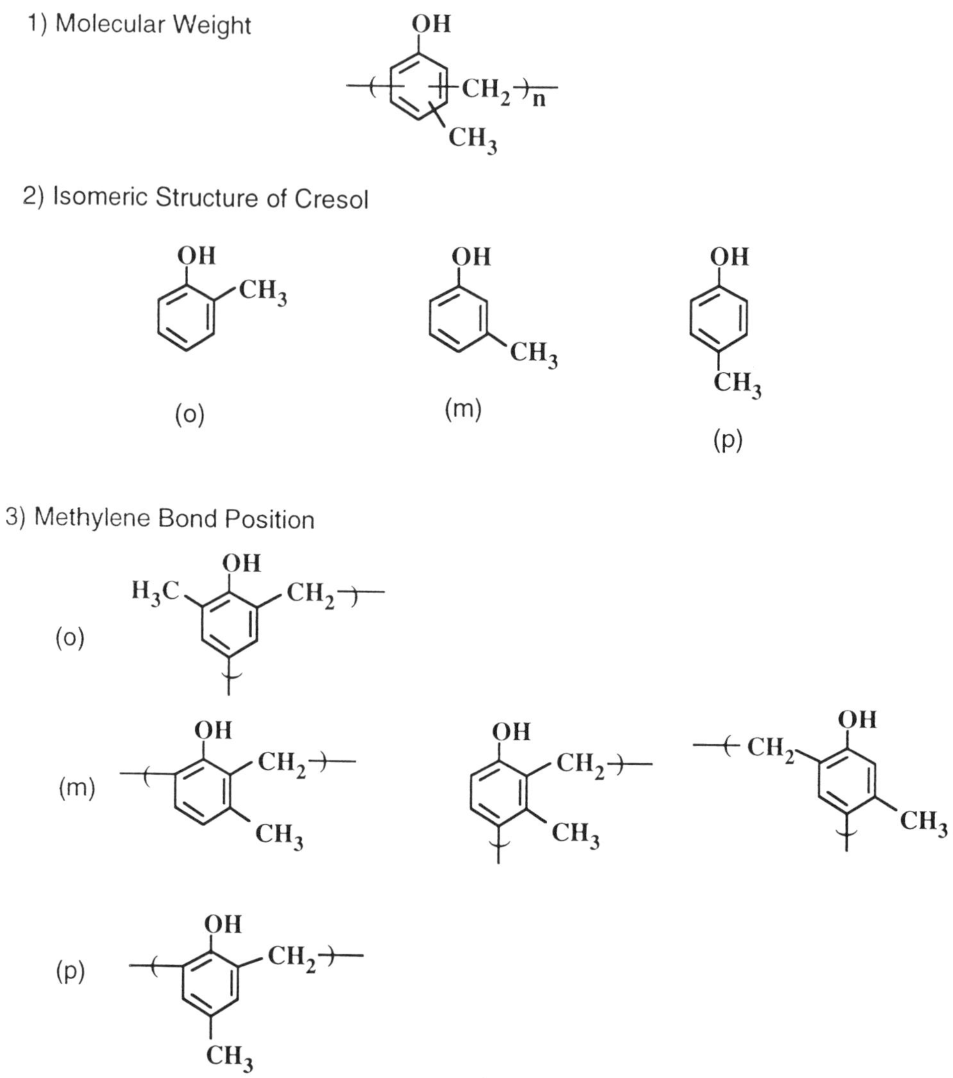

1) Molecular Weight

2) Isomeric Structure of Cresol

(o)   (m)   (p)

3) Methylene Bond Position

(o)

(m)

(p)

FIG. 4.14 Factors of novolak resins which influence resist characteristics.

polystyrene derivatives and cyclized rubber systems due to swelling.

Dissolution of polymers in a developer is schematically shown in Fig. 4.15. Developer molecules penetrate into the polymer matrix to form a penetration zone, which is a gel layer. Since the gel layer is a swelled polymer layer, the thickness of gel layer describes the degree of swelling. It is considered that the

Chemistry of Photoresist Materials

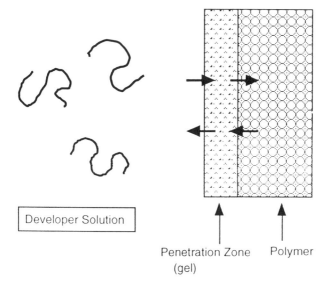

FIG. 4.15 Schematic representation during development.

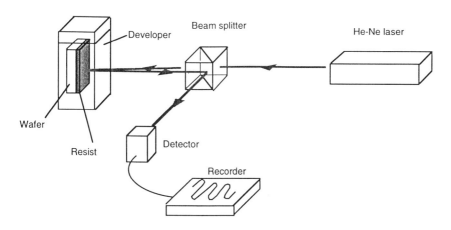

FIG. 4.16 An apparatus for dissolution rate measurement.

gel layer of a novolak resin in a aqueous base developer is negligibly small, which shows no evidence of swelling during the development. One can recognize the interference color change during the dissolution of a novolak film in an aqueous base, which indicates that the optical flat surface of the film is maintained during the dissolution to interfere the reflected light from the resist surface with that from the substrate. When a gel layer is negligible, one can measure the dissolution rate during the development. One typical apparatus for measuring the dissolution rate is shown in Fig. 4.16.

The most remarkable effect of a novolak resin on lithographic performance is the methylene bond position in the novolak resin. Fig. 4.17 shows the dependence of dissolution rate on the S_4 value of meta-cresol novolak resin. S_4 represents the ratio of "unsubstituted carbon-4 in benzene ring of cresol to carbon-5", which indicates a fraction of ortho-ortho methylene bonding (high ortho bonding) [18]. With increasing S_4 value, the content of type (B) structure increases in novolak resin. The dissolution rate of unexposed resist (R_0) shows a drastic decrease with increasing S_4 value as shown in Fig. 4.17, while the

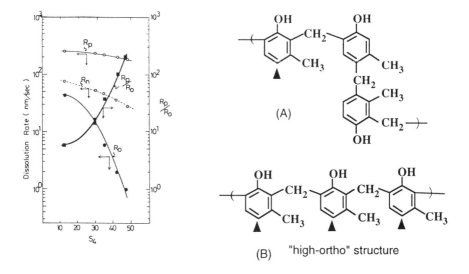

FIG. 4.17 Effect of content of "unsubstituted carbon-4" in benzene ring of cresol, S_4, in novolak resins synthesized from m-cresol on dissolution rates. The molecular weight of these novolak resins is almost the same. R_n is the dissolution rate of novolak resin, R_0 is the dissolution rate of unexposed resist, R_p is the dissolution rate of the exposed area. [Reprint with permission from *J. Vac. Sci. Technol.*, **B7**, 640(1989).]

dissolution rate of novolak resin (R_n) decreases slightly. It should be noted that the dissolution rate of the exposed area (R_p) remains constant for various S_4 values. Therefore, a high contrast resist is obtained without sensitivity loss using a novolak resin with high S_4 value. They explained this difference by an azo-coupling of the novolak resin with diazonaphthoquinone via base-catalytic reaction during development (Fig. 4.18). High-ortho (high S_4) novolak has more vacant para positions compared with a normal novolak resin, and these vacant positions enhance the electrophilic azo-coupling reaction.

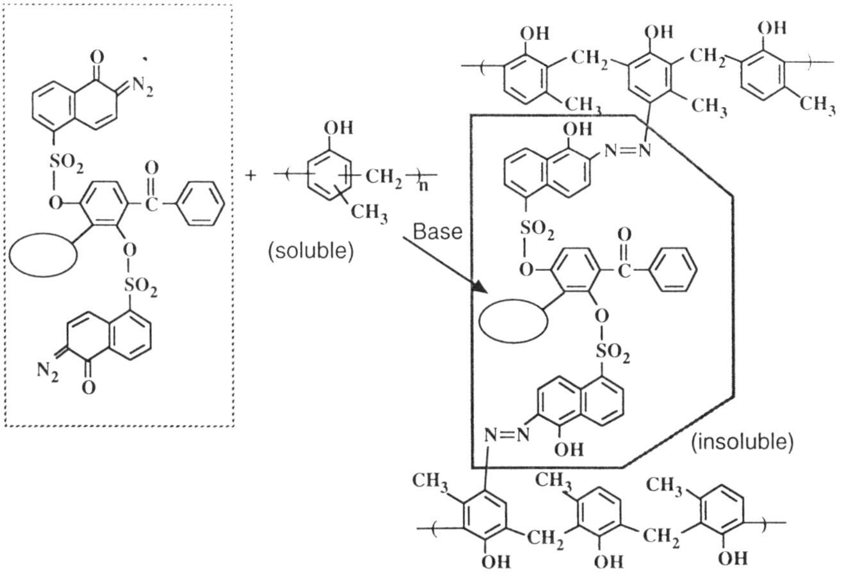

FIG. 4.18 Increase in molecular weight due to base catalyzed azo-coupling reaction of diazonaphthoquinone with a novolak resin as observed in GPC trace. [Reprint with permission from *J. Vac. Sci. Technol.*, **B7**, 640(1989).]

The effect of the molecular weight distribution of novolak resin on resist performance is shown in Fig.4.19 [18]. The dissolution rate of novolak resin increases with increasing molecular weight distribution (M_w/M_n). The

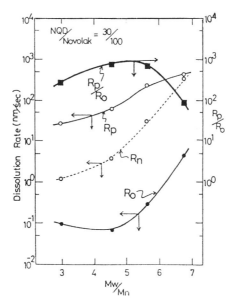

FIG. 4.19 Effect of molecular weight distribution of novolak resins on dissolution rate. The molecular weight of these novolak resins is almost the same. [Reprint with permission from *J. Vac. Sci. Technol.*, **B7**, 640(1989).]

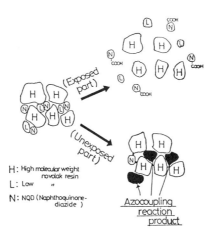

FIG. 4.20 "Stone wall" model for development of a positive photoresist. [Reprint with permission from *J. Vac. Sci. Technol.*, **B7**, 640(1989).]

discrimination between exposed and unexposed area is large at a certain M_w/M_n value, indicating that the optimum molecular weight distribution gives a high contrast resist.

On the basis of their systematic studies on novolak resins, Hanabata et al.[18] proposed the "stone-wall" model for positive photoresist with alkali development as shown in Fig. 4.20. In exposed parts, indenecarboxylic acid formed by exposure and low molecular weight novolak resin dissolves first into the developer. This increases the surface contact area of the high molecular weight novolak with the developer, leading to the enhancement of dissolution. In unexposed areas, an azocoupling reaction of low molecular weight novolak resin with DNQ retards the dissolution of low molecular weight resin. This stone wall model gave clues to design a high-performance positive photoresist in their following works.

The Sumitomo group also proposed a "tandem type novolak resin" which contains low-molecular weight (150-500) and high molecular weight novolak (>5000) as shown in Fig. 4.21 [19]. The advantage of the tandem type novolak

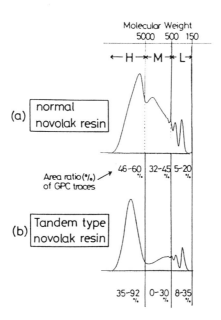

FIG. 4.21 Gel permeation chromatography traces of a normal novolak and "tandem type" novolak resin. [Reprint with permission from *Proc. SPIE*, **1466**, 132(1991).]

resins can be explained again by the stone-wall model. The low-molecular-weight novolaks and DNQ molecules are stacked between high molecular weight novolaks. In exposed areas, dissolution of indenecarboxylic acid and low-molecular-weight novolaks promote dissolution of high-molecular-weight novolaks due to an increase in surface contact to the developer. In unexposed areas, the azo-coupling reaction of DNQ compounds with low-molecular-weight novolaks retards the dissolution.

Studies of the effects of novolak molecular structures on resist performance have also been reported by several groups. Kajita et al. [20] of JSR investigated the effect of novolak structure on dissolution inhibition by DNQ. They found that the dissolution inhibition effect depends upon the structure of the novolak resin and that interaction between the naphthalene moiety and the novolak resin is important. They proposed a "host-guest complex" composed of a DNQ moiety and a cavity or channel formed with aggregation of several *ortho-ortho* linked units as shown in Fig.4.22, where the complex is formed via electrostatic interaction. The size of cavity is fit to that of the DNQ moiety.

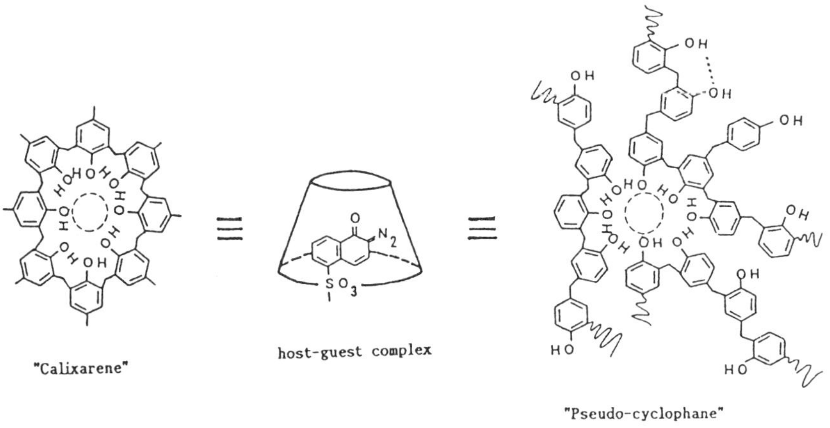

FIG. 4.22 "Host-guest complex" model for dissolution inhibition of a positive photoresist. [Reprint with permission from *Proc. SPIE*, **1466**, 161(1991).]

Honda et al. [21] proposed the dissolution inhibition mechanism called the "octopus-pot" model (FIG. 4.23) of novolak-PAC interaction and the relationship between novolak microstructure and DNQ inhibitor. The addition of DNQ to a novolak resin caused the OH band shift to a higher frequency (blue shift) in IR spectra, which suggests a disruption of the novolak hydrogen bonding by the inhibitor and concomitant hydrogen bonding with the inhibitor. The magnitude of the blue shift increases monotonously with the dissolution inhibition capability. The magnitude of the blue shift for *p*-cresol trimer was found to be dependent on the DNQ concentration and goes through a maximum at a mole ratio (*p*-cresol to DNQ) of 18. This suggests that a complex formation involving six units of p-cresol trimer and one molecule of DNQ is formed, probably through intermolecular hydrogen bonding leading to the "octopus-pot" model. These results suggest that dissolution inhibition in unexposed area is key the obtaining a high contrast resist.

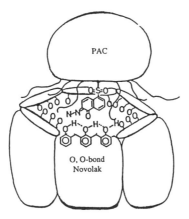

FIG. 4.23 Schematic of the "octopus-pot" model of a macromolecular complex of ortho-ortho bonded novolak microstructure with DNQ-PAC. [Reprint with permission from *Proc. SPIE*, **1262**, 493(1990).]

(2) DNQ Compounds

The chemical structure of DNQ inhibitors is another important factor which determines resist performance. The effect should be investigated in correlation with novolak structures. DNQ compounds are usually synthesized by the esterification reaction of phenol compounds with DNQ sulfonyl chloride. It is considered that DNQ-PAC (photoactive compound) derived from 2,3,4,4'-tetrahydroxybenzophenone is used for a g-line positive photoresist. Many DNQ-PACs have been associated with the change from g-line lithography to i-line lithography. When a positive photoresist for the g-line is used for i-line lithography, it suffers from the sloped side wall profile (low resolution) due to the absorbance of photoproducts of the benzophenone backbone structure. To avoid the absorbance of the backbone benzophenone structure, many DNQ-PACs have been synthesized.

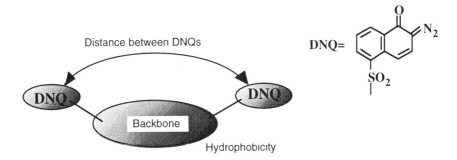

FIG. 4.24 Factors of DNQ-PAC which influence the resist performance.

Polyphotolysis model [22] for DNQ resists also stimulated study on PAC molecules. The model suggests that more DNQ groups in a single PAC molecule improves the resist contrast. Some results support the model, while some does not. It is not simply that the number of DNQ in a single molecule dictates the contrast. Based on the detail investigation of DNQ-PAC on dissolution characteristics, the distance between DNQ moieties in the PAC, the degree of dispersion of DNQ moieties in a resist film, and the hydrophobicity of a PAC are important (Fig. 4.24)[23,24].

The remained OH in PAC, which is selectively hindered for the esterification reaction, improve the solubilty and scum problems [25,26]. Several PACs are shown in Fig. 4.25. The OH in the figure can not be esterified due to steric hindrance. The OH group also promotes the dissolution in the exposed area, leading to high sensitivity.

The resolution capability of the DNQ-novolak resist has been remarkably improved for actual use in production. The resolution of the advanced resist has been reported to be 0.26 μm. However, this resist cannot be used for deep-UV lithography due to the high absorbance of a novolak resin and the high absorbance and non-bleaching DNQ-PAC at 248 nm and in the deep-UV region.

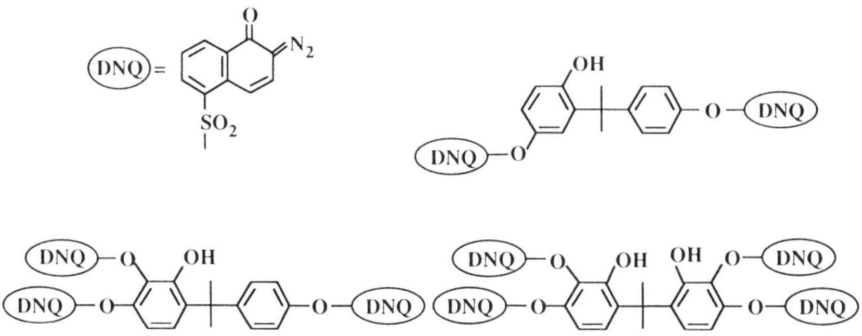

FIG. 4.25 DNQ-PACs with steric hindrance OH groups obtained by selective esterification of OH groups with DNQ-sulfonyl chlorides.

4.5.3 Dill's Model for Optical Characteristics of a Positive Photoresist

Dill's model is well-known for the analysis of optical properties of diazonaphthoquinone (DNQ)-novolak photoresist film [66]. Since this model has become the basis for the simulation of pattern formation in photolithography, it is of value to describe the formulations here. With light passing through the resist in the condition that the reflection of light can be neglected, one can use Lambert-Beer law to describe the optical absorption,

$$\frac{dI}{dx} = -I \Sigma a_i m_i \quad (4.3),$$

where I is light intensity, x the distance from the resist-air interface, m_i the molar concentration of the ith component, and a_i the molar absorption coefficient of the ith component.

For a positive photoresist, one needs to consider three absorbing species: a DNQ compound, a novolak resin, and the photochemical reaction product. Exposure with light converts the DNQ compound to reaction product of indenecarboxylic acid, reducing the total absorption of the film (bleaching). This change can be used to characterize the concentration of DNQ compound in the positive photoresist.

For a positive resist, the changes of light intensity and the concentration of the DNQ compound in a resist film can be described in the following equations,

$$\frac{\partial I(x,t)}{\partial x} = -I(x,t)[AM(x,t) + B] \qquad (4.4)$$

$$\frac{\partial M(x,t)}{\partial t} = -I(x,t) M(x,t) C \qquad (4.5)$$

where $I(x,t)$ is the light intensity at any depth (x) in the film and exposure time (t). $M(x,t)$ is DNQ concentration, normalized by initial concentration, at any depth (x) and exposure time (t). A corresponds to the difference of the absorbance between the initial and fully exposed resist film for 1μm thickness, B is the absorbance of the fully exposed film, and C is the fractional decay rate of DNQ

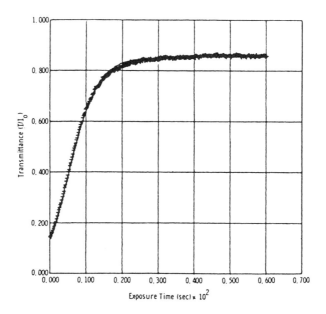

FIG. 4.26 Change of transmittance of a positive photoresist composed of diazonaphthoquinone (DNQ) with exposure time. [Reprint with permission from *IEEE Trans. Electron Devices*, **ED-22**, 445(1975).]

compound per unit light intensity. A, B, and C can be described by the following equations

$$A = (1/d) \ln[T(\infty)/T(0)] \qquad (4.6)$$

$$B = -(1/d)\ln T(\infty) \qquad (4.7)$$

$$C = \frac{(A+B)}{AI_0 T(0)\{1-T(0)\}} \cdot \frac{dT(0)}{dx} \qquad (4.8),$$

where $T(0)$ and $T(\infty)$ are the transmittance of initial and fully exposed film, respectively, d is the film thickness, and I_0 is the initial light intensity. The typical transmittance change with exposure time is shown in Fig. 4.26. A uniform monochromatic light beam is used to expose the resist film and the light intensity transmitted through the resist film and the substrate was measured using the detector.

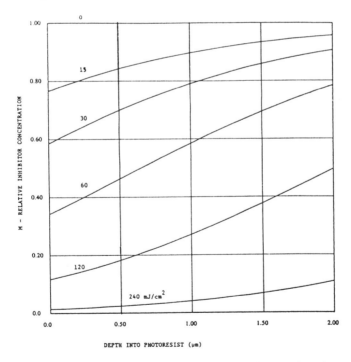

FIG. 4.27 Relation of normalized DNQ compound as a function of film thickness with increase in exposure time. [Reprint with permission from *IEEE Trans. Electron Devices*, **ED-22**, 445(1975).]

Once the values for A, B, and C are determined, the concentration distribution of DNQ compound in a resist film can be calculated. FIG.4.27 shows the normalized inhibitor concentration as a function of distance from resist-air substrate for different exposure energies. It is noted that the concentration of the DNQ compound changes along the film thickness once the exposure begins. With relatively thick films, a considerable range of inhibitor concentration distribution may be obtained.

Measurement of the dissolution rate of such a film as a function of depth into the film is necessary to characterize the development for that concentration range of DNQ compound. The development dissolution rate can be measured with in-situ resist thickness measurements, during the development, as described in Fig. 4.16. These analysis and measurements allow the determination of the relation of dissolution rate and M: relative inhibitor concentration. These relations are very important for the simulation of resist profiles for exposure systems.

4.6 CHEMICAL AMPLIFICATION RESIST SYSTEMS

No one denies that the lithography for the next generation is deep-UV lithography (among the several competing candidates) to obtain resolution below 0.30 μm. DNQ-based positive photoresists are not suitable for deep-UV lithography, since absorption of both novolak resins and DNQ-PACs is high and DNQ-PACs do not bleach at around 250 nm, resulting in resist profiles with severely sloping side walls. The light intensity at the wafer plane of deep-UV exposure tools (KrF excimer laser stepper and Micrascan step-and-scan system) is lower than that of conventional i-line steppers, which requires high sensitivity resists. Much attention has been focused on chemical amplification resist systems, especially for deep-UV lithography. These resists are advantageous because of high sensitivity and high transmittance in the deep-UV region. A chemical amplification system can also be used as electron beam and x-ray resists, which are strongly required to be highly sensitive.

In chemical amplified resist systems, a single photoevent initiates a cascade of subsequent chemical reactions. The resists are generally composed of an acid generator which produces acid upon exposure to radiation and acid-labile compounds or polymers which change the solubility in the developer by acid catalyzed reactions. As shown in Fig. 4.28, the photogenerated acid catalyzes the chemical reactions, which change the solubility in a developer. The change from insoluble to soluble is shown. The quantum yield for an acid catalyzed reaction is the product of the quantum efficiency of acid generation multiplied by the catalytic chain length. In chemically amplified resists, acid generation is the only photochemical event. Therefore, it is possible to design the acid-labile base polymer with high transmittance at deep-UV region. Since a small amount of acid can induce many chemical events, it is expected that yields of an acid catalyzed reaction along the film thickness can be alleviated even for concentration gradient of photogenerated acid along the film thickness. Another important aspect of chemically amplified resist systems is drastic polarity change by an acid catalyzed reaction, which can avoid the swelling during the development and give high contrast.

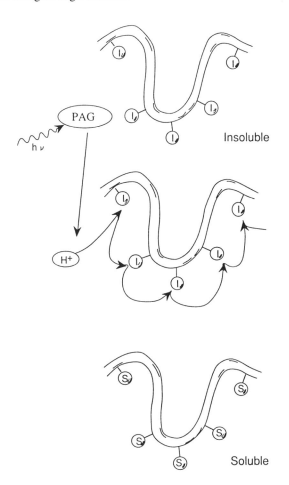

FIG. 4.28 Concept of a positive chemical amplification resist.

Many chemical amplification resist systems have been proposed since the report by the IBM group [27,28]. Most of these are based on acid-catalyzed reactions [29-34], though some with base-catalyzed reactions have been reported. Here, acid generators and acid-catalyzed reactions are discussed.

4.6.1 Acid Generators (Table 4.2)

4.6.1.1 Onium Salts

Most well-known acid generators are onium salts such as iodonium and sulfonium salts, which were invented by Crivello [35]. The photochemistry of diaryliodonium and triarylsulfonium salts has been studied in detail by Dektar and Hacker [36,37]. The products formed upon irradiation of diphenyliodonium salts and triphenyl sulfonium salts are shown in Fig. 4.29. Product formation from onium salts is complicated by in-cage reactions, cage-escape reactions and termination reactions. The efficiency of acid generation from onium salts in a polymer matrix depends on the polymer structure [38,39]. The sensitization of

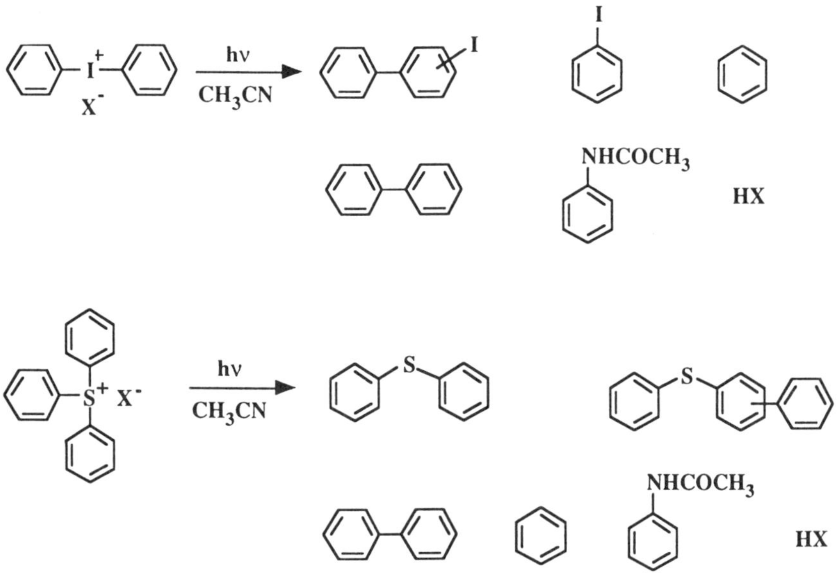

FIG. 4.29 Photoproducts from direct irradiation of diphenyliodonium salt and triphenylsulfonium salt.

Table 4.2 Acid Generators

Onium salts	$Ph_2I^+OTf^-$ [35]	$Ph_3S^+OTf^-$ [35]
Halogen compounds	XH_2CXHCH_2C–N(C=O)N(CH_2CHXCH_2X)C(=O)N(CH_2CHXCH_2X)C=O [41]	X_3C–(triazine)–CX_3, Ar-substituted [40]
Sulfonates	o-NO_2-C_6H_4-CH_2OSO_2Ar [42]; 1,2,3-tris(OSO_2CH_3)benzene [45]; 2-diazo-1-oxo-naphthalene-4-SO_2OR [48]; Ph-C(=O)-CH(CH_3)-OSO_2R [50]	p-NO_2-C_6H_4-CH_2OSO_2Ar [43]; phthalimide-N-OSO_3R [52]; Ph-C(=O)-C(OH)(Ph)-CH_2OSO_2R [49]; $Ar-SO_3-N=C(R')(R'')$ [51]
Sulfonyl compounds	$Ar-SO_2-C(=N_2)-SO_2-Ar$ [53]	$Ar-SO_2-SO_2-Ar$ [54]

onium salts by excited poly(t-butoxycarbonyloxy-styrene) (t-BOC-PHS) has been reported [38].

4.6.1.2 Halogen Compounds

Halogen compounds can be used as acid generators in chemical amplification systems [40]. These compounds, such as tricholoromethyl-s-triazene, have been known as free radical initiators for photopolymerization [29,40]. When exposed to halogen compounds with light, homolytic cleavage of the carbon-halogen bond produces a halogen atom radical followed by hydrogen abstraction, resulting in the formation of hydrogen halide acid, as shown in Fig. 4.30.

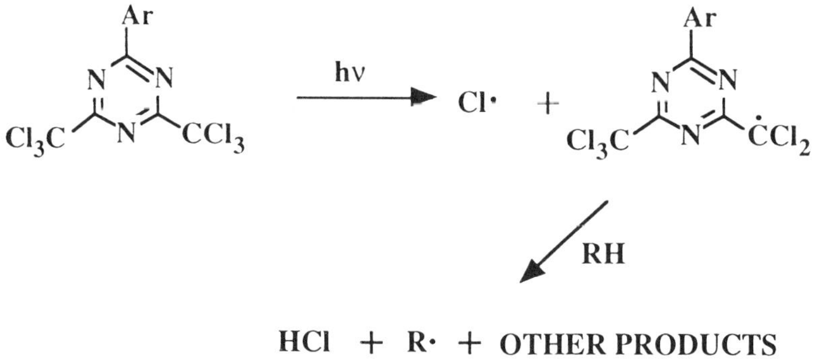

$$\text{Ar-triazine}(CCl_3)_2 \xrightarrow{h\nu} Cl\cdot + \text{Ar-triazine}(CCl_3)(\dot{C}Cl_2) \xrightarrow{RH}$$

HCl + R· + OTHER PRODUCTS

FIG. 4.30 Mechanism of acid formation from trichloromethyltriazene.

Calbrese et al. [41] proposed a mechanism for sensitization involving electron transfer from excited sensitizer to photoacid generators, since they found that the halogenated acid generator, 1,3,5-tris(2,3-dibromopropyl)-1,3,5-triazine-(1H,3H,5H)-trione, acts as an efficient acid generator in spite of low absorbance at 248 nm.

4.6.1.3 Sulfonic Acid Esters

Houlihan and co-workers have described acid generators based on 2-nitrobenzyl-sulfonic acid esters [42]. As shown in Fig. 4.31, the mechanism of the photoreaction of nitrobenzyl ester involves insertion of an excited nitro group oxygen into a benzylic carbon-hydrogen bond. Subsequent rearrangement and cleavage generates nitrosobenzaldehyde and sulfonic acid.

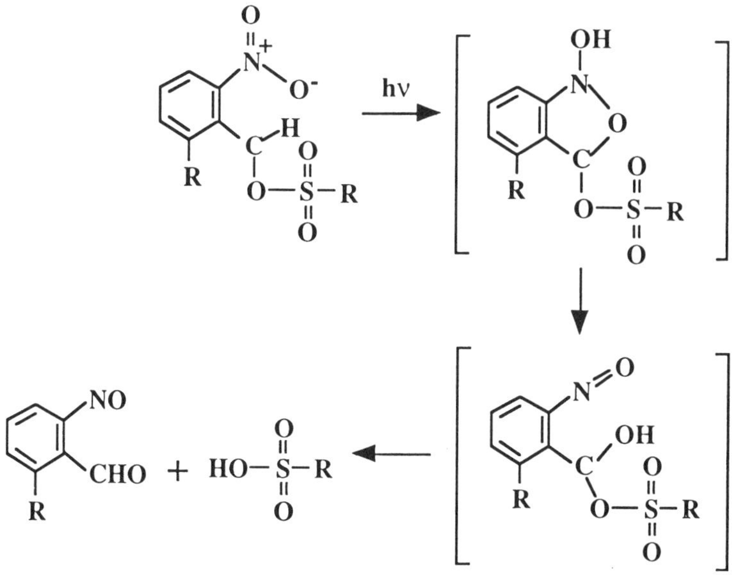

FIG. 4.31 Mechanism of acid formation from o-nitrobenzylesters.

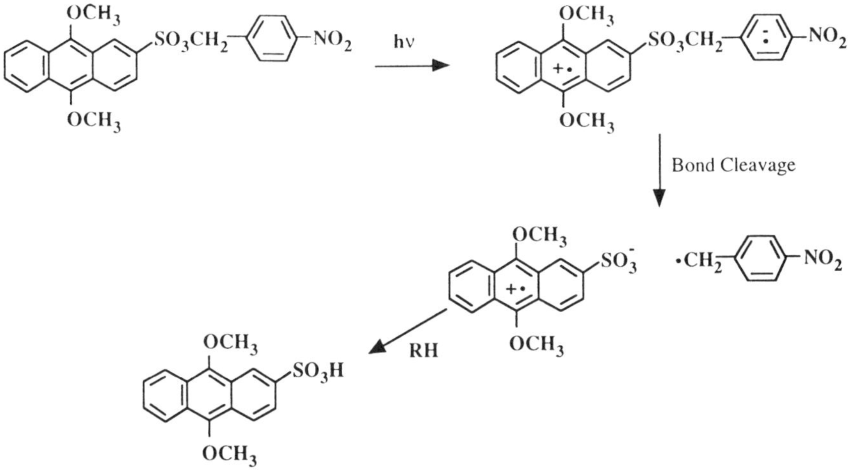

FIG. 4.32 Mechanism of acid formation from direct photolysis of p-nitrobenzyl-8,10-dimethoxyanthracene-3-sulfonate.

p-Nitrobenzylsulfonic acid esters such as *p*-nitrobenzyl-9,10-diethoxy-anthracene-2-sulfonate can act as a bleachable acid precursor [43]. Photodissociation of the *p*-nitrobenzyl ester proceeds via intramolecular electron transfer from the excited singlet state of 9,10-dimethoxyanthracene moiety to *p*-nitrobenzyl moiety followed by the heterolytic bond cleavage at the oxygen-carbon bond of sulfonyl ester as shown in Fig. 4.32 [44].

Simple alkylsulfonates such as tris(alkylsulfonyloxy)benzene were found to generate sulfonic acids upon deep-UV [45] and electron beam irradiation [46]. A high efficiency of acid generation (number of acid moieties generated per photon absorbed in a resist film) from these compounds in a novolak resin can be ascribed to a sensitization mechanism from excited novolak resin to the sulfonates, presumably via an electron transfer reaction as shown in Fig. 4.33 [47].

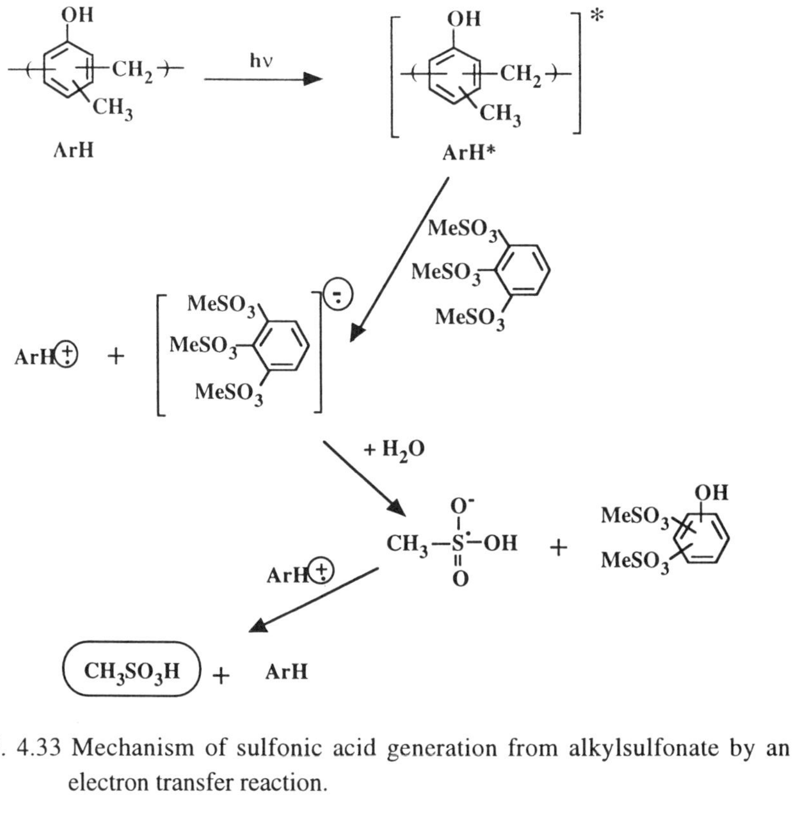

FIG. 4.33 Mechanism of sulfonic acid generation from alkylsulfonate by an electron transfer reaction.

It was reported that 1,2-diazonaphthoquinone-4-sulfonate (4-DNQ) can generate a sulfonic acid as well as carboxylic acid [48]. The photochemistry of 1,2-diazonaphthoquinone-4-sulfonate shown in Fig. 4.34 was proposed by Buhr et al. [48]. It is expected that the reaction follows its classical pathway starting from diazonaphthoquinone via the Wolff-rearranged ketene to the indene carboxylic acid. In polar media, possibly with proton catalysis, the phenol ester moiety can be eliminated leading to sulfene, which adds water to generate the sulfonic acid. 4-DNQ acid generators were used for an acid-catalyzed crosslinking of image-reversal resists and for an acid-catalyzed deprotection reaction.

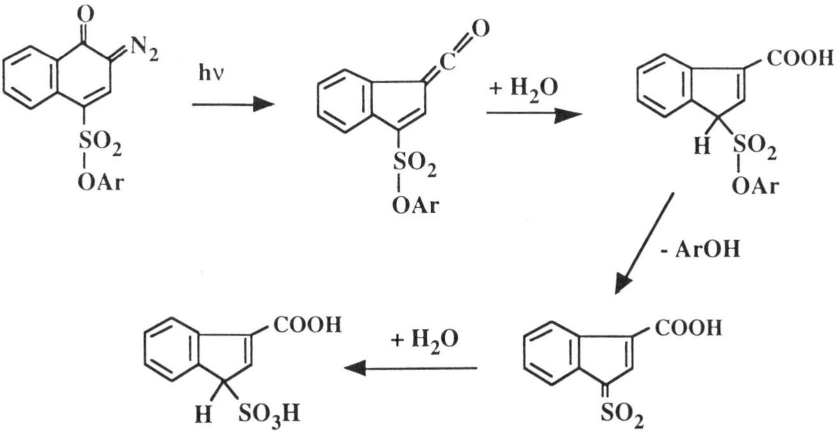

FIG. 4.34 Sulfonic acid generation from direct photolysis of DNQ-4-sulfonate.

α-Hydroxymethylbenzoin sulfonic acid esters [49], α-sulfonyloxyketones [50], iminosulfonates [51], and N-tosylphthalimide [52] were also reported as photoacid generators.

4.6.1.4 Sulfonyl Compounds

The proposed photochemical reaction mechanism for acid generation from α,α'-bisarylsulfonyl diazomethanes is shown in Fig. 4.35 [53]. The mechanism of acid generation is that the intermediate carbene formed during photolytic cleavage of nitrogen rearranges to a highly reactive sulfene, which then adds water present in the solvent mixture to give the sulfonic acid. The acid generation mechanism for disulfones is shown in Fig. 4.36 [54]. Homolytic cleavage of S-S bond is followed by hydrogen abstraction, yielding two equivalents of sulfinic acid.

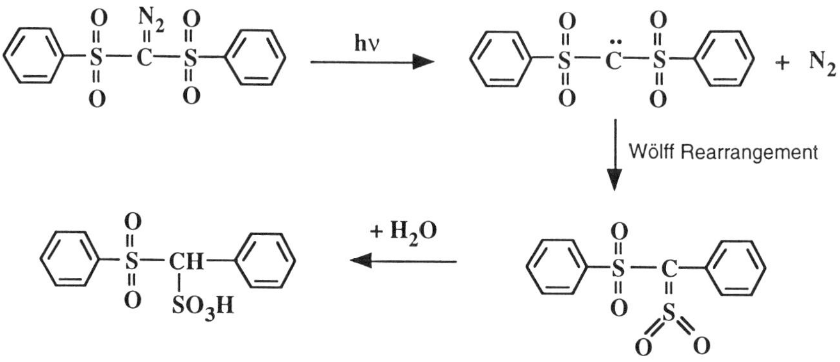

FIG. 4.35 Photodecomposition mechanism of α,α'-bisarylsulfonyl diazomethane.

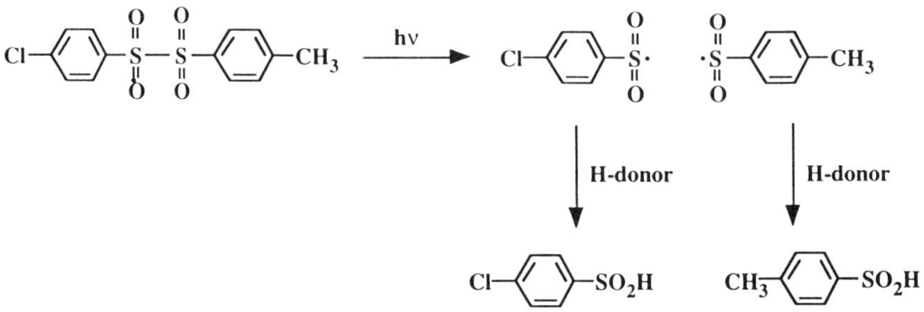

FIG. 4.36 Sulfinic acid generation mechanism from a disulfone.

4.6.2 Acid-Catalyzed Reaction

4.6.2.1 Deprotection Reaction

The acid catalyzed deprotection can be used as a positive resist, since a deprotected product, poly(hydroxystyrene) (PHS) is soluble in an aqueous base. Protection of reactive groups is a general method in synthesis [55]. There are several protecting groups for the hydroxy group of PHS as shown in Fig. 4.37. The chemical amplification resists are classified into two components and three component systems. The former consists of an acid labile protected PHS and a photoacid generator and the latter consists of an alkali soluble base polymer,

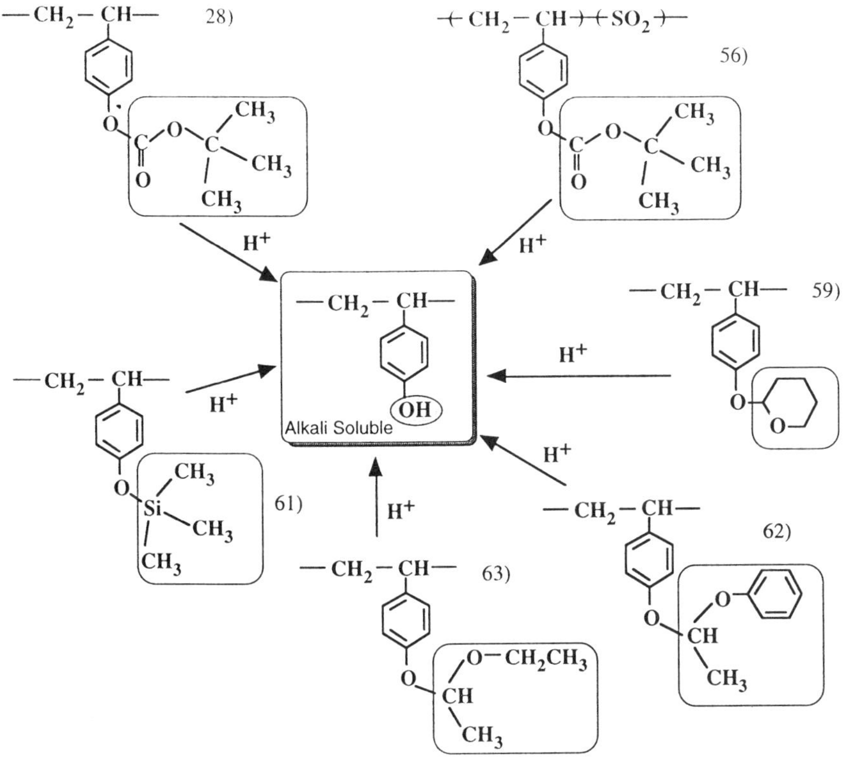

FIG. 4.37 Acid-catalyzed deprotection reactions which produce poly(hydroxystyrene) (PHS).

acid-labile protected compounds or polymers, and a photoacid generator.

(a) t-BOC Group

The *tert*-butoxycarbonyl (t-BOC) protecting group is the most well-known protecting group for acid-catalyzed thermolysis [27]. A resist formulated with poly(*p-tert*-butoxycarbonyloxystyrene) (tBOC-PHS) and an onium salt as a photoacid generator has been described by IBM workers [27,28]. The acid produced by photolysis of an onium salt catalyzes acidolysis of the t-BOC group, which converts t-BOC-PHS to poly(hydroxystyrene) (PHS) as shown in Fig. 4.38. Since the photogenerated acid is not consumed in the deprotection reaction, it serves only as a catalyst. This system is named "chemical amplification".

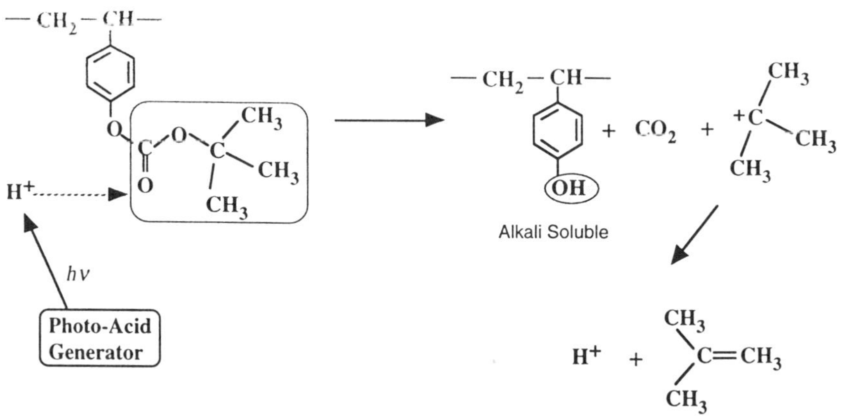

FIG. 4.38 Chemically amplified resist using acid-catalyzed t-BOC deprotection reaction.

The exposed part is converted to a polar polymer which is soluble in polar solvents such as alcohols or an aqueous base, whereas the unexposed area remains nonpolar. This large difference in polarity between exposed and unexposed areas gives a large dissolution contrast in the developer, which also allows negative or positive images depending on the developer polarity. Development with a polar solvent selectively dissolves the exposed area to give a positive tone image. Development with a nonpolar solvent dissolves the unexposed area to give a negative tone image. Since this resist is based on the polarity change, there is no evidence of swelling which is a severe problem in gel formation type resists such as cyclized rubber-based negative resists. The swelling phenomena causes pattern deformation and limits resolution capability.

There have been a lot of reports on t-BOC as a protecting group. Workers of AT&T proposed a series of copolymers of sulfur dioxide and *tert*-butoxycarbonyloxystyrene[56], which exhibits both improved sensitivity and high contrast. This is explained by the degradation of the matrix polymer. In three-component positive chemically amplified systems, t-BOC protected phenol derivatives as dissolution inhibitors for alkali developable base resins. These base resins such as partially protected PHS or poly(hydroxystyrene) derivatives should be transparent at 248 nm and have dissolution inhibition capability for t-BOC compounds.

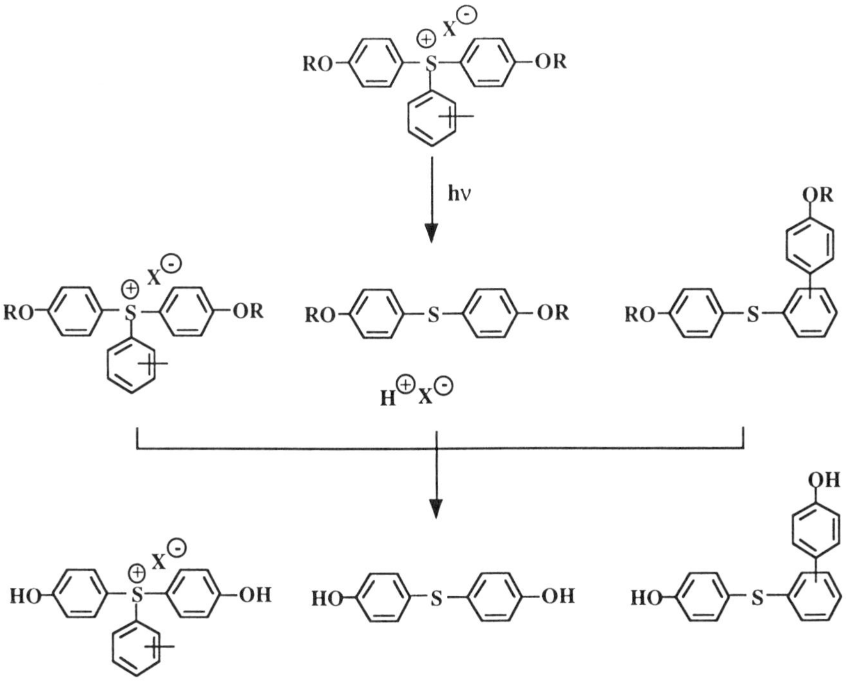

FIG. 4.39 Sulfonium compounds containing expellable sophisticated side groups (SUCCESS) concept: photoinduced complete solubility change.

Schwalm et al. [57] proposed a unique photoacid generator which completely converts to phenolic products after post-exposure-bake as shown in Fig. 4.39. Since one of the disadvantages of onium salts such as diaryliodonium and triarylsulfonium salts is strong dissolution inhibition for alkali development, sulfonium salts containing the acid-labile t-BOC protecting group in the same molecule were synthesized to improve the solubility of onium salts into the developer. After irradiation unchanged hydrophobic initiator and hydrophobic photoproducts are formed besides the acid, but upon thermal treatment, acid-catalyzed reaction converts all these compounds to phenolic products resulting in an enhanced solubility in aqueous base.

(b) Tetrahydropyranyl (THP) Group

The deprotection reaction mechanism for tetrahydropyranyl (THP) protected PHS is shown in Fig. 4.40 [58,59]. A fully THP protected PHS (THP-M) suffered from poor developability in aqueous base, when THP-M was used as a base polymer. The effect of the deprotection degree on dissolution rate was investigated by Hattori et al [60]. As shown in Fig. 4.41, a 30% protection degree is enough for negligible dissolution for alkali development. It should be noted that deprotection from 100% to 30% cannot induce a change in dissolution in 2.38% tetramethylammonium hydroxide solution. Optimization of the protection degree can provide alkali-developable two component resists for KrF lithography.

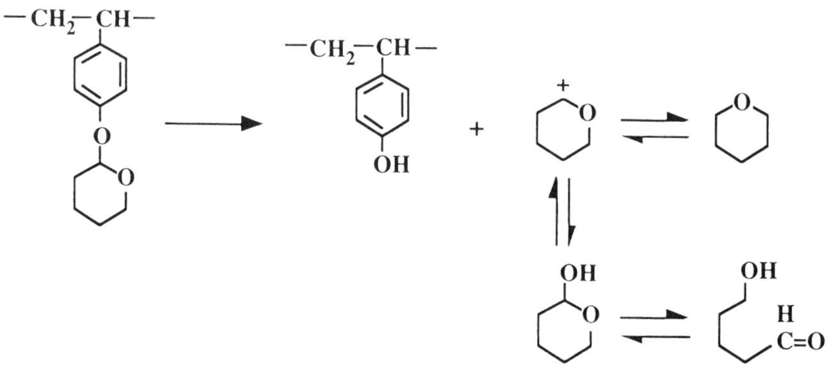

FIG. 4.40 Acid-catalyzed deprotection reaction of the tetrahydropyranyl (THP) group.

A three component system composed of THP-M as a dissolution inhibitor, novolak resin and acid generator has been applied to an electron beam resist (Fig. 4.42) [46]. A novolak resin can be used as a base polymer for electron beam resists because absorbance of resist in the deep-UV region has no effect on energy deposition for the electron beam.

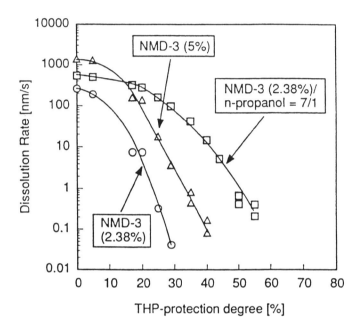

FIG. 4.41 Dissolution rate of THP (tetrahydropyranyl) protected poly(hydroxystyrene), THP, as a function of THP protection degree for various developers. NMD is tetramethylammonium hydroxide aqueous solution. [Reprint with permission from *Proc. SPIE*, **1925**, 146(1993).]

(c) Other Protecting Groups

Yamaoka et al. [43] have reported the trimethylsilyl group as a protecting group (Fig. 4.37) based on preliminary experiments on the rate of acid-catalyzed hydrolysis for a series of alkylsilylated and arylsilylated phenols. They reported a resist using trimethylsilyl protected PHS combined with p-nitrobenzylsulfonate as an acid generator [61].

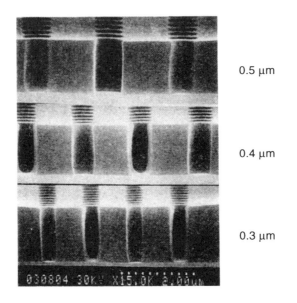

FIG. 4.42 Scanning electron micrographs of 0.3, 0.4, and 0.5 μm contact holes obtained by electron beam exposure at 50 kV using a resist composed of a novolak resin, THP-M and an acid generator. [Reprint with permission from *J. Vac. Sci. Technol.*, **B11**, 2812(1993).]

Jiang and Bassett [62] have proposed the phenoxyethyl group as a protecting group of PHS. It is expected that phenol is produced from the protection group via an acid-catalyzed cleavage. Since phenol is very soluble in an aqueous base, it acts as a dissolution promoter in exposed areas. Another acetal protecting group of ethoxyethyl group has been evaluated for KrF resists [63,64].

Onishi et al. [65] reported partially t-butoxycarbonylmethyl protected PHS which involves a methylene group between phenyl and carboxylic acid. The reaction product of this system is a dissolution promoter, carboxylic acid polymers, rather than the hydroxy group of phenol. This system can also avoid the high absorbance at 248 nm of poly(vinylbenzoic acid).

4.6.2.2 Depolymerzation

Ito and Willson [67] reported a positive resist acid-catalyzed depolymerization of polyphthalaldehyde (PPA) as a first stage of chemical amplification resists. Polyphthalaldehyde is classified as O,O-acetal. The polymerization of polyaldehyde is known to be an equilibrium process of quite low ceiling

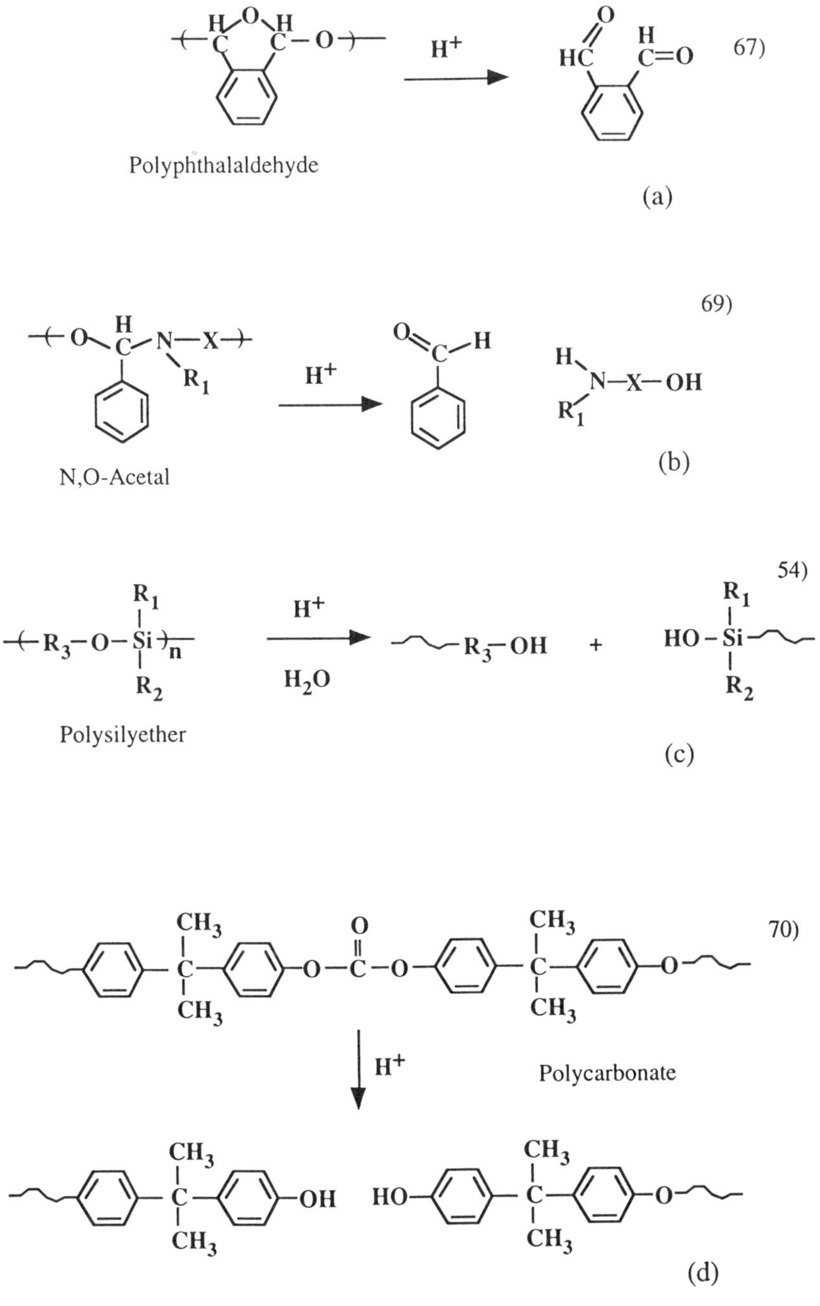

FIG. 4.43 Acid-catalyzed depolymerization.

temperature (T_c). Above T_c the monomer is more stable thermodynamically than its polymer. Therefore, once the bond is cleaved at above the ceiling temperature of about -40°C, PPA spontaneously depolymerizes to monomers.

The resist composed of PPA and an onium salt can give positive tone images without subsequent baking processes. This is called "self-development" imaging. As shown in Fig. 4.43(a) the acid generated from onium salts catalyzes the cleavage of the main chain acetal bond. PPA can also be used as a dissolution inhibitor of novolak resin in a three-component system. The photogenerated acid induces depolymerization of PPA, resulting in loss of dissolution inhibition capability to give a positive image.

Using acid-catalyzed depolymerization of O,O- and N,O-acetals, the workers of Höchst AG reported positive three-component chemical amplification resist systems [68,69]. The reaction mechanism is shown in Fig. 4.43(b). Poly-N,O-acetal is protonated at the oxygen atom and liberates an alcohol, XOH, which results in a decrease in molecular weight as well as dissolution promotion of cleavage products like alcohols and aldehydes. Therefore, acid-catalyzed depolymerzation leads to an increase in dissolution in aqueous base, when poly-N,O-acetal is used as a dissolution inhibitor for an alkali soluble phenolic resin.

The silicon polymer containing silylether groups in the main chain is hydrolyzed by acid (Fig. 4.43(c)) and degraded to low molecular weight compounds [54]. These polymers were used as dissolution inhibitors for three-component systems. Fréchet et al. prepared polymers of new imaging materials based on polycarbonates (Fig. 4.43(d)), polyethers and polyesters, which are all susceptible to acid catalyzed depolymerization [70]. The reaction is based on the protonation of the carbonyl group of a carbonate followed by cleavage of the adjacent carbon-oxygen bond to produce fragments to give a positive image.

4.6.2.3 Crosslinking and Condensation

(a) Acid Hardening of Melamine Derivatives

Negative tone resists based on acid hardening resin (AHR) chemistry are three-component systems composed of a novolak resin, a melamine crosslinking agent and a radiation sensitive acid generator [71]. The reaction leading to a decrease in dissolution rate in alkali aqueous solution via crosslinking is shown in Fig. 4.44. The protonated melamine liberates a molecule of alcohol upon heating to leave a nitrogen-stabilized carbonium ion. Alkylation of the novolak then occurs at either the phenolic oxygen (O-alkylation) or at a carbon on the aromatic ring (C-alkylation) and a proton is regenerated. There are several reactive sites on the melamine, allowing it to react more than once per molecule to give crosslinked polymer. The difference in dissolution rates between exposed and unexposed is quite high as reported by Liu et al.[72] (see the chapter on electron beam lithography). The acid-hardening type resists have been widely evaluated as resists for deep-UV and electron beam lithography.

Crosslinking of Novolak Resin

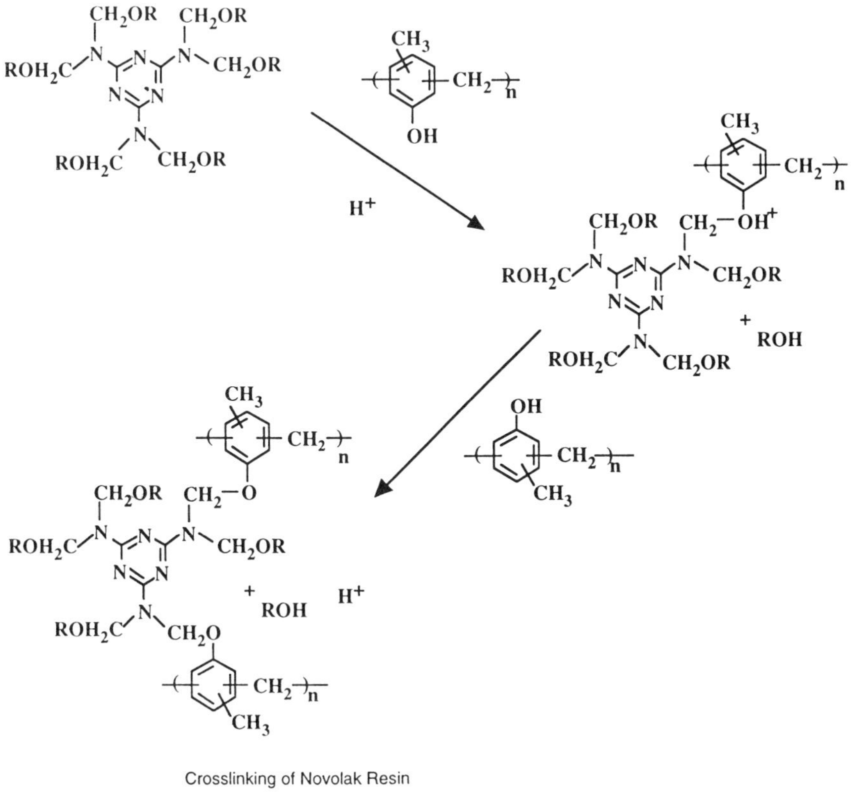

FIG. 4.44 Acid hardening resin of melamine.

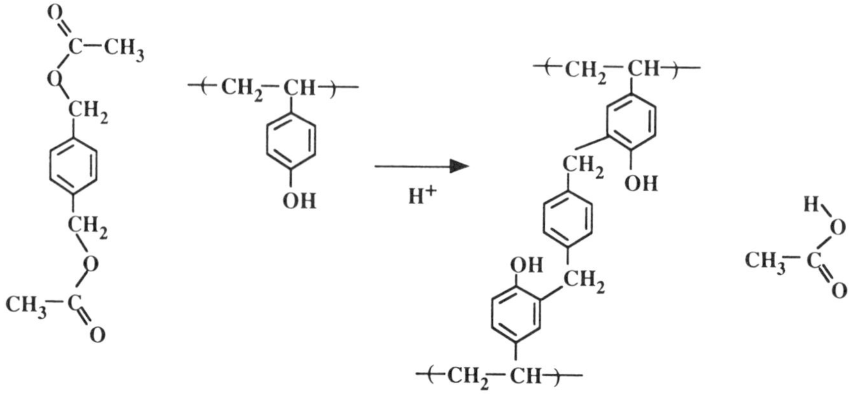

FIG. 4.45 Acid-catalyzed electrophilic aromatic substitution.

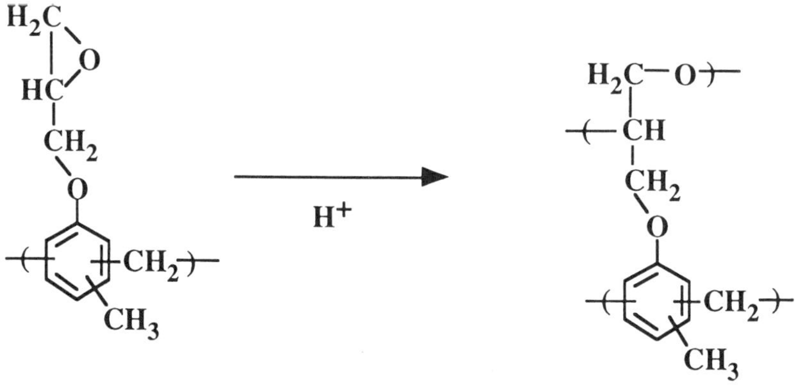

FIG. 4.46 Cationic ring-opening of epoxy group.

(b) Electrophilic Aromatic Substitution

Another type of acid-catalyzed electrophilic aromatic substitution reaction has been applied to negative resists [73]. The reaction mechanism is similar to acid hardening described above, as shown in Fig. 4.45. The photogenerated acid reacts with a latent electrophile, such as a substituted benzylacetate, to produce a carbocationic intermediate while acetic acid is liberated. The carbocationic intermediate then leads to an electrophilic reaction with neighboring aromatic moieties to produce crosslinking and a proton is regenerated.

Chemistry of Photoresist Materials 111

(c) Cationic Polymerization

Cationic ring-opening of the epoxy group leads to polymerization (crosslinking) as shown in Fig. 4.46, which can be applied to a negative resist [74]. Conley et al. [75] reported a three-component resist using certain classes of polyepoxide and monomeric epoxide compounds as a crosslinker for alkaline soluble phenolic resins to get a non-swelling resist. They showed that use of bis-

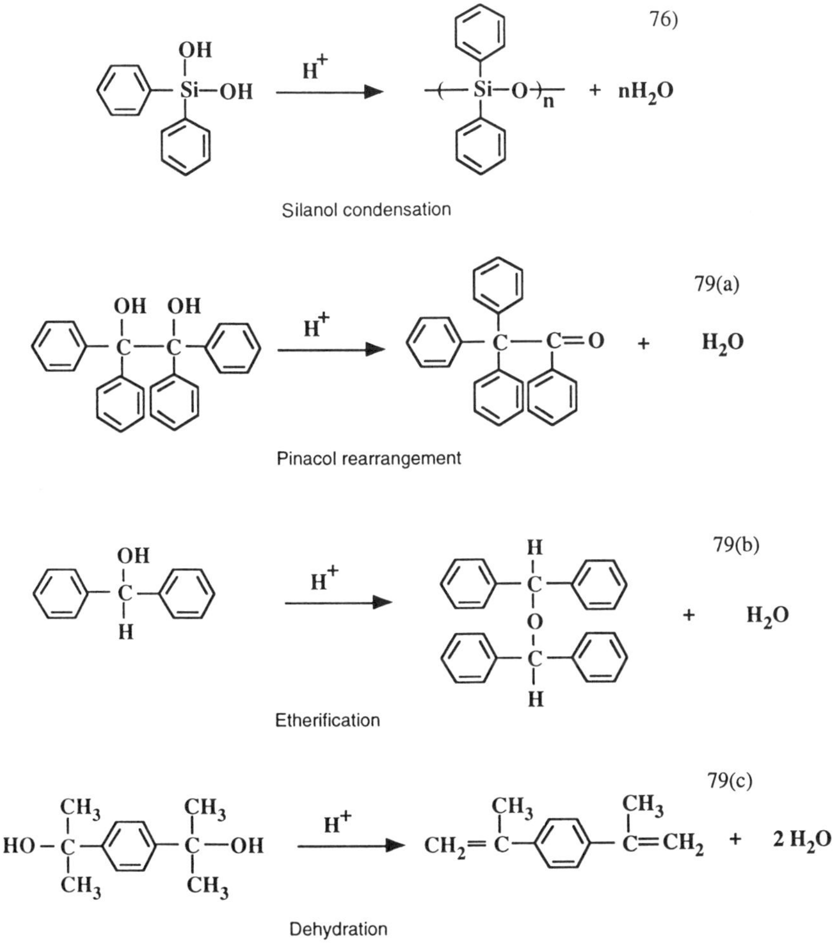

Silanol condensation — 76)

Pinacol rearrangement — 79(a)

Etherification — 79(b)

Dehydration — 79(c)

FIG. 4.47 Acid-catalyzed polarity change.

cyclohexane epoxide as a crosslinker shows high sensitivity, high resolution and excellent resistance to thermal ring-opening polymerization, while some epoxy compounds show sensitivity to the acidity of phenolic resin even at room temperature.

(d) Silanol Condensation

Silanol compounds undergo condensation to form siloxane in the presence of acid (Fig. 4.47). The acid catalyzed reaction produces a dissolution inhibitor, siloxane, which can be applied to a negative resist [76]. This acid catalyzed reaction can be applied to i-line phase-shifting lithography (Fig. 4.48) [77], KrF excimer laser lithography [76] and electron beam lithography [78].

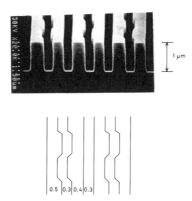

FIG. 4.48 Scanning micrograph of 0.3 μm space patterns produced by an i-line phase-shifting lithography. The resist is composed of a novolak and a silanol compound and a photoacid generator. [Reprint with permission from *Jpn. J. Appl. Phys.* **29**, 2632(1990).]

4.6.2.4 Polarity Change

Although the resist based on silanol condensation shows high resolution capability and high sensitivity, it causes the problem of SiO_x residue formation by oxygen plasma removal processes. Uchino et al. [79] designed and tested several types of acid-catalyzed reactions, such as a pinacol rearrangement [79a], etherification of carbinols [79b] and intramolecular dehydration [79c] of the α-hydroxypropyl group that can be applied to negative resists as shown in

Fig.4.47. The basic concept is the same as that of silanol condensation: a carbinol acts as a dissolution promoter in a novolak resin while the acid-catalyzed reaction products act as a dissolution inhibitor. These systems are also called "reverse polarity" [80].

4.6.3 Route for Actual Use of Chemically Amplified Resists

Some chemical amplification resists have been extensively evaluated for actual use in deep-UV lithography including KrF excimer laser lithography. It is not an easy task to select resists from the commercially available resists, since many things have to be taken into consideration, such as sensitivity, resolution, depth-of focus, heat resistance, dry-etch resistance, shelf life, impurity, etc. In addition, the composition of commercially available resists are not disclosed. For positive resists, however, a resist using the acid-catalyzed deprotection reaction is one of the candidates. Especially there have been many reports on the resists using the t-BOC protecting group. As for negative resists, resists using acid hardening of melamines from Shipley Co. have been widely accepted.

At present, it is said that optimum sensitivity is in the range of 20 to 50 mJ/cm^2 for a KrF excimer laser stepper. The sensitivity required for KrF lithography is not a critical issue, as far as acid catalyzed reactions are used. Resolution limitation of chemically amplified resists used in KrF lithography is usually better than that of DNQ-novolak resists used for i-line lithography, since a shorter wavelength is used. However, more depth-of-focus (DOF) latitude is required. The dry-etch resistance of chemically amplified resists is similar to

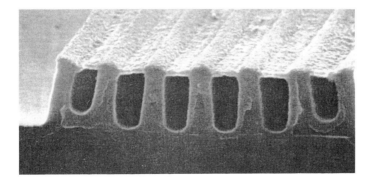

FIG. 4.49 Scanning micrograph of a T-shape isolubilizing layer occurred in a positive chemical amplification resist.

that of DNQ-novolak resists, as a phenolic resin is used as a base polymer. An acid-hardening type negative resist is expected to show better dry-etch resistance, as exposed areas are crosslinked.

The main issues for chemically amplified resists are some intrinsic problems associated with acid-catalyzed reactions. Chemically amplified resists are susceptible to process conditions such as airborne contamination, baking condition, and delay time between exposure and post-exposure-baking (post-exposure delay). Underlying substrates sometimes influence the resist profile. The shelf life is also a critical issue compared with DNQ based resists. Most positive chemically amplified resists suffer from line width shift and/or formation of an insolubilization layer or "T-top" profiles (Fig. 4.49) depending on the post-exposure delay. Improvement in process stability is a route for actual use of chemical amplification resists in a production line.

4.6.4 Improvement in Process Stability

Some additives to solve the delay problems were investigated by BASF workers [81]. Reactions of additives such as sulfonic acid esters and disulfones with airborne contamination, usually base, are expected to alleviate the delay effect at the expense of sensitivity. Przybilla and coworkers of Höchst proposed a unique idea for process stability using a photosensitive base which loses basity upon irradiation [82]. In the exposed region, acid is generated and at the same time a base is decomposed. When the acid diffuses into unexposed areas, it is neutralized by a photosensitive base. It was reported that this improves the line width stability for process condition.

Since the acid-catalyzed reaction is sensitive to the impurity, one should pay attention to the polymer end group derived from polymerization catalyst. Ito et al. [83] observed the difference in t-BOC deprotection yield for different molecular weights. This difference may be ascribed to the different concentration of poisoning CN end group derived from 2,2-azobis(isobutyronitrile) (AIBN): as the molecular weight becomes lower, the end group concentration becomes higher. When t-BOC-PHS prepared by radical polymerization using benzoyl peroxide (BPO) or living anionic polymerization with *sec*-butyllithium was used, a higher extent of acid catalyzed deprotection reaction was observed compared with the polymer obtained by radical polymerization with AIBN initiator. Polymer end groups are important to maximize the performance of chemical amplification resists.

Ito and coworkers [84] proposed the design concept of ESCAP

(environmentally stable chemically amplification positive resist), composed of a thermally stable resin and a thermally stable photoacid generator: a copolymer of 4-hydroxystyrene (HOST) with *tert*-butylacrylate (TBA) and camphorsulfonyloxynaphthalimide (CSN) as an organic nonionic acid generator (Fig. 4.50). This ESCAP can employ a bake temperature of 150°C or above. This resist showed a 2 hour stability for postexposure delay.

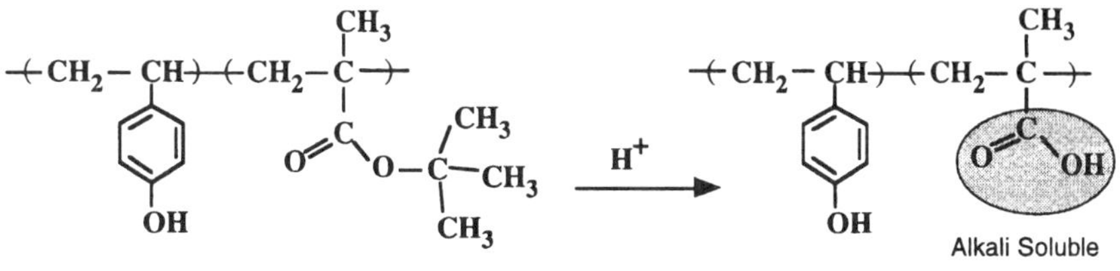

FIG. 4.50 Environmentally stable chemically amplified positive resist (ESCAP).

4.7 RESISTS FOR ArF LITHOGRAPHY

For 193 nm lithography, deep-UV resists currently formulated cannot be used. The aromatic polymers are not suitable due to their high absorption coefficients at 193 nm. Several approaches to obtain resists with improved transmittance at 193 nm have been reported. Resists using acrylic polymers as base resins combined with an acid catalyzed reaction have been reported.

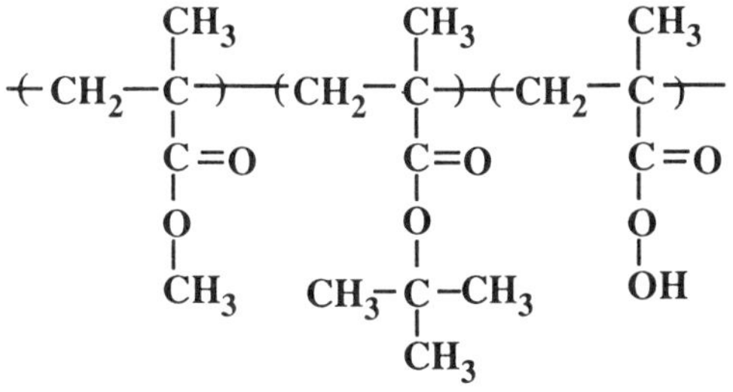

85)

FIG. 4.51 Acrylic polymer of chemical amplification resists for ArF lithography.

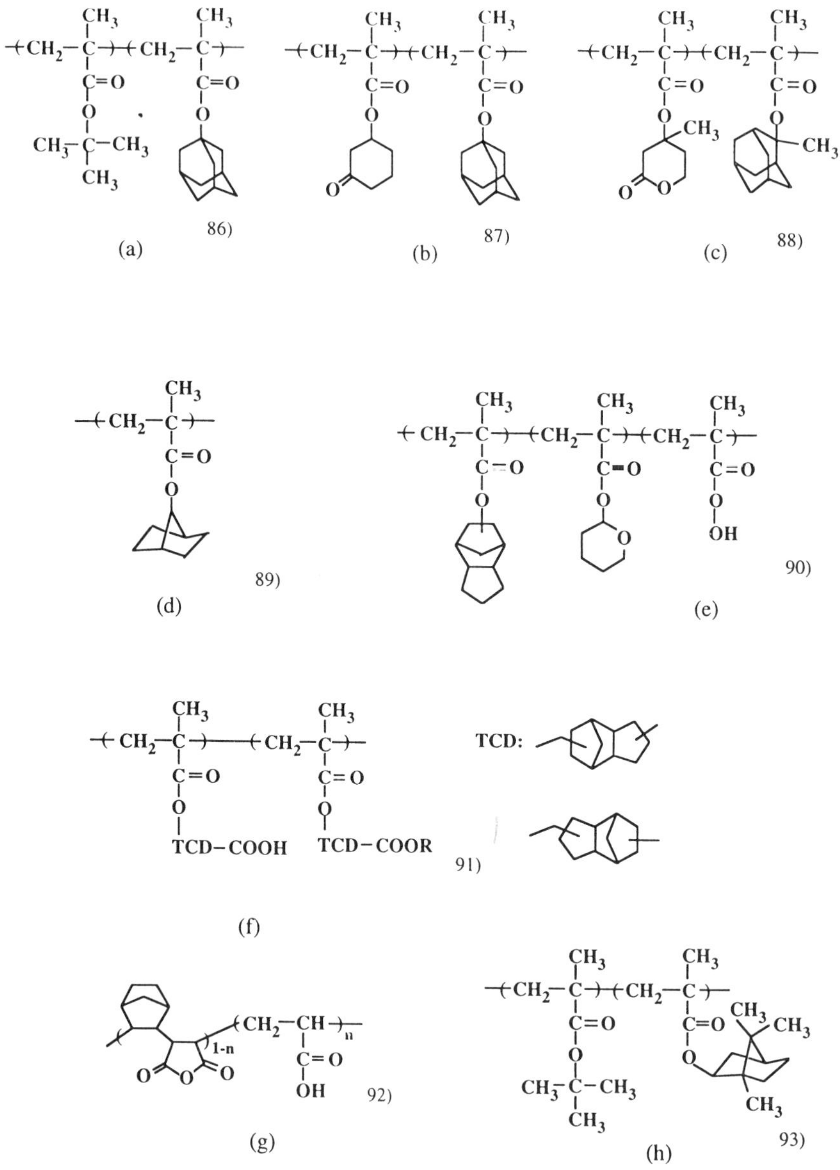

FIG. 4.52 Chemical amplification resists for ArF lithography.

Allen et al. [85] demonstrated patterning with 193 nm exposure using acrylic polymers and a photoacid generator, which was initially developed for printed circuit board technology. The acrylic polymers are terpolymers of methylmethacrylate, *tert*-butylmethacrylate and methacrylic acid (Fig. 4.51). The photogenerated acid induces acid catalyzed deprotection of the *tert*-butyl group to yield polymethacrylic acid. Carboxylic acid polymer is soluble in aqueous base, resulting in a positive image.

One disadvantage of acrylic polymers is poor dry-etch resistance. In order to improve the dry-etch resistance, alicyclic polymers containing adamantane or norbornane groups have been reported (Fig. 4.52a, b, c, d) [86-93]. The alicyclic component without a conjugated double bond is desirable for transmittance at 193 nm and improved dry etch resistance compared with an aliphatic methacrylate such as poly(methylmethacrylate) (PMMA) or poly(*tert*-butylmethacrylate). Although it was reported that the dry-etch resistance of these polymers is compatible with that of DNQ-novolak resists, it is to be noted that the dry-etch resistance is always dependent on etching conditions, such as etching gas, pressure and flow rate of the etching gas, apparatus, power, shape and size of the electrodes, pumping rate, etc.

Nakano et al. of NEC [90] reported a methacrylate terpolymer, poly(tricyclodecanylmethacrylate-co-tetrahydropyranylmethacrylate-co-methacrylic acid) (Fig. 4.52e) as a base polymer for 193 nm resist. They have also described polymers

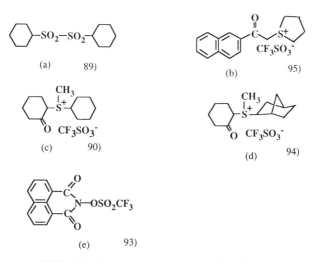

FIG. 4.53 Acid generators for ArF resist.

with tricycledecanyl groups which have carboxylic acid and carboxylic acid protecting groups (Fig. 4.52f). The dissolution behavior of these polymers are similar to poly(hydroxystyrene). Workers of Bell laboratories [92] have reported polymers containing an alicyclic group in the main chain, cycloolefin-maleic anhydride alternating copolymers as shown in Fig. 4.52g.

Acid generators used are summarized in Fig. 4.53. Nakano et al. of NEC [90,94] reported an new alkylsulfonium salt, cyclohexylmethyl-(2-oxocyclohexyl)sulfonium triflate and norbornyl-(2-oxocyclohexyl)sulfonium triflate (Fig. 4.53c,d) as an acid generator, which shows high transmittance at 193 nm.

Workers of Toshiba [95] found that the compounds containing the naphthalene moiety afford lower absorption coefficient at 193 nm than those with the phenyl group (Fig. 4.54). This is ascribed to a shift of λ_{max} (wavelength of absorption maximum) induced by conjugation extension from benzene to naphthalene. This finding leads to resists composed of polymers and compounds containing the naphthalene moiety rather than phenyl groups.

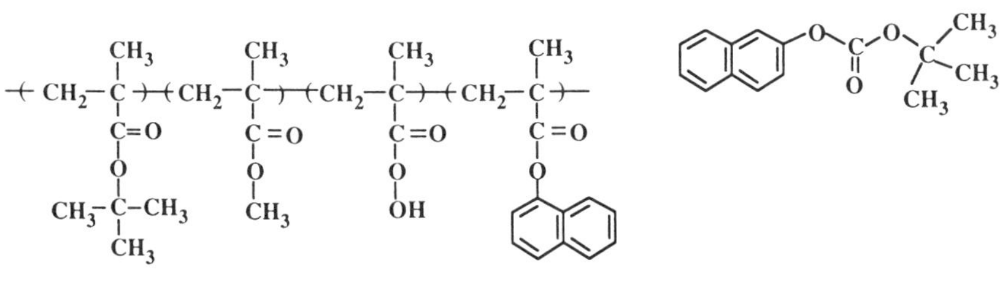

FIG. 4.54 ArF resist using naphthalene moiety.

Using the high absorbance of phenolic polymers at 193 nm, Hartney et al. [96] described surface imaging based on silylation. Irradiation of 193 nm produces direct crosslinking in phenolic polymers near the surface (Fig. 4.55). The crosslinking prevents diffusion of organosilicon reagent to generate silylation selectivity. Subsequent oxygen plasma treatment forms an etch-resistant silicon-oxide mask in the silylated area and etches the unsilylated areas to give a positive tone image.

Chemistry of Photoresist Materials 119

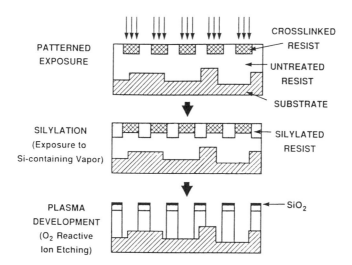

FIG. 4.55 Schematic process flow for 193 nm silylation process. [Reprint with permission from *Proc. SPIE*, **1466**, 238(1991).]

4.8 NEW APPROACH OF CONTRAST ENHANCEMENT DURING DEVELOPMENT

Increased demand for high resolution requires new types of resists of high resolution capability. Several attempts to use base-catalyzed reaction have been reported to improve resolution capability. As described in DNQ-novolak resists, a base-catalyzed reaction of azoxy and azocoupling formation during development have been reported. This enhances the dissolution inhibition in the exposed area.

When a DNQ compound derived from phenolphthalein is used as a photoactive compound, the resist showed improved behavior of scumming. The patterns did not web or scum as is usually observed [97]. This can be explained by a base-catalyzed hydrolysis of a lactone ring which leads to a more soluble photoproduct as shown in Fig. 4.56.

The base-catalyzed hydrolysis of the lactone ring has been utilized for three-component chemically amplified resists. Phenolphthalein and cresolphthalein protected with *tert*-butoxycarbonyl group have been used as dissolution inhibitor [98-100]. These inhibitors were decomposed by acid-catalyzed thermal reaction then the lactone ring was cleaved by a base-catalyzed reaction in the developer, which may enhance the dissolution rate of the exposed area (Fig. 4.57).

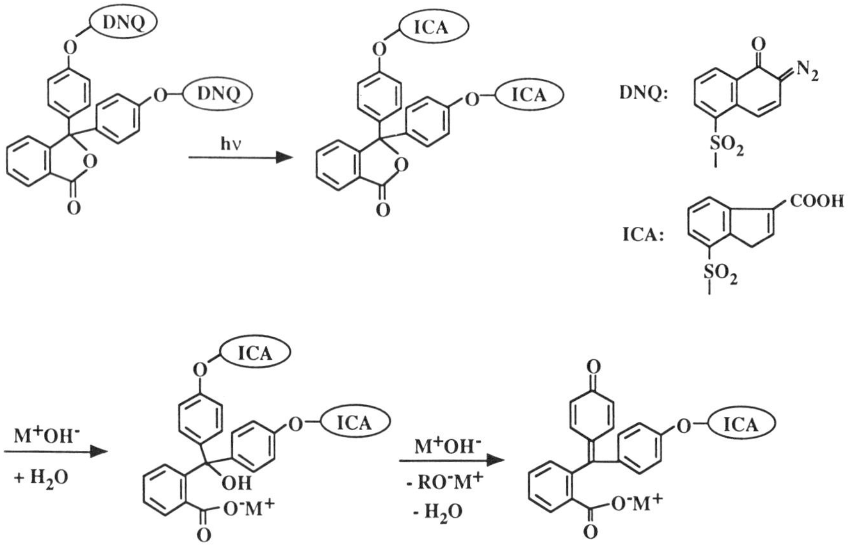

FIG. 4.56 Reaction mechanism of DNQ compound of phenolphthalein in an aqueous base.

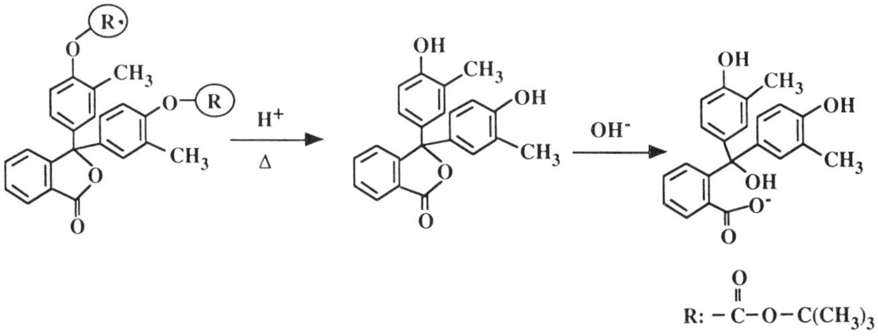

FIG. 4.57 Acid-catalyzed deprotection of t-BOC compound of cresolphthalein and reaction in an aqueous base.

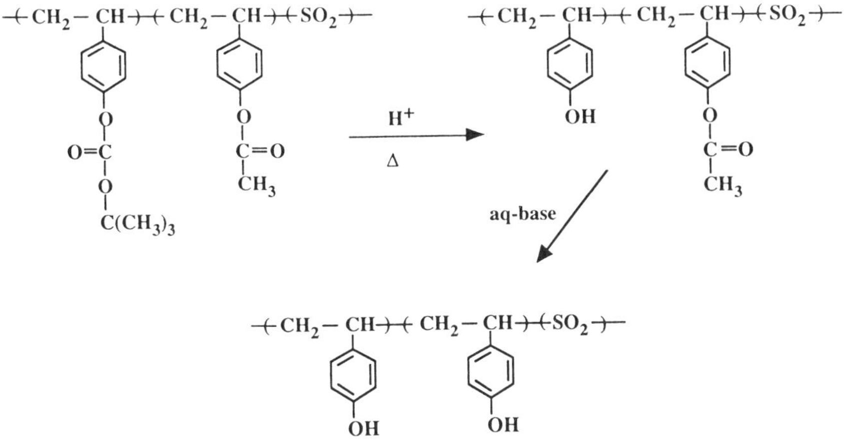

FIG. 4.58 Acid-catalyzed deprotection reaction during post-exposure-bake and base-catalyzed cleavage during development.

Workers of AT&T prepared a new terpolymer, poly[(*tert*-butoxycarbonyloxy)-styrene-co-acetoxystyrene-co-sulfone] as a base polymer for a chemically amplified resist. The acetoxy group can be cleaved from acetoxystyrene monomer in an aqueous base solution as shown in Fig. 4.58 [101]. This cleavage occurs only when sulfone is incorporated into the copolymer in the appropriate amounts. It is expected that this base-catalyzed cleavage during development enhances solubility of the exposed area where the t-BOC group is removed by acid.

A new contrast enhancement method called contrast boosted resists (CBRs) as shown in Fig. 4.59 [102] has been proposed. The resists consist of a phenolic resin, a photoactive compound, and a base-labile compound. The negative CBR offers an enhanced difference in dissolution rate between exposed and unexposed area by the reaction of base-labile water-repellant compound in the aqueous base. In the exposed area, the photochemically produced hydrophobic compounds work together with the water-repellent compound to retard permeability of the base developer penetrating into the resist. Thus, the exposed area of the CBR is completely insolubilized in the developer. On the other hand, the base developer permeates into the unexposed area of CBR and converts the water-repellent

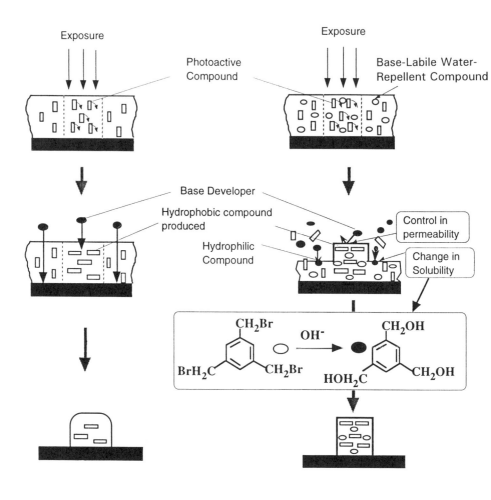

FIG. 4.59 Schematic representation of contrast enhancement during development for negative CBR (contrast boosted resist).

compound into hydrophilic compound that gives complete loss of dissolution inhibition in a phenolic resin. Therefore, CBRs exhibit a large contrast compared to conventional resists. When tris(bromoacetyl)benzene is added to a resist composed of azide-phenolic resin, the compound converts to tris(hydoroxyacetyl)benzene during the development in an aqueous base, which is the change from water-repellent to hydrophilic compounds, which gives high resolution patterns (Fig. 4.60).

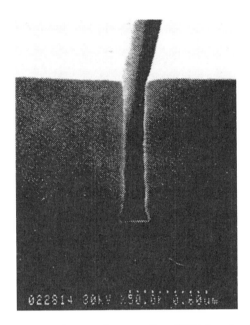

FIG. 4.60 Fine groove patterns of a negative CBR (contrast boosted resist) delineated using i-line stepper in conjunction with an edge-type phase-shift mask. [Reprint with permission from *Proc. 10th International Conference Photopolymers*, p.306(1994).]

4.9 CONCLUSION

The resists for present and future lithography have been described. The strong demand for reduction of minimum feature size will replace the DNQ-novolak positive resists with chemical amplification resists for deep-UV lithography. This demand also requires more precise line width control and higher aspect ratios over substrate topography. More precise control of process and environmental conditions are necessary. Although some of the chemical amplification resists described above show promising results, continuous effort to improve the process stability and shelf life is necessary in cooperation with engineers in the process field. In ArF excimer laser lithography, the surface imaging and a single-layer process using chemical amplification resists will be the candidates for development. In the the surface imaging processes, line width control associated with surface functionalization and dry development will be a critical issue. To use chemical amplification resists for a single-layer process, the transmittance of the resists and dry etching durability will be a trade-off.

REFERENCES

1. T. Ueno, H. Shiraishi, S. Uchino, T. Sakamizu, and T. Hattori, Chemical amplification resists for future lithography, *J. Photopolym. Sci. Technol.*, **7**, 397(1994).
2. S. Okazaki, Lithographic technologies for future ULSI, *Appl. Surf. Sci.*, **70/71**, 603(1993).
3. N. J. Turro, Modern Molecular Photochemistry, University Science Book, Mill Valley, 1991.
4. W.S. DeForest, Photoresist Materials and Processes, McGraw Hill, New York(1975).
5. T. Iwayanagi, T. Kohashi, S. Nonogaki, T. Matsuzawa, K. Douta, and H. Yanazawa, Azide-phenolic resin photoresists for deep UV lithography, *IEEE Trans.Electron Devices*, **ED-28**, 1306(1981); T. Matsuzawa, A. Kishimoto, and H. Tomioka, Profile simulation of negative resist MRS using the SAMPLE photolithographic simulator, *IEEE Electron Device Lett.*, **EDL-3**, 58(1982).
6. M. Hashimoto, T. Iwayanagi, H. Shiraishi, and S. Nonogaki, Photochemistry of azide-phenolic resin photoresists, *Polym. Eng. Sci.*, **26**, 1090(1986).
7. (a) S. Uchino, T. Tanaka, T. Ueno, and T. Iwayanagi, Azide-novolak resin negative photoresists for i-line phase-shifting lithography, *J. Vac. Sci. Technol.*, **B9**, 3162(1991); (b) K. T. Hattori, T. Hattori, S. Uchino, T. Ueno, N. Hayashi, S. Shirai, N. Moriuchi, and M. Morita, Bisazidobiphenyl / novolak resin negative resist systems for i-line phase-shifting lithography, *Jpn. J. Appl. Phys.*, **31**, 4307(1992); (c) N. Asai, A. Imai, T. Ueno, Y. Azuma, T. Miyazaki, T. Tanaka, and S. Okazaki, KrF trilayer resist system using azide-phenol resin resist, *J. Photopolym. Sci. Technol.*, **7**, 23(1994).
8. R. Dammel, Diazonaphthoquinone-based resists, SPIE tutorial text TT 11, SPIE Optical Engineering Press (1993).
9. O. Süs, *Liebigs Ann. Chem.*, **556**, 65(1944).
10. J. Packansky and J. R. Lyerla, Photochemical decomposition mechanism for AZ-type photoresists, *IBM J. Res. Develop.*, **23**, 42(1979); J. Packansky, Recent advances in the photodecomposition mechanism of diazo-oxides, *Polym. Eng. Sci.*, **20**, 1049(1980).
11. K. Nakamura, S. Udagawa, and K. Honda, Studies of photosensitive resins V: The photodecomposition mechanism of *o*-naphthoquinonediazides,

Chem. Lett., **1972**, 763.
12. T. Shibata, K. Koseki, T. Yamaoka, M. Yoshizawa, H. Uchiki, and T. Kobayashi, Mechanism of photochemical conversion of 1,2-naphthoquinonediazide in solution, *J. Phys. Chem.*, **92**, 6269(1988).
13. M. Barra, T. A. Fisher, G. J. Cernigliaro, R. Sinta, and J. C. Scaiano, On the photodecomposition mechanism of *o*-diazonaphthoquinones, *J. Am. Chem. Soc.*, **114**, 2680(1992).
14. J. Andraos, Y. Chiang, C.-G. Huang, A. J. Kresge, and J. C. Scaiano, Flash photolysis generation and study of ketene and carboxylic acid enol intermediates formed by the photolysis of diazonaphthoquinones in aqueous solution. *J. Am. Chem. Soc.*, **115**, 10605(1993).
15. K. Tanigaki and T. W. Ebbsen, Dynamics of the Wölff rearrangement: Spectroscopic evidence of oxirene intermediate, *J. Amer. Chem. Soc.*, **109**, 5883(1989); K. Tanigaki and T. W. Ebbsen, Dynamics of the Wölff rearrangement of six-membered ring *o*-diazido ketones by laser flash photolysis, *J. Phys. Chem.*, **93**, 4531(1989).
16. F. A. Vollenbroek, W. P. M. Nijssen, C. M. J. Mutsaers, M. J. H. J. Geomini, M. E. Reuhman, and R. J. Visser, The chemistry of g-line photoresist process, *Polym. Eng. Sci.*, **29**, 928(1989).
17. J. Sheats, Reciprocity failure in novolak/ diazoquinone photoresist with 364-nm exposure, *IEEE Trans. Electron Devices*, **ED-35**, 129(1988).
18. M. Hanabata, Y. Uetani, and A. Furuta, Design concept for a high-performance positive photoresist, *J. Vac. Sci. Technol.*, **B7**, 640(1989).
19. M. Hanabata, F. Oi, and A. Furuta, Novolak design for high resolution positive photoresists (IV), Tandem type novolak resin for high performance positive photoresists, *Proc. SPIE*, **1466**, 132(1991).
20. T. Kajita, T. Ota, H. Nemoto, Y. Yumoto, and T. Miura, Novel novolak resins using substituted phenols for high performance positive photoresists, *Proc.SPIE*, **1466**, 161(1991).
21. K. Honda, B. T. Beauchemin, Jr., R. J. Hurditch, A. J. Blakeney, K. Kawabe, T. Kokubo, Studies of the molecular mechanism of dissolution inhibition of positive photoresists based on novolak-DNQ, *Proc. SPIE*, **1262**, 493(1990).
22. P. Trefonas III and B. K. Daniels, New principle for image enhancement in single layer positive photoresists, *Proc. SPIE*, **771**, 194(1987).
23. H. Nemoto, K. Inomata, T. Ota, Y. Yumoto, T. Miura, and H. Chaanya, Structural effects of NQD-PAC and novolak resin on resist performance, *Proc. SPIE*, **1672**, 305(1992).

24. K. Uenishi, Y. Kawabe, T. Kokubo, and A. Blakeney, Structural effects of DNQ-PAC backbone on resist lithographic properties, *Proc. SPIE*, **1466**, 102(1991).
25. K. Uenishi, S. Sakaguchi, Y. Kawabe, T. Kokubo, M. A. Toukhy, A. T. Jeffries,III, S. G. Slater, and R. J. Hurditch, Selectively DNQ-esterified PAC for high performance positive photoresists, *Proc. SPIE*, **1672**, 262(1992).
26. R. Hanawa, Y. Uetani, and M. Hanabata, Design of PACs for high-performance photoresists (I) Role of di-esterified PACs having hindered -OH groups, *Proc. SPIE*, **1672**, 231(1992); R. Hanawa, Y. Uetani, and M. Hanabata, Design of PACs for high-performance photoresists (II) Effects of number and orientation of DNQs and -OH of PACs on lithographic performance, *Proc. SPIE*, **1925**, 227(1993).
27. H. Ito, C. G. Willson, and J. M. J. Fréchet, Paper presented at the 1982 Symposium on VLSI Technology, Oiso, Japan, Sept. 1982; H. Ito, C. G. Willson, and J. M. J. Fréchet, U. S. Pat. 4,491,628 (1985).
28. H. Ito and C. G. Willson, Applications of photoinitiators to the design of resists for semiconductor manufacturing, *ACS Symp.Ser.*, **242**, 11(1983).
29. H. Steppan, G. Buhr, and H. Vollmann, The resist technique - A chemical contribution to electronics, *Angew. Chem. Int. Ed. Engl.*, **21**, 455(1982).
30. T. Iwayanagi, T. Ueno, S. Nonogaki, H. Ito, and C. G. Willson, Materials and processes for deep-UV lithography, *Adv. Chem. Ser.*, **218**, ch.3 (1988).
31. J. Lingnau, R. Dammel, and J. Theis, Recent trends in x-ray resists: Part I, *Solid State Technol.*, 32(9), 105(1989); J. Lingnau, R. Dammel, and J. Theis, Recent trends in x-ray resists: Part II, *Solid State Technol.*, **32**(10), 107(1989).
32. J. R. Sheats, Photoresists for deep UV lithography, *Solid State Technol.*, **33**(6), 79(1990).
33. A. A. Lamola, C. R. Szmanda, and J. W. Thackeray, Chemically amplified resists, *Solid State Technol.*, **34**(8), 53(1990).
34. E. Reichmanis, F. M. Houlihan, O. Nalamasu, and T. X. Neenan, Chemical amplification mechanism for microlithography, *Chem. Mater.*, **3**, 394(1991).
35. J. V. Crivello, Possibility for photoimaging using onium salts, *Polym. Eng. Sci.*, **23**, 953(1983); J. V. Crivello, Cationic polymerization - Iodinium and sulfonium salt photoinitiators, *Adv. Polym. Sci.*, **62**, 1(1984).

36. J. L. Dektar and N. P. Hacker, Photochemistry of diaryliodonium salts, *J. Org. Chem.*, **55**, 639(1990).
37. J. L. Dektar and N. P. Hacker, Photochemistry of triarylsulfonium salts, *J. Amer. Chem. Soc.*, **112**, 6004(1990).
38. N. P. Hacker and K. M. Welsh, Photochemistry of triarylsulfonium salts in poly[4-[(*tert*-butoxycarbonyl)oxy]styrene]: evidence for a dual photoinitiation process, *Macromolecules*, **24**, 2137(1991).
39. D. R. Mckean, U. Schaedeli, and S. A. MacDonald, Brönsted acid generation from triphenylsulfonium salts in acid-catalyzed photoresist films, *ACS Symp. Ser.*, **412**, 27(1989).
40. G. Buhr, R. Dammel, and C. Lindley, Non-ionic photoacid generating compounds, *ACS Polym. Mater. Sci. Eng.*, **61**, 269(1989).
41. G. Calabrese, A. Lamola, R. Sinta and J. Thackeray, Electron transfer mechanism for photocatalyst generation in some chemically-amplified resists, *Polymer for Microelectronics-Science and Technology*, Ed. by Y. Tabata, I. Mita, S. Nonogaki, K. Horie, and S. Tagawa, Kodansha, Tokyo, p.435(1990).
42. F. M. Houlihan, A. Schugard, R. Gooden, and E. Reichmanis, Nitrobenzyl ester chemistry for polymer processes involving chemical amplification, *Macromolecules*, **21**, 2001(1988).
43. T. Yamaoka, M. Nishiki, K. Koseki, and M. Koshiba, A novel positive resists for deep UV lithography, *Polym. Eng. Sci.*, **29**, 856(1989).
44. K. Naitoh, K. Yoneyama, and T. Yamaoka, Photoinduced intramolecular electron transfer mechanism for photochemical dissociation of *para*-substituted benzyl 9,10-dimethoxyanthracene-2-sulfonates, *J. Phys. Chem.*, **96**, 238(1992).
45. T. Ueno, H. Shiraishi, L. Schlegel, N. Hayashi, and T. Iwayanagi, Chemical amplification positive resist systems using novel sulfonates as acid generators, *Polymers for Microelectronics--Science and Technology*, Ed. by Y. Tabata, I. Mita, S. Nonogaki, K. Horie, and S. Tagawa, Kodansha, Tokyo, p.413(1990).
46. H. Shiraishi, N. Hayashi, T. Ueno, T. Sakamizu, and F. Murai, Novolak resin-based positive electron-beam resist system utilizing acid-sensitive polymeric dissolution inhibitor with solubility reversal reactivity, *J. Vac. Sci. Technol.*, **B9**, 3343(1991).
47. T. Sakamizu, H. Yamaguchi, H. Shiraishi, F. Murai, and T. Ueno, Development of positive electron-beam resist for 50 kV electron-beam direct-writing lithography, *J. Vac. Sci. Technol.*, **B11**, 2812(1993); L.

Schlegel, T. Ueno, H. Shiraishi, N. Hayashi, and T. Iwayanagi, Acid formation and deprotection reaction by novel sulfonates in a chemical amplification positive photoresist, *Chem. Mater.*, **2**, 299(1990).
48. G. Buhr, H. Lenz, and S. Scheler, Image reversal resist for g-line exposure: chemistry and lithographic evaluation, *Proc. SPIE*, **1086**, 117(1989).
49. H. Röschert, Ch. Eckes, and G. Pawlowski, α-Hydroxymethylbenzoin sulfonic acid esters: A versatile class of photoacid generating compounds for chemically amplified deep-UV photoresists, *Proc. SPIE*, **1925**, 342(1993).
50. Y. Onishi, H. Niki, Y. Kobayashi, R. H. Hayase, and N. Oyasato, Acid catalyzed resists for KrF excimer laser lithography, *J. Photopolym. Sci. Technol.*, **4**, 337(1991).
51. M. Shirai, T. Masuda, M. Tsunooka, and M. Tanaka, Photo-crosslinking of poly(2,3-epoxypropyl methacrylate) with imino sulfonates, *Makromol. Chem., Rapid Commun.*, **5**, 689(1984).
52. C. A. Renner, US Patent 4,371,605 (1980); W. Brunsvold, R. Kwong, W. Montgomery, W. Moreau, H. Sachdev, and K. Welsh, Sensitivity enhancement for chemically amplified resists, *Proc. SPIE*, **1262**, 162 (1990).
53. G. Pawlowski, R. Dammel, C. Lindley, H.-J. Merrem, H. Röschert, and J. Lingau, Chemically amplified DUV photoresists using a new class of photoacid generating compounds, *Proc. SPIE*, **1262**, 16(1990).
54. T. Aoai, A. Umehara, A. Kamiya, N. Matsuda, and Y. Aotani, Application of silicon polymer as positive photosensitive materials, *Polym. Eng. Sci.*, **29**, 887(1989).
55. T. W. Green, *Protective Groups in Organic Synthesis*, Wiley, New York (1981).
56. R. G. Tarascon, E. Reichmanis, F. Houlihan, A. Schugard, and L. Thompson, Poly(t-BOC-styrene sulfone)-based chemically amplified resists for deep-UV lithography, *Polym. Eng. Sci.*, **29**, 850(1989).
57. R. Schwalm, SUCESS: A novel concept regarding photoactive compounds, *ACS Polym. Mater. Sci. Eng.*, **61**, 278(1989).
58. G. H. Smith and J. A. Bonham, US patent 3,779,778 (1973).
59. S. A. M. Hesp, N. Hayashi, and T. Ueno, Tetrahydropyranyl- and furanyl-protected polyhydroxystyrene in a chemical amplification systems, *J. Appl. Polym. Sci.*, **42**, 877(1991); T. Sakamizu, H. Shiraishi, H. Yamaguchi, T. Ueno, and N. Hayashi, Acid-catalyzed reactions of tetrahydropyranyl-protected polyvinylphenol in a novolak-resin-based positive resist, *Jpn. J. Appl. Phys.*, **31**, 4288(1992).

60. T. Hattori, L. Schlegel, A. Imai, N. Hayashi, and T. Ueno, Chemical amplification positive deep ultraviolet resist by means of partially tetrahydropyranyl-protected polyvinylphenol, *Opt. Eng.*, **32**, 2368(1993).
61. M. Murata, T. Takahashi, M. Koshiba, S. Kawamura, and T. Yamaoka, An aqueous base developable novel deep-UV resist for KrF excimer, *Proc. SPIE*, **1262**, 8(1990).
62. Y. Jiang and D. Bassett, Chemically Amplified Deep UV Photoresists Based on Acetal-protected Poly(vinylphenols), *ACS Polym. Mater. Sci. Eng.*, **66**, 41(1992).
63. R. Schwalm, H. Binder, T. Fisher, D. Funhoff, M. Goethals, A. Grassmann, H. Moritz, P.Paniez, M. Reuhman-Huisken, F. Vinet, H. Dijkstra, and A. Krause, A robust and environmentally stable deep UV positive resist: Optimisation of SUCCESS ST2, *Proc. SPIE*, **2195**, 2(1994).
64. T. Hattori, A. Imai, R. Yamanaka, T. Ueno, and H. Shiraishi, Delay-free deprotection approach to robust chemically amplified resist, *J. Photopolymer Sci. Technol.*, **9**, 611(1996).
65. Y. Onishi, N. Oyasato, H. Niki, R. H. Hayase, Y. Kobayashi, K. Sato, and M. Miyamura, Acid catalyzed resist for KrF excimer laser lithography (2) Polymer dissolution inhibitors, *J. Photopolym. Sci. Technol.*, **5**, 47(1992).
66. F. H. Dill, W. P. Hornbergner, P. S. Hauge, and J. M. Shaw, Characterization of positive photoresist, *IEEE Trans. Electron Devices*, **ED-22**, 445(1975).
67. H. Ito and C. G. Willson, Chemical amplification in the design of dry developing resist materials, *Polym. Eng. Sci.*, **23**, 1012(1983).
68. J. Lingnau, R. Dammel, and J. Theis, Highly sensitive novolak-based x-ray positive resist, *Polym. Eng. Sci.*, **29**, 874(1989); G. Pawlowski, K.-J. Przybilla, W. Spiess, H. Wengenroth and H. Röchert, Chemical amplification & dissolution inhibition: A novel high performance positive tone deep UV resist, *J. Photopolym. Sci. Technol.*, **5**, 55(1992).
69. H. Röschert, K.-J. Przybilla, W. Spiess, H. Wegenroth, and G. Pawlowski, Critical process parameters of an acetal based deep UV photoresist, *Proc. SPIE*, **1672**, 33(1992).
70. J. M. J. Fréchet, F. Bouchard, F. Houlihan, B. Kryczka, and C. G. Willson, Polycarbonates derived from *o*-nitrobenzyl glycidyl ether: synthesis and radiation sensitivity, *ACS Polym. Mater. Sci. Eng.*, **53**, 263(1985).

71. W. E. Feely, J. C. Imhof, and C. M. Stein, The role of the latent image in a new dual image, aqueous developable thermally stable photoresist, *Polym. Eng. Sci.*, **26**, 1101(1986).
72. H.-Y. Liu, M. P. de Grandpre, and W. E. Feely, Characterization of a high-resolution novolak based negative electron-beam resist with 4μC/cm^2 sensitivity, *J. Vac. Sci. Technol.*, **B6**, 379(1988).
73. J. M. J. Fréchet, S. Matuszczak, H. D. H. Stoever, C. G. Willson, and B. Reck, Nonswelling negative resists incorporating chemical amplification: The electrophilic aromatic substitution approach, *ACS Symp. Ser.*, **412**, 74(1989).
74. H. Ito and C. G. Willson, Application of photoinitiators to the design of resists for semiconductor manufacturing, *ACS Symp. Ser.*, **242**, 11(1983).
75. W. Conley, W. Moreau, S. Perreault, G. Spinillo, R. Wood, J. Gelorme, and R. Martino, Negative tone aqueous developable resist for photon, electron and x-ray lithography, *Proc. SPIE*, **1262**, 49(1990).
76. T. Ueno, H. Shiraishi, N. Hayashi, K. Tadano, E. Fukuma, and T. Iwayanagi, Chemical amplification negative resist systems composed of novolak, silanols, and acid generators, *Proc. SPIE*, **1262**, 26(1990).
77. N. Hayashi, K. Tadano, T. Tanaka, H. Shiraishi, T. Ueno, and T. Iwayanagi, Negative resist for i-line lithography utilizing acid-catalyzed silanol-condensation reaction, *Jpn. J. Appl. Phys.*, **29**, 2632(1990).
78. H. Shiraishi, E. Fukuma, N. Hayashi, K. Tadano, and T.Ueno, Insolubilization mechanism of a chemical amplification negative-resist system utilizing an acid-catalyzed silanol condensation reaction, *Chem. Mater.*, **3**, 621(1991).
79. (a) S. Uchino, T. Iwayanagi, T. Ueno, and N. Hayashi, Negative resist systems using acid-catalyzed pinacol rearrangement reaction in a phenolic resin matrix, *Proc. SPIE*, **1466**, 429(1991); S. Uchino and C. W. Frank, Mechanistic study on chemically amplified resist systems using pinacol rearrangement in phenolic resin, *Polym. Eng. Sci.*, **32**, 1530(1992); (b) S. Uchino, M. Katoh, T. Sakamizu, and M. Hashimoto, Chemical amplified negative resists using acid-catalyzed etherification of carbinol, *Microelectronic Eng.*, **18**, 341(1992); (c) T. Ueno, S. Uchino, K. T. Hattori, T. Onozuka, S. Shirai, N. Moriuchi, M. Hashimoto, and S. Koibuchi, Negative resists for i-line lithography utilizing acid catalyzed intramolecular dehydration reaction, *Proc SPIE*, **2195**, 173(1994).
80. R. Sooriyakumaran, H. Ito, and E. A. Mash, Acid-catalyzed pinacol rearrangement: chemically amplified reverse polarity, *Proc. SPIE*, **1466**,

419(1991).
81. D. J. H. Funhoff, H. Binder, and R. Schwalm, Deep-UV resists with improved delay capabilities, *Proc. SPIE*, **1672**, 46(1992).
82. K.-J. Przybilla, Y. Kinoshita, S. Masuda, T. Kudo, N. Suehira, H. Okazaki, G. Pawlowski, M. Padmanabam, H. Röchert, and W. Spiess, Delay time stable chemically amplified deep UV resist, *Proc. SPIE*, **1925**, 76(1993).
83. H. Ito, W. P. England, and S. B. Lundmark, Effect of polymer end groups on chemical amplification, *Proc. SPIE*, **1672**, 2(1992).
84. H. Ito, G. Breyta, D. Hofer, R. Sooriyakumaran, K. Petrillo, and D. Seeger, Environmentally stable chemical amplification positive resist: principle, chemistry, contamination resistance, and lithographic feasibility, *J. Photopolym. Sci. Technol.*, **7**, 433(1994).
85. R. D. Allen, G. M. Wallraff, W. D. Hinsberg, and L. L. Simpson, High performance acrylic polymers for chemically amplified photoresist application, *J. Vac. Sci. Technol.*, **B9**, 3357(1991).
86. Y. Kaimoto, K. Nozaki, S. Takechi, and N. Abe, Alicyclic polymer for ArF and KrF excimer resist based on chemical amplification, *Proc. SPIE*, **1672**, 66(1992).
87. K. Nozaki, Y. Kaimoto, M. Takahashi, S. Takechi, and N. Abe, Molecular design and synthesis of 3-oxocyclohexyl methacrylate for ArF and KrF excimer laser resist, *Chem. Mater.*, **6**, 1492(1994).
88. S. Takechi, M. Takahashi, A. Kotachi, K. Nozaki, E. Yano, and I. Hanyu, Impact of 2-methyl-2-adamantyl group used for 193-nm single-layer resist, *J. Photopolymer Sci. Technol.*, **9**, 475(1996); K. Nozaki, K. Watanabe, E. Yano, A. Kotachi, S. Takechi, and I. Hanyu, A novel polymer for a 193-nm resist, *J. Photopolymer Sci. Technol.*, **9**, 509(1996).
89. K. Yamashita, M. Endo, M. Sasago, N. Nomura, H. Nagano, S. Mizuguchi, T. Ono, and T. Sato, Performance of 0.2 μm optical lithography using KrF and ArF excimer laser sources, *J. Vac. Sci. Technol.*, **B11**, 2692(1993).
90. K. Nakano, K. Maeda, S. Iwasa, J. Yano, Y. Ogura, and E. Hasegawa, Transparent photoacid generator (ALS) for ArF excimer laser lithography and chemically amplified resist, *Proc. SPIE*, **2195**, 194(1994).
91. K. Maeda, K. Nakano, T. Ohfuji, and E. Hasegawa, Novel alkaline-soluble alicyclic polymer poly(TCDMACOOH) for ArF chemical amplified positive resists, *Proc. SPIE*, **2724**, 377(1996); T. Ohfuji, K. Maeda, K. Nakano, and E. Hasegawa, Dissolution behavior of alycyclic polymers

designed for ArF excimer laser lithography, *Proc. SPIE*, **2724**, 386(1996).
92. T. I. Wallow, F. M. Houlihan, O. Nalamasu, E. Chandross, T. Neenan, and E. Reichmanis, Evaluation of cycloolefin-maleic anhydride alternating copolymers as single-layer photoresists for 193 nm photolithography, *Proc. SPIE*, **2724**, 355(1996).
93. R. D. Allen, G. M. Wallraff, R. A. DiPietro, D. C. Hofer, and R. R. Kunz, Single layer resists with enhanced etch resistance for 193 nm lithography, *J. Photopolym. Sci. Technol.*, **7**, 507(1994).
94. K. Nakano, K. Maeda, S. Iwasa, T. Ohfuji, and E. Hasegawa, Positive chemically amplified resist fro ArF excimer laser lithography composed of a novel transparent photoacid generator and alycyclic terpolymer, *Proc. SPIE*, **2438**, 433(1995).
95. T. Naito, K. Asakawa, N. Shida, T. Ushirogouchi, and M. Nakase, Highly transparent chemically amplified ArF excimer laser resists by absorption band shift for 193nm, *Jpn. J. Appl. Phys.*, **33**, 7028(1994).
96. M. A. Hartney, D. W. Johnson, and A. C. Spencer, Evaluation of phenolic resists for 193 nm surface imaging, *Proc. SPIE*, **1466**, 238(1991).
97. W. Brunsvold, N. Eib, C. Lyons, S. Miura, M. Plat, R. Dammel, O. Evans, M. D. Rahman, S. Jain, P. Lu, and S. Ficner, Novel DNQ PACs for high-resolution i-line lithography, *Proc. SPIE*, **1672**, 273(1992).
98. N. Kihara, T. Ushirogouchi, T. Tada, T. Naitoh, S. Saitoh, and O. Sasaki, Novel chemical amplification positive resist materials for EB lithography, *Proc. SPIE*, **1672**, 194(1992).
99. H. Koyanagi, S. Umeda, S. Fukunaga, T. Kitaori, and K. Nagasawa, A new positive-acting chemically amplified resist system for electron-beam lithography, *Proc. SPIE*, **1672**, 125(1992).
100. H. Ban, J. Nakamura, K. Deguchi, and A. Tanaka, High-speed positive x-ray resist suitable for precise replication of sub-0.25-µm features, *J. Vac. Sci. Technol.*, **B12**, 3905(1994).
101. J. M. Kometani, M. E. Galvin, S. A. Heffner, F. M. Houlihan, O. Nalamasu, E. Chin, and E. Reichmanis, A novel approach to inducing aqueous base solubility in substituted styrene-sulfone polymers, *Macromolecules*, **26**, 2165(1993).
102. S. Uchino, T. Ueno, S. Migitaka, T. Tanaka, K. Kojima, T. Onozuka, N. Moriuchi, and M. Hashimoto, Contrast boosted resist (CBR) using base-labile water-repellent compounds, *Proc. Reg. Tech. Conf. Photopolymers*, SPE, Ellenville, New York, p.306(1994).

5
Practical Processes in Microlithography

5.1 ADHESION ENHANCEMENT OF SUBSTRATE

Silicon dioxide is the most commonly used insulator in semiconductor devices. It is formed on the surface of the silicon wafer by thermal oxidation of the surface with oxygen or water vapor at a temperature between 1000 and 1200°C. The oxide layer is also deposited onto a substrate which is not necessarily silicon from the vapor phase by the oxidation of silane with oxygen at a temperature between 400 and 500°C.

Although a fresh surface of a thermally grown silicon dioxide is hydrophobic, it reacts with water vapor in the atmosphere to form silanol (Si - OH) and becomes gradually hydrophilic. In the chemical vapor deposition of silicon dioxide, only silanolated surface is formed. Lack of adhesion of resist material to these silanolated surfaces is often observed in the course of development or wet etching. Therefore, a surface treatment to enhance the adhesion is necessary before the resist film deposition.

Collins and Deverse [1] invented a very effective process to improve the adhesion to the silicon dioxide surface. One of the simplest ways they have described is to subject the surface to an atmosphere containing the vapor of hexamethyldisilazane [$(CH_3)_3$Si-NH-Si$(CH_3)_3$] for a period of time at a temperature sufficient for the surface to react with this reagent. Hexamethyldisilazane (HMDS) is a colorless liquid which boils at 125°C. In a practical application, the surface of oxide layer is exposed the vapor of HMDS about one minute at a temperature between 60 and 70°C.

Collins and Deverse, the inventors of this surface treatment, described HMDS as an adhesive. However, HMDS is now regarded as a silane-coupling agent which converts hydrophilic silanol groups on the surface into hydrophobic siloxanes by the reaction illustrated in Fig. 5.1.

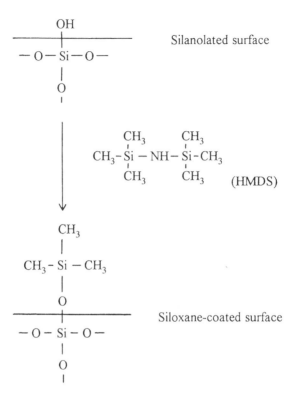

Fig. 5.1 Reaction of HMDS with a silanolated surface of silicon dioxide.

Yanazawa carried out detailed experiments regarding the effect of silane-coupling agents including HMDS on several kinds of surfaces relevant to semiconductor devices [2]. Although his work was focused on the surfaces of silicon dioxide and silicon nitride, other surfaces were also dealt with there. A summary of his results is shown in Table 5.1, where the adhesion of positive photoresist (AZ 1350J) to each surface is indicated quantitatively and related to the adhesion factor f defined by

Table 5.1. Observed photoresist adhesion characteristics and calculated adhesion factor f for surfaces relevant to semiconductor devices. (H. Yanazawa [2])

Substrate	Adhesion	f defined by eq. (5.1)
SiO_2		
As grown	Good	0.8
Etched with aq. HF	Poor	1.2
HMDS-treated	Good	0.25
SiO_2-P_2O_5		
As deposited	Poor	1.35
HMDS-treated	Good	0.5
Polycrystalline Si		
As deposited	Good	1.0
HMDS-treated	Good	0.75
Si_3N_4		
As deposited	Poor	1.35
HMDS-treated	Good	0.35
Al		
As deposited	Good	0.52

$$f = \frac{\gamma_L (\cos\theta_{L/P} + \cos\theta_{L/S})}{2(\gamma_P^d \gamma_S^d)^{1/2}} . \qquad (5.1)$$

Here, γ_L is the surface tension of water, $\theta_{L/P}$ is the contact angle of water on the photoresist and $\theta_{L/S}$ is that of water on the substrate, and γ^d_P and γ^d_S are the dispersion components of surface free energy of photoresist and substrate, respectively. The dispersion component of surface free energy of a solid material γ^d was estimated by

$$\gamma^d = \frac{\gamma_L (1+\cos\theta)^2}{4} , \qquad (5.2)$$

where γ_L and θ are the surface tension and contact angle of a nonpolar liquid on the surface of the solid material. The adhesion factor f is considered to be a balancing factor between adhering and liquid-inserting forces as illustrated in Fig. 5.2.

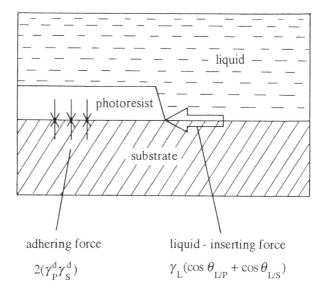

Fig. 5.2. Model for adhesion of photoresist to substrate immersed in a liquid. (H. Yanazawa [2])

As seen in Table 5.1, for the surface on which the adhesion of photoresist is poor, f exceeds unity and vice versa, indicating that f can be used as a criterion of resist adhesion in the processes of development and etching.

5.2 RESIST COATING PROCESS

5.2.1 Dynamics of Spin Coating

Spin coating is the most common method to coat a substrate with a resist solution in microlithography. The solution is spread onto the substrate which is held on a vacuum spindle. The substrate is then instantly accelerated up to a constant rotating speed, which is held for 30 to 60 seconds. The film of solution is thinned rapidly by centrifugal force and dried to form a solid film.

The first mathematical description of spin-coating process was presented by Emslie and others [3]. In their work, a Newtonian fluid (with a linear relationship between shear stress and shear rate), was assumed to be flowing on a rotating infinite plane. In this subsection, we follow their work to understand why a uniform film thickness is achieved by spin coating.

We consider a liquid flow on a rotating disk as shown in Fig. 3.5 (A).

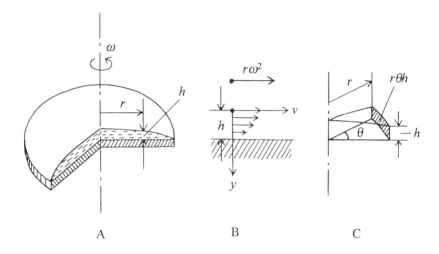

Fig. 5.3. Liquid flow on a rotating disk.

The centrifugal force at a distance r [cm] from the rotation axis is $mr\omega^2$ [dyne], where m [g] is the mass of the liquid under consideration and ω [1/s] is the angular velocity of the rotating disk. By this force, there arises a shear stress within the liquid, which is expressed as $\rho r\omega^2 y$ [dyne/cm^2], where ρ [g/cm^3] is the density of the liquid and y [cm] is the depth from the surface, as illustrated in Fig. 5.3 (B). If the liquid has a Newtonian flow characteristic, the radial velocity v [cm/s] of the liquid at the depth y must satisfy the equations,

$$\eta \frac{\partial v}{\partial y} = -\rho r\omega^2 y , \qquad (5.3)$$

$$v = 0 \quad \text{at} \quad y = h , \tag{5.4}$$

where η [poise=gcm^{-1}s^{-1}] is the viscosity of the liquid and h [cm] is the thickness of the liquid film at the distance r.

Integration of eq. (5.3) with regard to y under the boundary condition (5.4) results in

$$v = r\omega^2 \frac{\rho}{2\eta}(h^2 - y^2) . \tag{5.5}$$

If we use the kinetic viscosity, $\kappa = \eta/\rho$ [cm^2/s], eq. (5.5) becomes

$$v = r\omega^2 \frac{1}{2\kappa}(h^2 - y^2) . \tag{5.6}$$

Under the assumption that the rotational velocity is constant and the Coriolis force is neglected, the flow of liquid is confined within a wedge-shaped portion shown in Fig. 5.3 (C). The space velocity V [cm^3/s] of the liquid passing through the cross-section of the wedge at the distance r, indicated by hatching in Fig. 5.3 (C), is calculated from eq. (5.6) as follows.

$$V = r\theta \int_0^h v\,dy = r^2\theta \frac{\omega^2}{3\kappa} h^3 , \tag{5.7}$$

where θ is the wedge angle.

The equation of volume balance in the flow is written as

$$r\theta \cdot dr \cdot \frac{\partial h}{\partial t} = -\frac{\partial V}{\partial r} \cdot dr , \tag{5.8}$$

where t is the time.

Using eqs. (5.7) and (5.8), we obtain a fundamental equation in the spin-coating process,

$$\frac{\partial h}{\partial t} = -\frac{\omega^2}{3\kappa}(2h^3 + 3rh^2 \frac{\partial h}{\partial r}) . \tag{5.9}$$

By solving eq. (5.9) under a given initial condition, we obtain the liquid film thickness h at a distant r from the rotation axis at a time t.

First, we consider the case where the initial film thickness is uniform, that is,

$$h = h_0 = \text{constant} \quad \text{at } t = 0. \tag{5.10}$$

In this case, the last term in the right-hand side of eq. (5.9) vanishes, and the equation becomes independent of r, giving an analytical solution

$$\frac{1}{h^2} = \frac{1}{h_0^2} + \frac{4\omega^2}{3\kappa}t. \tag{5.11}$$

Equation (5.11) shows that the film thickness decreases uniformly with increasing time. An example of this time dependence of film thickness is shown in Fig. 5.4, where the kinetic viscosity and the spin speed are assumed to be 0.3 cm²/s and 3000 rpm (revolutions per minute), respectively. The initial film thickness was varied from 0.5 to 2 mm. However, the difference in the calculated film thickness decreased very rapidly with the increasing time, and disappeared in the plotting in Fig. 5.4.

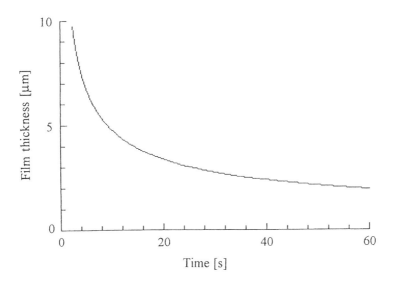

Fig. 5.4. Thickness of a liquid film as a function of spinning time. The kinetic viscosity and the spin speed are 0.3 cm²/s and 3000 revolutions per minute, respectively.

Secondly, we consider a general case where the initial film thickness varies along the radial direction. In this case, eq. (5.9) can not be solved analytically. However, numerical calculation shows that the solution approaches very rapidly to a solution with uniform film thickness. An example of the result of numerical calculation is shown in Fig. 5.5, where a parabolic profile of initial film thickness is assumed, and κ and ω are assumed to be the same as used in the calculation described above.

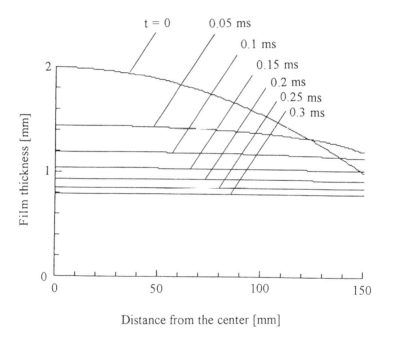

Fig. 5.5. Film thickness profile as a function of spinning time.

As shown in Fig. 5.5, the radial change in film thickness, which is initially 2 to 1 mm, becomes indiscernible in the figure only after 0.25 ms of spinning. By combining this result with that in the case of uniform film thickness, we conclude that a spin-coating process using a Newtonian resist solution gives a uniform film thickness regardless of the initial film thickness profile.

In the discussion described above, we neglected the Coriolis force which is perpendicular to the flow of liquid. The ratio of its strength to that of centrifugal force reaches $\omega h^2/\kappa$. Therefore, in the early stage of spinning where h is relatively large, we must consider a tangentially distorted wedge instead of that shown in Fig. 5.3 (C). However, as the spinning continues, h decreases very rapidly, as indicated in Fig. 5.5, and the effect of the Coriolis force becomes small enough to be neglected.

5.2.2 Thickness-Spin Speed Correlation

In a spin-coating process, the flow of coating solution under centrifugal force and the evaporation of solvent take place simultaneously, making it difficult to correlate the final solid film thickness with the rotational velocity.

An example of an experimentally obtained solid film thickness-spin speed correlation is shown in Fig. 5.6, where silicon wafers are coated with 20% of poly(vinylphenol) with cyclohexanone at different spin speeds. The acceleration time and spinning time in each coating process were 0.1 and 60 s, respectively.

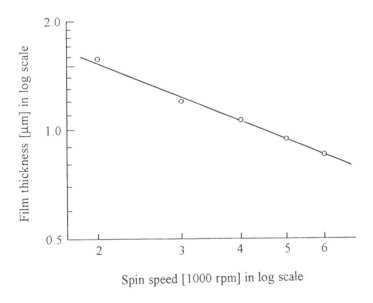

Fig. 5.6. Final solid film thickness as a function of spin speed. A 20% solution of poly(vinylphenol) with cyclohexanone is spin-spread onto a silicon wafer.

As seen in Fig. 5.6, the solid film thickness in this case is nearly proportional to $f^{-0.56}$, where f is the rotational velocity.

Meyerhofer reported that the final solid film thickness of a positive photoresist is inversely proportional to the square root of rotational velocity [4]. He also developed a model similar to that considered in the foregoing subsection but allowed the solvent to evaporate during the spinning process. For the purpose of calculation, he assumed a constant rate of evaporation and the viscosity-concentration relationship expressed by

$$\eta = \eta_{\text{solvent}} + \eta_{\text{solid}} c^{\gamma}, \tag{5.12}$$

where c is the concentration of solid, and γ has a value around 2.5. The calculated results on the relationship between solid film thickness and rotational velocity were in good agreement with the experimental results.

5.2.3 Prebaking

After a resist film is formed by a spin-coating process, the film is usually baked. The purpose of the baking is to evaporate the solvent remaining in the film and enhance the adhesion to the substrate. The apparatus used there is an air oven, where the resist-coated wafer is placed on a metal plate kept at oven temperature. In some cases, the wafer is pressed onto the metal plate by air suction to enhance the thermal conduction. The temperature and time of prebaking are usually 60 to 100°C and 1 to 20 minutes, respectively. In the case of a specially designed chemical amplification resist, the baking temperature is elevated to ensure the complete evaporation of coating solvent and to decrease the free volume in the resist film [5].

5.3 EXPOSURE OF PHOTORESIST

5.3.1 Contact Printing

After the prebaking, the film of photoresist is exposed to visible or ultraviolet light. In the case of contact printing, the photoresist-coated wafer is fixed onto a flat face of mobile stage by air suction and brought under an optical mask usually called a "photomask". After being aligned with the mask pattern by using an optical microscope, the wafer is pressed to the photomask by vertical movement of the stage. The film of photoresist is then exposed to a vertically incident flux of light emitted from a high-pressure mercury lamp.

Silver halide photographic plates have been most commonly used as the photomasks in photolithography. However, in the microfabrication of semiconductor devices, they have been almost completely replaced by chrome masks because of their insufficient resolution and mechanical hardness. A cross-section of a chrome mask is illustrated in Fig. 5.7.

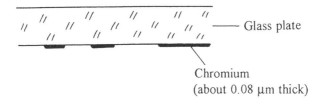

Fig. 5.7. Cross-sectional view of a chrome mask.

In the preparation of a chrome mask, chromium is evaporated onto a glass plate to a thickness of about 0.08 μm and then patterned by electron-beam lithography.

A negative type of photoresist composed of cyclized polyisoprene and aromatic bisazide is most commonly used in contact printing for the microfabrication of semiconductor devices. The resist is suited for this purpose because it is mechanically strong and not damaged by contact with the photomask.

5.3.2 Projection Printing

Practical projection exposure is carried out by using a very sophisticated exposure system, where all the functions including loading and unloading of wafers, focusing, alignment and exposure time control are automated. The most commonly used type of projection exposure system is the reduction projection type, where the pattern image with designed dimensions is projected through the lens from an enlarged mask pattern. The reduction ratio is usually 1/5.

The light used to expose the resist film in lens-projection printing is monochromatized by an interference filter to avoid the effect of chromatic aberration. The lights commonly used in this type of exposure system are g-line (with the wavelength of 0.436 μm) and i-line (0.365 μm), both emitted from a high-pressure mercury lamp. The same type of exposure system using the light from a KrF excimer laser (oscillating at the wavelength of 0.248 μm) has been recently developed.

A positive type of photoresist is now being used, most commonly in the projection printing, mainly because of its high resolution capability. The resist is not suited for contact printing because it is mechanically fragile and apt to be damaged by contact with the photomask. However, the fragility of the resist sets no problem in projection printing. The phenolic resin-based negative type of photoresists and chemical amplification photoresists are also used in projection printing.

5.3.3 Phase-Shifting Masks

The optical mask used in reduction projection printing is called the "reticle". The mask pattern on the reticle is enlarged from the pattern with designed dimensions by a factor m, where $1/m$ is the reduction ratio of the optical system. At present, most reticles are prepared from chrome mask blanks by using electron-beam lithography.

In the phase-shifting projection printing, the reticle has phase-shifting areas which shift the phase of light waves by 180°. In the case of an additive type of phase-shifting mask, a patterned layer of phase shifter is formed on the surface of mask, as already shown in Fig. 1.2 (B). The material of the phase shifter is conveniently deposited by spin coating. The material used there is called spin-on glass, which is tetraethoxysilane dissolved in ethanol. When deposited by spin coating and then baked, it gives a layer of silicon dioxide. Afterwards, the layer is patterned by using electron-beam lithography.

Several types of phase-shifting printing have been proposed, and some of them, including the original one shown in Fig. 1.2 (B), have already been applied to practical microfabrication. Among them, the so-called half-tone phase-shifting printing is most frequently used in the practical application. An example of the mask used in this type of printing is shown in Fig. 5.8.

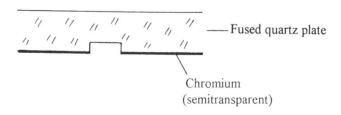

Fig. 5.8. Cross-sectional view of a half-tone phase-shifting mask.

The mask shown in Fig. 5.8 is a subtractive type, where the phase shift arises from the etched depth in the transparent areas. The light intensity distribution of the image of this type of phase-shifting mask can be calculated in the same manner as described in Section 3.3. An example of the calculated result is shown in Fig. 5.9, where the light intensity distribution in the vicinity of the image projected from an isolated slit pattern of a conventional mask is compared with that from a phase-shifted pattern with the same dimension.

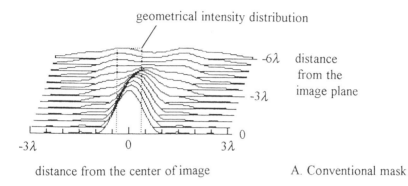

A. Conventional mask

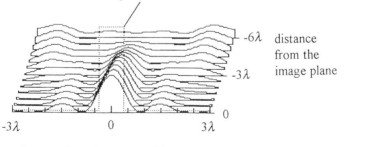

B. Half-tone phase-shifting mask

Fig. 5.9. Calculated light intensity in the vicinity of the image of isolated slit projected from conventional mask (A) and that from phase-shifting mask (B), where the optical transmittance of background is 2.5 %. The width of the geometrical image is 0.75λ, where λ is the wavelength of light. The NA and σ of the optical system are assumed to be 0.5 and 0, respectively.

As seen in Fig. 5.9, the light intensity profile is thinned by the use of the phase-shifting mask, resulting in the increase in resolution of pattern transfer. The width of the geometrical line image which is transferred from the phase-shifting mask pattern to the resist pattern within the deviation range of ±10% by four times the threshold exposure dose is calculated and plotted against the absolute distance from the image plane in Fig. 5.10.

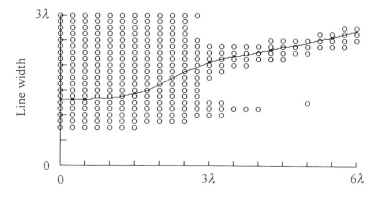

Distance from the image plane

Fig. 5.10. Width of geometrical bright line image transferred from half-tone phase-shifting mask pattern to resist pattern within the deviation range of ±10% by four times the threshold exposure dose as a function of distance from the image plane. The solid curve indicates the minimum line width of the geometrical image transferred from a conventional mask under the same condition. The optical transmittance of the background of the phase-shifting mask is 2.5 %. The NA and σ of the optical system are 0.5 and 0, respectively.

As seen in Fig. 5.10, a higher resolution than that obtained by the conventional method (indicated by the solid curve in the figure) is achieved by using the phase-shifting mask. There is no theoretical basis of four times the threshold exposure dose being optimum in the case of phase-shifting exposure. However, it may be reasonable in practical applications to control the exposure dose to be that value, especially in the case of hybrid use of conventional and phase-shifting mask patterns.

Practical Processes in Microlithography 147

A very thin black line image is obtained by using a transparent phase-shifting mask. The mask used to produce such an image is illustrated in Fig. 5.11. The half-wavelength phase shift arises from the step formed on the mask.

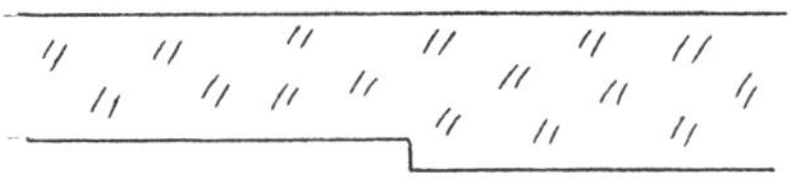

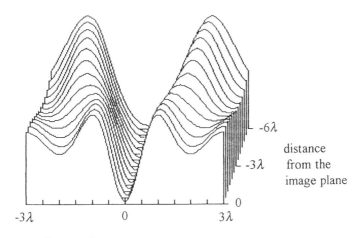

Fig. 5.11. Phase-shifting mask to produce a thin black line image.

An example of the calculated light intensity distribution in the vicinity of the black line image projected by using this type of phase-shifting mask is shown in Fig. 5.12.

Fig. 5.12. Calculated light intensity in the vicinity of the image projected from the step on a transparent phase-shifting mask. The NA and σ of the optical system are 0.5 and 0, respectively.

The width of black line image projected from the transparent phase-shifting mask is dependent on the NA of the optical system. The width of resist line pattern obtained by four times the threshold exposure dose is calculated and plotted against the absolute distance from the image plane for varied NA in Fig. 5.13.

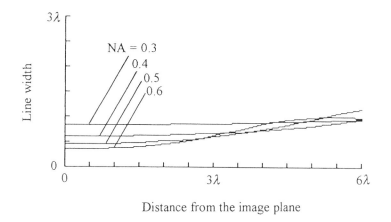

Fig. 5.13. Line width of resist pattern transferred from a step on a transparent phase-shifting mask by four times the threshold exposure dose as a function of distance from the image plane and the NA of the optical system.

As seen in Fig. 5.13, a resist line pattern with a width smaller than half wavelength is obtained by using this type of phase-shifting mask and an optical system with an NA not smaller than 0.5. Similarly to the case of half-tone phase-shifting mask, there is no theoretical basis of four times the threshold exposure dose being optimum in the present case. As indicated in Fig. 5.12, a thinner resist pattern can be obtained by lowering the exposure dose.

5.3.4 Post-Exposure Baking

In the case where a chemically amplified resist is used, post-exposure baking is generally necessary to promote the chemical amplification. The baking process is carried out in the same manner as that of the prebaking process. In some cases where the chemical amplification takes place very rapidly at room temperature, the process is not necessary [6].

Practical Processes in Microlithography 149

5.4 DEVELOPMENT OF RESIST

5.4.1 Development Process

After the exposure or post-exposure baking, the resist film is developed in an automated development apparatus. The apparatus has a vacuum spindle, on which the wafer is fixed by air suction and then the resist film on it is developed with liquid developer. In most cases, the development is carried out in puddle style as illustrated in Fig. 5.14.

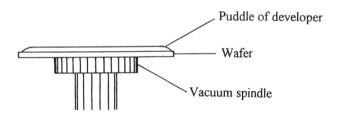

Fig. 5.14. Puddle-style development.

In the puddle-style development, the developer is poured onto the resist-coated surface of the wafer fixed on the still spindle to make a puddle. After a predetermined period of time, the developer on the wafer driven out by spinning the spindle. After that, the wafer on the rotating spindle is washed with an organic solvent or pure water and then dried by fast spinning.

5.4.2 Chemical Development

There are two types of development process, which are chemical development and physical development. In the case of chemical development, the main component of resist system dissolves in the developer by a chemical reaction. A typical example of chemical development is that for a positive photoresist composed of phenolic resin and naphthoquinonediazide. The resist is developed with the aqueous solution of a strong base such as trisodium phosphate or tetramethylammonium hydroxide. The phenolic resin in the resist dissolves in the developer by the chemical reaction expressed by

$$P\text{-}OH + M^+ + OH^- = P\text{-}O^- + M^+ + H_2O,$$

where P-OH, P-O$^-$ and M$^+$ stand for phenolic resin, phenolate ion, and alkalimetal or tetraalkylammonium ion, respectively.
The reaction does not occur on the unexposed areas because of the dissolution-inhibiting effect of naphthoquinonediazide which remains unchanged there. Therefore, the unexposed resist film does not either dissolve or swell in the developer. This nonswelling type of development is one of the main reasons why a positive photoresist has a high resolution capability.

5.4.3 Physical Development

A typical example of physical development is that for a negative photoresist composed of cyclized polyisoprene and aromatic bisazide. An unexposed film of the resist is developed with an organic solvent such as xylene. No chemical reactions take place during the development process. The resist film on the exposed areas has been insolubilized by the photocrosslinking reaction, and, therefore, does not dissolve in the developing solvent. The insolubilized resist film, however, swells in the developing solvent. This swelling lowers the resolution capability of the resist significantly.
  In the case where a chemically amplified resist is developed in a nonpolar solvent to give a negative pattern, the resist is also physically developed [7].

5.4.4 Postbaking

The resist pattern obtained by the development is usually baked to enhance both the adhesion to the substrate and resistance to etching. The baking process is carried out in the same manner as that of prebaking. The baking temperature is between 90 and 180°C. In the case of a phenolic resin-based resist, the temperature is usually kept below 150°C in order to prevent the thermal deformation of the resist pattern profile.

5.4.5 Deep UV Hardening

The pattern of a phenolic resin-based resist is hardened by exposure to short-wavelength ultraviolet light [8]. This hardening process is useful to prevent the thermal deformation of the resist pattern profile during the dry-etching process, where the temperature of the wafer often exceeds the deformation temperature of an untreated resist. A low-pressure mercury lamp is normally used as a source of short-wavelength ultraviolet light.

5.5 ETCHING

5.5.1 Wet Etching

Before the dry etching technique was developed, all of the etching processes in semiconductor device fabrication have been carried out by wet etching where the substrate was etched with an etchant solution. The process is comprised of dipping the wafer in the etchant solution and subsequent cleaning with pure water. The etch rate is mainly dependent on the composition of etchant solution and the temperature. Since these two factors and the etching time are all easily controllable, the reproducibility of the result is generally good.

Silicon dioxide is etched with a mixed aqueous solution of hydrogen fluoride (HF) and ammonium fluoride (NH_4F). A detailed study on the etch rate in this case has been published by Parisi and others [9]. They have presented a simple equation,

$$\gamma_d = 4.5 \times 10^9 \, [HF] \, \exp(-4980/T), \qquad (5.13)$$

where γ_d = etch rate in angstroms per minute, [HF] = concentration of hydrogen fluoride in moles per liter, and T = absolute temperature in degrees Kelvin. This expression is valid for the dissolution of thermally grown silicon oxide within the range of parameter levels, viz., etchant solutions between [HF] = 0.9 and 3.8 and temperature range from 25 to 55°C. For example, γ_d for [HF] = 3.8 and T = 298 (25°C) is calculated to be 950, which is very much close to the observed value.

Silicon is etched with a mixed aqueous solution of hydrogen fluoride (HF) and nitric acid (HNO_3). Detailed experiments on this etching process have been carried out by Schwartz and Robbins [10]. According to the data they have obtained, the etching proceeds by a sequential oxidation-followed-by-dissolution process. In those composition regions where the solution is very low in HNO_3 and rich in HF, the rate-limiting process is the oxidation step.

Consequently, the electron concentration, crystal defects, and catalysis by lower oxides of nitrogen play an important role. In those composition ranges where HF is in limited supply, dissolution of the formed oxide is the rate-controlling step and the diffusion of fluoride species is the important factor. The latter case seems to be suited for practical applications because the etch rate is mainly dependent on the concentration of HF. To give an example, the etch rate of the (111) plane of 2 Ω-cm n-type silicon in 7.9% HF-58.4% HNO_3 solution is measured to be 25 micrometers per minute.

Aluminum is etched with many kinds of etchant. However, the etchants used in microlithography must satisfy the condition that they etch aluminum without evolving bubbles of hydrogen which hinder the progress of etching locally. An example of such an etchant solution is composed of phosphoric acid, nitric acid, acetic acid and water in volume ratio of 16:1:2:1. The evolution of hydrogen is eliminated by the oxidative reaction of nitric acid.

Chromium is etched by an aqueous acidic solution of ceric salt. An example of etchant solution is prepared by mixing 17g of ceric nitrate, 5ml of perchloric acid and 100ml of water. Most chrome masks are now being fabricated by using this type of etchant solution.

5.5.2 Plasma Etching

The first dry etching technique introduced in the microfabrication of semiconductor devices was called plasma etching. As the etching in this technique proceeds isotropically, the term "plasma etching" at present implies an isotropic etching of the substrate. The main part of a barrel-type plasma etching apparatus is illustrated in Fig. 5.15.

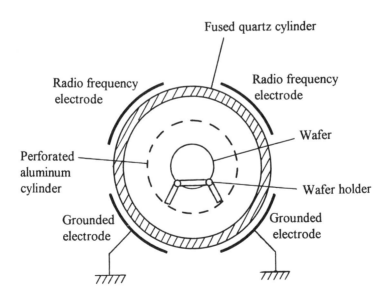

Fig. 5.15. Barrel-type plasma etching apparatus.

The wafers are coaxially held by a holder made of fused quartz and placed at the center of a barrel-type chamber. After the chamber has been evacuated, gaseous carbon tetrafluoride (CF_4) is introduced into the chamber to a pressure of about 1 torr (133 Pa), and then a 13.75 MHz electric power is applied to external electrodes to produce a gaseous plasma in the chamber. A variety of ionic and neutral fragments including C^+, F^+, CF^+, CF_2^+, CF_3^+, C, F, CF, CF_2 and CF_3 has been detected in such plasma [11].

As the wafers are shielded from the ionic fragments by a perforated aluminum cylinder, the etching proceeds only by the reaction of the fluorine radical (F) with the substrate. Substrate materials such as silicon, silicon dioxide and silicon nitride are etched by this technique. The etch rates are in the order as silicon dioxide < silicon nitride < silicon. The etch rate of silicon nitride in practical applications is between 10 and 80 nm per minute. Since the etching proceeds isotropically, this type of dry etching is not suited for precise pattern transfer, and consequently, has been replaced mostly with an anisotropic type called reactive ion etching.

5.5.3 Reactive Ion Etching

Reaction ion etching (RIE) is the most frequently used dry etching technique in the microfabrication of semiconductor devices. It is essential for this technique that the wafers are electrically contacted with one of the internal electrodes, and the inner pressure of the chamber is low (below 10 Pa). A typical example of RIE apparatus is illustrated in Fig. 5.16.

A volume of plasma is produced in the space between the two electrodes by radio frequency gas discharge. As the wafers are contacted with one of the electrodes, the ambient condition on the surface of the wafer is the same as that of the electrode. Because of a large difference in mobility between electrons and ions, an ion-rich layer called an ion sheath is formed on both of the surfaces. At the time when the surfaces of wafer and electrode are negatively charged, a large difference in electric potential is arises across the ion sheath as illustrated in Fig. 5.17.

By the gradient of potential, ions in the ion sheath are accelerated perpendicularly to the surface of the wafer. In the case where the pressure is relatively low, and consequently, the mean free path of ions is comparable with the thickness of the ion sheath, the ions are incident on the surface nearly perpendicularly. As the etching reaction is significantly accelerated by the impact of incident ions, the etching proceeds aniotropically.

The gases commonly used in RIE are shown in Table 5.2.

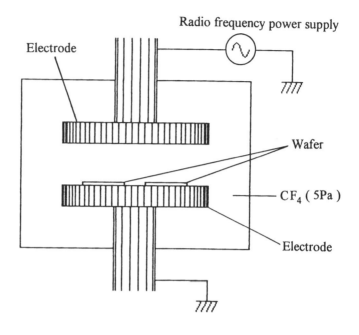

Fig. 5.16. Reactive ion etching apparatus.

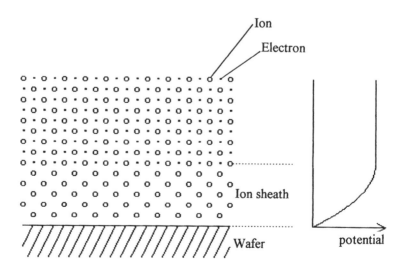

Fig. 5.17. Ion sheath and the accelerating potential in it.

Table 5.2. Gases used in reactive ion etching.

substrate material	gases
silicon	CF_4, SF_6, $CF_4 + O_2$, CCl_4
silicon dioxide	$CF_4 + H_2$, C_2F_6, C_3F_8, C_4F_8
silicon nitride	$CF_4 + H_2$, CH_2F_2, CH_3F
aluminum	CCl_4, BCl_3, Cl_2
refractory metals and refractory silicides	CF_4, SF_6, CCL_4, Cl_2, $Cl_2 + O_2$

Silicon is etched by RIE using a fluorine-containing gaseous compound such as carbon tetrafluoride (CF_4). The addition of oxygen to the gas composition significantly increases the etch rate of silicon. The increase in the etch rate is attributed to the increase in concentration of the fluorine radical (F) in the plasma. Since the etch rate of the resist material is also increased by the addition of oxygen, the concentration of oxygen in the gas composition is practically limited within the range of 10%. Silicon is also etched by RIE using a chlorine-containing gaseous compound such as carbon tetrachloride (CCl_4). Although the etch rate is relatively low, nearly vertical side walls of the etched pattern are formed by this type of etching.

Silicon dioxide is etched by RIE using a fluorine-containing gaseous compound. In most cases of practical application, the layer of silicon dioxide is formed or deposited on the surface of silicon and patterned by a microlithographic process. Therefore, there must be a large difference in etch rate between silicon dioxide and silicon in order that the etching is easily stopped at the interface of oxide and silicon. All of the gases shown as the etchant gases for silicon dioxide in Table 5.2 are suited for this purpose. Ephrath has reported that silicon dioxide-to-silicon etch rate ratios as high as 35 to 1 have been measured and silicon dioxide-to-resist etch rate ratios have been found to exceed 10 to 1 in the case where CF_4-H_2 is used as an etchant gas mixture [12]. He explained the observed large difference in etch rate as the result of etching retardation by polymeric films formed on silicon and resist surfaces.

Silicon nitride is etched by RIE using a fluorine-containing gaseous compound. In the case where the layer of nitride is deposited on the surface of silicon, the etch rate of silicon substrate must be relatively low compared with that of nitride by the same reason as described above.

Aluminum is etched by RIE using a chlorine-containing gaseous compound such as carbon tetrachloride (CCL_4). Fluorine-containing gaseous compounds can not be used for this purpose because of the formation of nonvolatile aluminum fluoride film on the surface of aluminum.

Refractory metals and refractory silicides such as tungsten (W) and tungsten silicide (WSi_2) are etched by RIE using fluorine-containing or chlorine-containing gaseous compounds such as carbon tetrafluoride (CF_4) or carbon tetrachloride (CCl_4). In the case where these materials are deposited on the surface of silicon dioxide, the etch rate of the oxide substrate is required to be much lower than that of the deposited material to prevent the etching of the substrate. The use of a chlorine-containing etchant gas such as CCl_4 meets this requirement.

5.6 RESIST REMOVAL

After the etching process, the resist pattern remaining on the surface of wafer is removed by either a wet or dry process.

In the wet process, the resist is removed by using a commercially available resist remover which is a mixture of strong solvents. The wafer is dipped in the remover heated at about 100°C for 10 to 20 minutes, washed with a solvent such as methanol, washed again with pure water, and finally dried.

Oxygen plasma is most frequently used in the dry process of resist removal. As all of the practical resist materials are composed of organic compounds, they are removed from the substrate by the oxidative reaction of oxygen plasma. The apparatus used for this process is similar to that shown in Fig. 5.15 or Fig. 5.16. The pressure of oxygen is about 1 torr (133 Pa).

REFERENCES

1. R. H. Collins and F. T. Deverse, Process for improving photoresist adhesion, *U. S. Patent* 3,549,368 (1970).
2. H. Yanazawa, Adhesion model and experimental verification for polymer-SiO_2 system, *Colloids and Surfaces*, **9**, 133-145 (1984).
3. A. G. Emslie, F. T. Bonner and L. G. Peck, Flow of a viscous liquid on a rotating disk, *J. Appl. Phys.*, **29**, 858-862 (1958).

4. D. Meyerhofer, Characteristics of resist films produced by spinning, *J. Appl. Phys.*, **49**, 3993-3997 (1978).
5. H. Ito, G. Breyta, D. Hofer, R. Sooriyakumaran, K. Petrillo and D. Seeger, Environmentally stable chemical amplification positive resist: Principle, chemistry, contamination resistance and lithographic feasibility, *J. Photopolymer Sci. Technol.*, **7**, 433-447 (1994).
6. T. Hattori, A. Imai, R. Yamanaka, T. Ueno and H. Shiraishi, Delay-free deprotection approach to robust chemically amplified resist, *J. Photopolymer Sci. Technol.*, **9**, 611-618 (1996).
7. H. Ito and C. G. Willson, Chemical amplification in the design of dry developing resist materials, Polymer Eng. Sci., 23, 1012-1018 (1983).
8. H. Hiraoka and J. Pakansky, UV hardening of photo- and electron beam resist patterns, J. Vac. Sci. Technol., 19, 1132-1135 (1981).
9. G. I. Parisi, S. E. Haszko and G. A. Rozgonyi, Tapered windows in SiO_2: The effect of NH_4F:HF dilution and etching temperature, *J. Electrochem. Soc.*, **124**, 917-921 (1977).
10. B. Schwartz and H. Robbins, Chemical etching of silicon IV. Etching technology, *J. Electrochem. Soc.*, **123**, 1903-1909 (1976).
11. T. M. Mayer and R. A. Barker, Reactive ion beam etching with CF_4: Characterization of a Kaufman ion source and details of SiO_2 etching, *J. Electrochem. Soc.*, **129**, 585-591 (1982).
12. L. H. Ephrath, Selective etching of silicon dioxide using reactive ion etching with CF_4-H_2, *J. Electrochem. Soc.*, **126**, 1419-1421 (1979).

6

X-ray Lithography

6.1 INTRODUCTION

Far shorter in wavelength of electromagnetic waves is the x-ray. It has passed almost a quarter century since x-ray lithography was described in 1972 by Spears and Smith [1]. A lot of research institutes have been concerned with x-ray lithography. However, the application of x-ray lithography to real production environments has not yet materialized. The main reason for this delay is that optical lithography has been doing much better than expected a decade ago. When x-ray lithography was proposed, the resolution limit of optical lithography was considered to be 2-3 µm. Today one does not have much difficulty in fabricating half-submicron patterns with a current i-line reduction projection aligner. Attempts to develop the exposure machines with large NA lens and for short-wavelength irradiation have been continued to obtain better resolution. Developments in resist processes and resist materials for high resolution have extended the use of optical lithography.

X-ray lithography is a novel technology requiring an entirely new combination of source, mask, resist and alignment systems. It is said that "quantum jumps" are required on each front for practical use of x-ray lithography [2]. Therefore, there has been reluctance to consider x-ray lithography as an alternative to optical lithography. In view of these difficulties and continued encroachment by optical lithography, it is not unreasonable to ask if x-ray lithography still has a future [3].

Optical lithography has been extending the resolution. However, it is difficult to obtain smaller patterns than the irradiation wavelength with a certain depth of focus, unless such an exotic technique as wavefront engineering including interference lithography [4,5] is utilized. No one denies that it is necessary to continue research on x-ray lithography to prepare for the end of the optical lithography era.

The x-ray lithography system proposed by Spears and Smith [1] is proximity

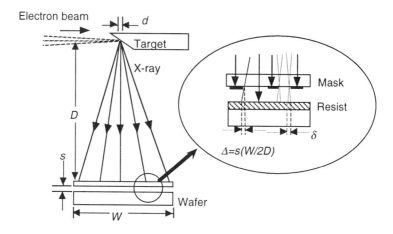

FIG. 6.1 Schematic diagram of an x-ray lithography system. [Reprint with permission from *Solid State Technol.*, 15(7), 21(1972).]

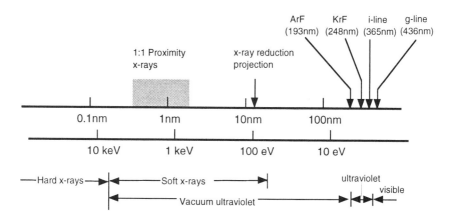

FIG. 6.2 Wavelength of x-rays.

X-ray Lithography

printing as shown in Fig. 6.1. Although there have been reports on reduction projection x-ray lithography (see 6.7), more focus on x-ray proximity printing is discussed here.

6.2 SELECTION OF X-RAY WAVELENGTH

It is generally accepted that the x-ray wavelength range lies from several 10 nm to 0.1 nm as shown in Fig. 6.2. X-ray overlaps with the ultraviolet region in the longer wavelength region and with γ-rays in the shorter wavelength region. Although x-rays cover a wide range of wavelengths, the region of wavelengths for x-ray lithography is rather limited.

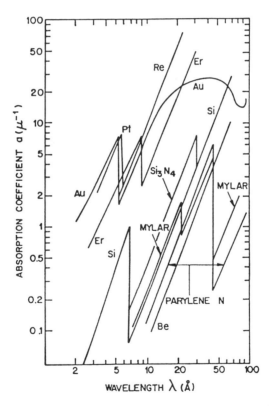

FIG. 6.3 Absorption coefficient of some of the most absorbing and most transparent materials for x-rays. [Reprint with permission from *Polm. Eng. Sci.*, **17**, 385(1977).]

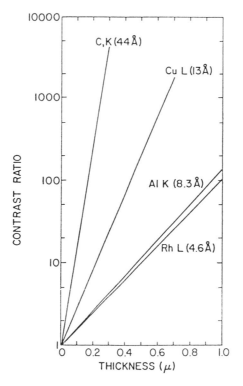

FIG. 6.4 X-ray contrast of Au as a function of Au thickness for four different wavelength. (Bremstrahlung neglected). [Reprint with permission from *Polm. Eng. Sci.*, **17**, 385(1977).]

The choice of materials for masking determines the wavelength range for x-ray lithography. The absorption coefficient for mask materials as a function of wavelength is shown in Fig. 6.3 [6]. A mask with a reasonable contrast ratio is necessary to obtain good definition of x-ray images. According to Spears and Smith [1], the wavelength must be longer than 0.4 nm to obtain 90% x-ray absorption using the most highly absorbing materials (Au, Pt, Ta, W etc.) with 0.5 μm thickness. On the other hand, substrate materials (Be, Si, SiC, Si_3N_4, BN and organic polymers) restrict the usable wavelength to less than 2 nm to transmit more than 25% of the incident x-rays (see Fig. 6.3). Thus, the wavelength range is limited to $0.4 < \lambda < 2$ nm. A plot of contrast ratios (or MTF of the mask) as a function of Au thickness for four different wavelengths is shown in Fig. 6.4 [6].

Since x-rays are generated under vacuum, a "vacuum window" separating the vacuum from exposing room is needed. The material (Be) for this window also dictates the wavelength range of the x-ray. Although exposure systems under vacuum have been proposed, wafer handling would be complicated.

As will be discussed in the next section, the range of photoelectrons and diffraction effect limit the resolution [7]. The effects of photoelectron range and diffraction on resolution are depicted in Fig. 6.5. The range of photoelectrons is smaller for longer wavelengths, while the diffraction effect is smaller for shorter wavelengths. The optimum wavelength for x-ray lithography seems to fall into ~1 nm.

It is known that absorption coefficient varies strongly with wavelength. Organic resist materials mainly consist of C, H, and O and absorption coefficient for these elements are higher in the longer wavelength region and lower in the shorter one. Therefore, the longer wavelength is desirable for higher resist sensitivity. The amount of the absorption energy is closely related to

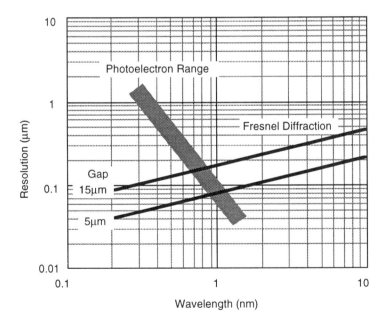

FIG. 6.5 Relation between resolution and x-ray wavelength. The effect of photoelectron range and mask-wafer gap on resolution as a function of the wavelength is also shown.

resist sensitivity, which will be described in 6.6.1.

In summary, the materials for the x-ray mask are of utmost importance in determining the wavelength range 0.4< λ< 2 nm.

6.3 RESOLUTION OF X-RAY LITHOGRPHY

6.3.1 Geometrical Factors

A typical exposure system for x-ray lithography using an electron-beam bombardment source is schematically depicted in Fig. 6.1. The opaque part of the mask cast shadows onto the wafer below. The edge of the shadow is not absolutely sharp because of the finite size of the x-ray source d (diameter of focal spot of electrons on the anode) at a distance D from the mask. If the gap between the mask and wafer is called s, the penumbral blur is given by

$$\delta = s(d/D) \tag{6.1}$$

Smaller gap and smaller x-ray source size lead to smaller penumbra. Small size as well as high intensity are important factors for developing x-ray sources. The incident angle of x-rays on the wafer varies from 90 degrees at the center of the wafer to $tan^{-1}(2D/W)$ at the edge of the wafer diameter W. The shadows are slightly longer at the edge by the amount Δ that is given by

$$\Delta = s(W/2D) \tag{6.2}$$

The smaller gap and larger D gives the smaller Δ. A full wafer exposure system can be adopted when this geometrical distortion is acceptable. Otherwise, the step-and-repeat exposure mode is necessary.

6.3.2 Effect of Secondary Electrons

Three processes are involved in the absorption of x-rays: Compton scattering, the generation of photoelectrons and the formation of electron-positron pairs. The magnitude of these three processes strongly depends on x-ray energy and the atomic number as shown in Fig. 6.6 [8].

The electron-positron pair formation is possible at x-ray energies above 1.02 MeV. The energies of the x-ray quanta in x-ray lithography are lower than 10 keV. For such energies, the cross section for the photoelectric effect is about

100 times larger than that for the Compton scattering. Therefore, only the effect of the photoelectrons needs to be considered.

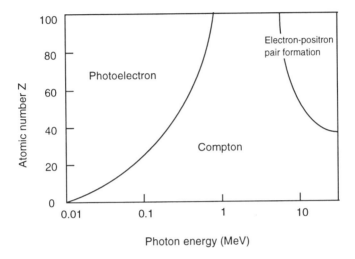

FIG. 6.6 Main contribution of photon-atom interaction as a function of atomic number and photon energy. [Reprint with permission from *Radiation Effect on Atoms and Molecules*, S. Shida Ed. Kyoritsu Pub. p. 30 (1966).]

The absorption of an x-ray by an atom is one of innershell excitation followed by emission of a photoelectron as shown in FIG. 6.7. When an x-ray photon is absorbed, a photoelectron is generated with kinetic energy E_p where

$$E_p = E_x - E_b. \tag{6.3}$$

Here E_x is the energy of x-ray photon and E_b is the binding energy required to release an electron from an atom. The vacancy that is created in the atom is filled quickly and the energy E_b is distributed to the surroundings via Auger electrons or fluorescent radiation as also shown in Fig. 6.7. Usually the vacancy is filled by an electron from the next higher level and the energy released is a fraction smaller than E_b. X-ray fluorescence occurs at a wavelength slightly longer than the wavelength corresponding to the absorption edge.

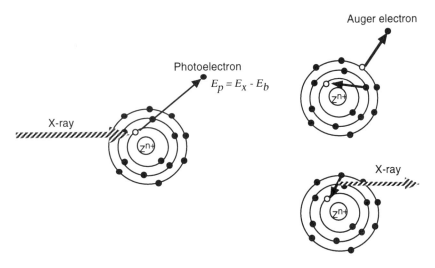

FIG. 6.7 Interaction of x-ray with atoms.

The ratio of the probability for Auger emission to fluorescence for a light element is 9:1. Hence for light elements 90% of the binding energy is transferred to the surroundings by Auger electrons. These electrons, produced by photoelectric effect and Auger process, cause the excitation and ionization of atoms and molecules in the resist film. The excitation and ionization cause a complicated process described in electron beam lithography, leading to chemical changes in polymers. Hence the range of these electrons relates to the resolution of exposed patterns.

The range of electrons in PMMA films as a function of electron energy is shown in Fig. 6.8 [1]. The energies of the characteristic copper, aluminum and molybdenum x-rays are indicated for comparison.

Measurements of the maximum penetration depth of electrons from a heavy metal layer into PMMA resists were carried out by Feder et al. [6]. The effective range was determined by the experimental arrangement shown in Fig. 6.9. The erbium film evaporated on the resist film acted as an x-ray absorber and electron generator to expose the resist. After the exposure the erbium film was removed and the resist was developed. In Fig. 6.9 a plot of the change in resist film thickness as a function of development time is also shown. In all cases the initial stages of development showed a rapid decrease in thickness followed by a normal development curve representative of PMMA. They regarded that extrapolation of the normal part of the curve as the effective range of the electrons.

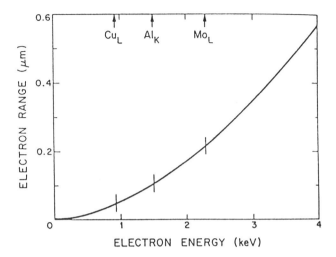

FIG. 6.8 Characteristic electron range as a function of electron energy for a typical polymer film ($\rho=1\text{g/cm}^3$). [Reprint with permission from *Solid State Technol.*, 15(7), 21(1972).]

It is clear from this figure that effective range of the electrons depends on the wavelength of the x-rays: the range is smaller for longer wavelengths. These values are much smaller than the expected values shown in Fig. 6.8, though the energy of photoelectrons from Er would be different from those from C and O. Experiments by Early et al. [9] and Deguchi et al. [10] also clarified that the maximum range of photoelectrons does not determine the resolution of x-ray lithography.

The energy delivered by photo- and Auger-electrons near the interface of resist and silicon substrate should be taken into account. Ticher and Hundt [11] have reported the depth profile of deposited energy of photo- and Auger electrons generated in the resist and silicon by Al_K and Rh_L x-rays. The angular distribution of photo- and Auger electrons is important in calculating the distribution of energy transferred. The Auger electrons have a spherical or isotropic distribution from their starting point, while the photoelectrons (generated by x-rays with energies below 10 keV) are preferentially emitted perpendicular to the impinging x-rays. While for Al_K radiation the electrons from the silicon have only minor influence on the deposited energy profile, the electrons generated by Rh_L in the silicon can give a high contribution to the energy density at the resist/silicon interface even in the unexposed area as shown in Fig. 6.10. This difference was attributed to the larger electron range for Rh_L

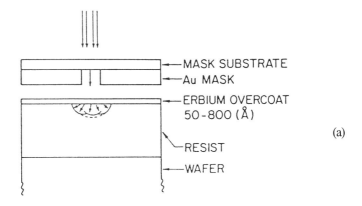

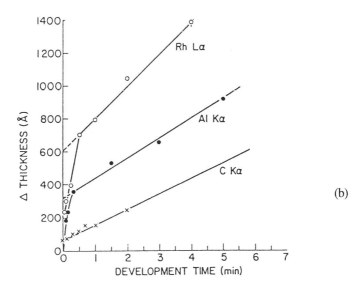

FIG. 6.9 Schematic showing for measurement of effective range of electrons generated by x-ray exposure (a). The depth of the developed exposed area in resist plotted as a function of development time (b). The intercepts on the vertical axis represent the maximum penetration depth of electrons as measured in the developed resist. [Reprint with permission from *Polm. Eng. Sci.*, **17**, 385(1977).]

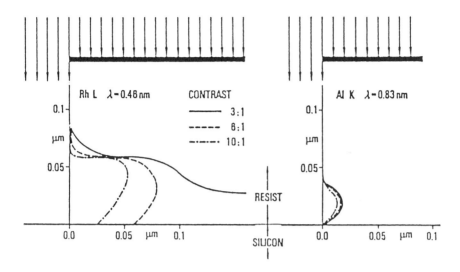

FIG. 6.10 Effect of the secondary (photo and Auger) electrons on a deposition energy in a resist film near the interface of the resist/substrate. Equi energy density curves ($D(x,z)/D_0=0.5$) near the bottom of the resist layer for various contrast values for Rh_L and Al_K radiation. [Reprint with permission from *Proc. Symp. 8th Electron Ion Beam Sci. Technol.*, vol.**78-5**, p.444(1987).]

than that for Al_K radiation. The calculated energy density curves for Rh_L and Al_K radiation are in good agreement with the resist profile after development.

Recently Ogawa et al. [12] have investigated the effect of secondary electrons from an Si substrate in synchrotron radiation x-ray lithography. The secondary electrons produce an undercut of 0.1 μm in the replicated resist patterns. To avoid larger energy deposition at the resist/silicon interface, it is important to expose them with a wavelength above the absorption edge of silicon: $\lambda > 0.7$ nm.

Photoelectrons are also generated from the absorber on the mask by the absorption of the incident radiation as shown in Fig. 6.11 [13]. These electrons cause unwanted exposure on the resist which deteriorates the mask contrast and system resolution. These photoelectrons can be eliminated by coating the mask with an organic layer.

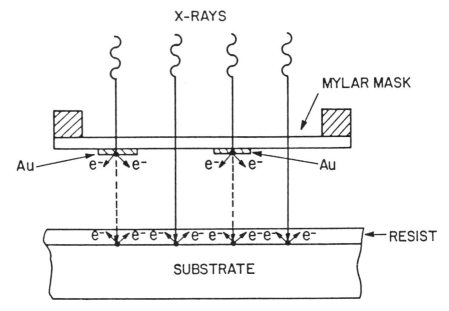

FIG. 6.11 Schematic representation of spurious exposure effects affecting the x-ray exposure. [Reprint with permission from *J. Vac. Sci. Technol.*, **12**, 1329(1975).]

6.3.3 Diffraction Effect

As described above, one has to take into account the effect of diffraction of x-rays on resolution, even though this effect in x-ray lithography is smaller than in optical lithography. Atoda et al.[14] have reported the diffraction effect on pattern replication with synchrotron radiation. In Fig. 6.5 the effect of diffraction for various gaps between mask and wafer is shown as a function of x-ray wavelength. X-ray intensity distribution calculated for line and space with infinite length is shown in Fig. 6.12 [14].

They discussed the pattern formation with parameter U_0 given by the following relation

$$W = U_0(\lambda s/2)^{1/2} \qquad (6.4)$$

where W is the pattern width, s the mask-to-wafer distance and λ the wavelength. From the observation of pattern transfer for various values of

U_0, it clearly demonstrated that the diffraction effect becomes serious for small patterns and large gaps (large U_0 in Fig. 6.12). The value of U_0 should be larger than 3 for satisfactory pattern replication. Similar analysis has also been reported by Smith's group [15].

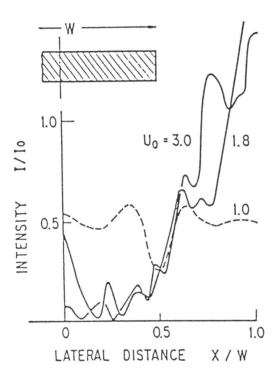

FIG. 6.12 Calculated intensity (I/I_0) distribution in direction across line and space pattern with finite length and period of $2W$. U_0 is $W/(\lambda s/2)^{1/2}$, where W is pattern width, s the mask-to-wafer gap, λ the x-ray wavelength. U_0 should be larger than 3 for satisfactory pattern replication [14]. [Reprint with permission from *J. Vac. Sci. Technol.*, **B1**, 1267(1983).]

6.4 X-RAY SOURCES

6.4.1 Electron Beam (EB) Bombardment X-ray Sources

X-ray radiation produced by bombardment of material with accelerated electrons has been utilized as an x-ray source since the discovery of x-rays. When an accelerated electron (several 10 keV) is bombarded at the target, two types of radiations are produced as shown in Fig. 6.13. One is continuous radiation primarily produced from interactions with nuclei; the electron sometimes emits energy when it experiences the strong electric field near the nucleus (bremsstrahlung). The other is characteristic radiation (x-ray fluorescence); when

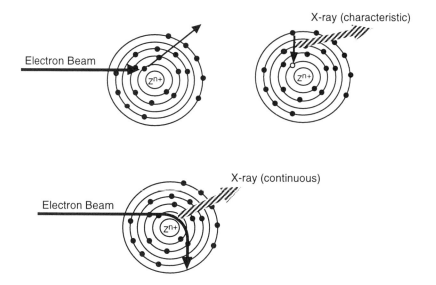

FIG. 6.13 Generation of x-ray with electron beam bombardment.

the incident electron energy is high enough, it can cause innershell excitation followed by the emission of characteristic lines. A typical x-ray spectra generated by EB bombardment is shown in Fig. 6.14 [16], which shows a sharp high intensity characteristic radiation and broad low intensity bremsstrahlung. Although the characteristic radiation is mainly used for x-ray lithography, the influence of the continuous radiation on exposure dose cannot be neglected. The energy of the characteristic line is determined by the materials used as a target.

A Gaines-type x-ray source is shown in Fig. 6.15 [17]. This source has an inverted cone geometry providing a large surface area with a minimum (and symmetric) project spot. At the same time, the inverted cone acts as an excellent black body absorber for electrons. The cathode of the electron gun is

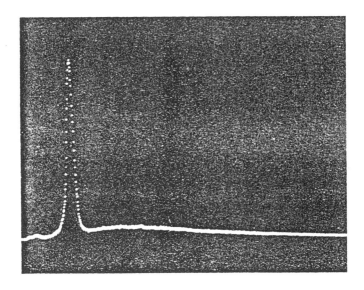

FIG. 6.14 X-ray emission spectra from Pd with electron beam bombardment. [Reprint with permission from *J. Vac. Sci. Technol.*, **B1**, 1251(1983).]

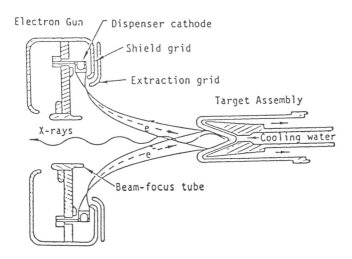

FIG. 6.15 Schematic view of electron gun and target assembly for x-ray generation. [Reprint with permission from *Nucl. Instrum. Method*, **126**, 99(1975).]

ring-shaped and masked from the view of the target, since it is necessary to prevent evaporated cathode material from being deposited on the targets. The target is cooled by high velocity water flow.

6.4.2 Synchrotron Orbit Radiation (SOR)

Synchrotron orbit radiation (SOR) is a very intense and well-collimated x-ray source. SOR is emitted when a relativistic electron experiences an acceleration perpendicular to its direction of motion: usually when relativistic electrons are bent round a radius by a magnetic field as shown in Fig. 6.16. It has received a

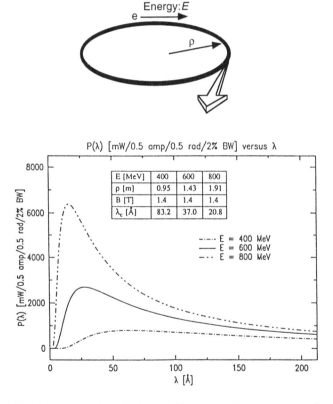

FIG. 6.16 Synchrotron orbit radiation and its spectral power output for different electron energies. [Reprint with permission from *Jpn. J. Appl. Phys.*, **28**, 2074(1989).]

great attention as a source for x-ray lithography since the report by Spiller et al. [18].

The characteristics of SOR are summarized as follows [18]:

(1) The radiation is a broad continuum, spanning the infrared through the x-ray range.
(2) The intensity of the flux is several orders of magnitude larger than that of conventional EB bombardment sources.
(3) The radiation is collimated vertically and its divergence is small.
(4) The radiation is horizontally polarized in the orbital plane (electron trajectory plane).
(5) The source is clean in a high vacuum.
(6) The radiation can be thought of as pulsed, since the burst of radiation can be seen under circular motion.

The characteristics of (2)and (3) are utilized in x-ray lithography. The high intensity of x-ray radiation can reduce the exposure time. The small divergence of SOR essentially eliminates the problem of geometrical distortion, which imposes severe constrains on mask to wafer positioning with conventional x-ray sources.

The spectral distribution and intensity of synchrotron radiation is dependent on electron energy, magnetic field, and orbital radius of the deflection magnet. The SOR power output at a given wavelength λ, in a band width $\Delta\lambda/\lambda$, and integrated over all vertical angles is given by [19]

$$\text{Power}(\lambda) \text{ (mW)} = 8.73 \times 10^3 \, E^4 \, \{\text{GeV}\} \, I \, (\text{amp})$$
$$\times \Theta \, (\text{mrad})(\Delta\lambda/\lambda)G_2(y)/\rho \quad (6.5)$$

where $y=\lambda_c/\lambda$; $G_2(y)$ is given by

$$G_2(y) = y^2 \int_y^\infty K_{5/3}(x)dx \quad (6.6)$$

$K_{5/3}(x)$ is the Bessel function, and λ_c is the critical wavelength given by

$$\lambda_c(A) = 6.59\rho(\text{m}) / E^3(\text{GeV}). \quad (6.7)$$

As described in 6.2, the wavelength region for x-ray lithography is determined by the mask contrast and mask substrate. Therefore, the electron energy and magnetic field of SOR should be optimized to offer desirable spectral distribution and intensity for x-ray lithography.

Although SOR is a collimated beam and the effective source size is small, the x-ray beam is rectangular or slit-like in shape at some distance from the source. The emitted radiation is horizontally uniform but very ununiform vertically. To get enough exposure area, several methods have been reported [19]: a wafer was moved with a mask during exposure; a mirror which scans the reflected light vertically was oscillated; and the electron was oscillated in the storage ring.

Other disadvantages of synchrotron radiation are large physical dimension and high cost. Although several attempts for compact SOR were reported to reduce the construction cost [56], it still requires one billion dollars for a system. The cost for a beam port can be reduced by using a multiport system, which could compete with optical lithography. However, at least two SOR sources are necessary in case of shut-down.

6.4.3 Plasma X-ray Sources

Several attempts to obtain high intensity x-ray emission from extremely high-energy plasma have been made. Devices capable of producing such plasma rely on the ability to deliver energy to a target more rapidly than it can be carried away by loss processes. Several devices capable of producing dense high temperature plasma by electrical discharge and laser pulse irradiation have been reported.

Economou and Flanders [20] reviewed the gas-puff configuration reported by Stalling et al. [21], which is shown in Fig. 6.17. This consists of a fast valve and supersonic nozzle. When the valve is fired, the gas expands through the nozzle and forms a hollow cylinder at the nozzle exit. A high, pulsed current is driven through the gas cylinder, causing the gas to ionize. The magnetic pressure induced by the current results in plasma cylinder onto axes, forming a dense, hot plasma which is a strong x-ray source.

Either an IR or UV laser is used with pulses varying from 50 psec to 10 nsec for these types of sources (Fig. 6.18). The beam focused on the target (10^{14} W/cm^2 is needed) creates a plasma of high enough temperature to produce black body radiation [22,23]. The conversion efficiency from laser energy to x-ray photon is higher than that of electron impact excitation, but the conversion efficiency from electrical energy to laser is low. The advantage of this approach is that the power supply can be positioned away from the aligner, preventing electromagnetic interference.

X-ray Lithography

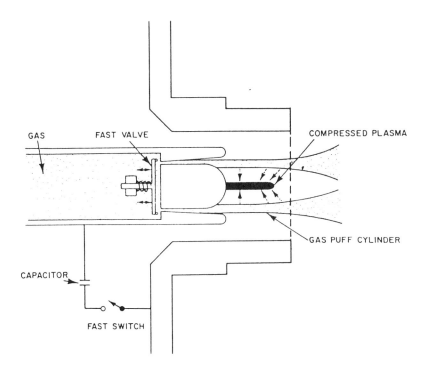

FIG. 6.17 Schematic representation of the gas puff configuration. The fast acting gas release valve and the shaped nozzle form a cylinder of gas. A discharge current through the gas causes the cylinder collapse, forming the plasma x-ray source. [Reprint with permission from *J. Vac. Sci. Technol.*, **19**, 868(1981).]

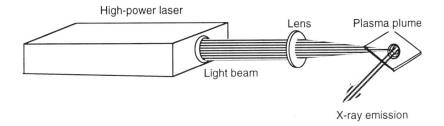

FIG. 6.18 Schematic representation of a laser plasma x-ray source.

6.5 X-RAY MASK

In proximity x-ray lithography, production, inspection, and repair of x-ray masks is the most problematical aspect. Schemes of x-ray mask fabrication for the Au additive process and Ta subtractive process are shown in Fig. 6.19 [24]. The additive process indicates the plating of an x-ray absorber on the resist patterned membrane, while the subtractive process indicates etching the x-ray absorber using the resist patterns as an etching mask. These resist patterns are fabricated by electron beam lithography. The x-ray mask uses a very thin membrane as a substrate instead of the glass substrate used for the photomask. Materials currently investigated as a membrane include SiC, SiN, and Si. X-ray absorber materials are mostly Au, Ta, and W, which define the circuit pattern on the membrane.

X-ray mask fabrication using an additive method includes the deposition of a membrane film on a silicon wafer, back-etching the silicon to the membrane film, glass frame attachment, deposition of Cr for plating base, resist coating, pattern formation by electron beam lithography, Au plating (additive process), and finally resist removal. In the subtractive method, after deposition of a SiN film for membrane and Ta film for absorber, resist patterns are formed on Ta film by electron beam lithography. The patterns are first transferred to SiO_2, then SiO_2 patterns are transferred to Ta film by dry etching. The reason for use of SiO_2 as an etch mask is due to low dry-etching selectivity of a resist to Ta. Finally silicon substrate is etched from the backside to the membrane. One of the issues for both additive and subtractive processes in x-ray mask fabrication is to relax the stress of the membrane.

Optical lithography has extended its resolution capability down to 0.25 µm, and the use of an ArF excimer laser (193 nm) as a light source and wavefront engineering may have the potential of resolution near the 0.1 µm level. X-ray lithography for this generation requires that absorber patterns below 0.2 µm should be fabricated on the mask membrane in a 1:1 dimension precisely, which are usually made by electron beam lithography. However, most of the electron beam exposure machines currently used are unable to meet the requirement for very accurate pattern size and beam placement needed for pattern features below 0.2µm. This difficulty can be understood if one considers the difficulty in photomask fabrication processes currently used in spite of 5 times as large as actual pattern size on the chip. The photomask for a 0.30 µm process would require a precise 1.5 µm pattern on the photomask. In subtractive method, another proximity problem arises. Resist pattern formation should be carried

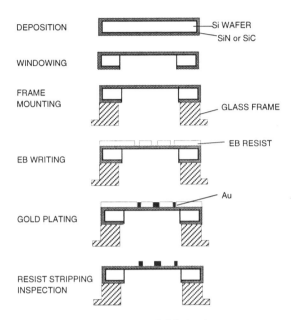

(a) Additive x-ray mask fabrication process

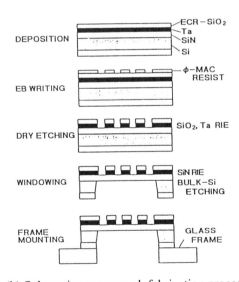

(b) Subtractive x-ray mask fabrication process

FIG. 6.19 Mask fabrication processes.[Reprint with permission from *Jpn. J. Appl. Phys.*, **28**, 2074(1989).]

out on the x-ray absorber materials of high-atomic number, such as Ta and W, which usually show higher back scattering effect than those of low-atomic number. The back-scattering causes electron energy deposition in unwanted areas, the proximity effect, resulting in pattern size variation. The proximity correction during electron beam exposure makes the electron beam exposure more complicated.

X-ray wavelength ~1 nm is used as described in "Section 2. Selection of x-ray wavelength". The difference in the absorption coefficient of the material on x-ray mask can provide image contrast. To obtain appropriate mask contrast, absorber thickness, 0.5 to 1 µm is needed as described before. These structures become more difficult to fabricate due to the high aspect ratio as the minimum feature size becomes smaller. In addition these precise high-aspect-ratio patterns should be maintained with low distortion on the thin film membrane. However, high-aspect-ratio problem can be alleviated, if one can use thinner absorber films with expectation of the phase shift effect [25,26]. It was reported that a thinner absorber (0.3 to 0.35 µm) can improve the image quality by letting some of the x-ray radiation pass through in the same manner as a "leaky chrome" optical phase shift mask. Smith et al. also demonstrated thinner absorber for x-ray lithography using a relatively longer wavelength (Cu_K: 1.3 nm) [27].

In photolithography, chromium on the photomask reflects exposure light, though some percentage of exposure light is absorbed by Cr. On the other hand, in x-ray lithography, the difference in the absorption coefficient is used for pattern exposure on the wafer. Absorption of x-rays by the absorber results in temperature rise of the mask which may cause distortion of the x-ray mask. In addition, continuous exposure of the mask may cause radiation damage to the membrane. Inspection and repair are other issues to be solved. Although much effort has been devoted to x-ray mask issues, it is still considered that the difficulty in mask fabrication is an obstacle to actual use in x-ray lithography.

6.6 X-RAY RESIST MATERIALS

Requirements for x-ray resists strongly depend on x-ray sources and lithographic processes. A variety of multi-layer resist systems as well as the conventional single layer resist process has been extensively studied [28]. A simple single layer resist is desirable for practical use of x-ray lithography to be compatible with the existing process for optical lithography. A single layer resist process requires a resist with high resolution as well as high dry-etch resistance.

6.6.1 Factors Determining Sensitivity of X-ray Resists

The absorption of a photon is the first step of photochemical or radiation chemical reactions in the resist film. Photo- and Auger electrons produced following absorption of x-rays in a resist film cause ionization and excitation of atoms and molecules in a resist film. The ionization and excitation induce chemical changes leading to a differential solubility behavior for the developer. Therefore, chemical reactions induced by x-ray irradiation are similar to those exposed by electron beam irradiation. Most of electron beam resists can be used as x-ray resists. In addition, x-ray sensitivity shows a linear relation with electron beam sensitivity, as shown in Fig. 6.20 [29].

The amount of x-ray energy absorbed determines the x-ray resist sensitivity. X-ray energy absorption is given by Beer's law

$$I = I_0 exp(-\mu_m \rho l) \qquad (6.8)$$

where I_0 is the intensity of incident x-rays, I the intensity of x-ray after

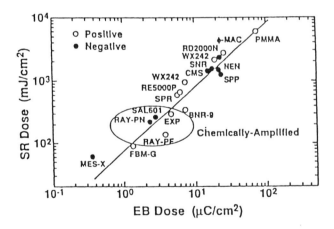

FIG. 6.20 Relation of x-ray sensitivity with electron sensitivity for various resists. [Reprint with permission from Innovation of ULSI lithography), p.245, Science Forum (1994).]

penetration through a thickness l of a homogeneous material having a mass absorption coefficient μ_m for x-ray, and ρ the bulk density of the material.

Therefore, the mass absorption coefficient for a polymer, μ_{mp}, is given by

$$\mu_{mp} = \Sigma A_i \mu_{mi} / \Sigma A_i \qquad (6.9)$$

where A_i and μ_{mi} are the atomic weights and mass absorption coefficient, respectively. The percent of x-ray absorbed in a polymer can be calculated from the relation

$$\%absorbed = (I_0\text{-}I)/I_0 = 1 - exp(-\mu_{mp}\rho l) \qquad (6.10)$$

For these calculations, values for μ_m in Table 6.1 [28] can be used. For example, absorption fractions of x-ray energy by a high resolution electron beam resist poly(methyl methacrylate) (PMMA) with 1 µm thickness are only 3.1% and 1.7 % for Mo(5.14A) and Pd(4.36A) x-rays, respectively.

Improvement in the sensitivity of an x-ray resist is not an easy task because of the low deposit energy of an x-ray in resist film. As can be seen in Table 6.1, the mass absorption coefficients for halogen atoms and metals have larger values for 4~15 A x-rays. Therefore, incorporation of these atoms into polymers is an effective approach to increase x-ray absorption. Since the Cl atom strongly absorbs the Pd(4.36A) x-ray, there was a report on Cl containing acrylate polymers, poly(2,3-dichloro-1-propylacrylate), DCPA by Bell Laboratories workers [30]. The absorption fraction for DCPA is at most 9.9% for the Pd(4.36A) x-ray. On the contrary, absorption fractions of 1 µm AZ-1350J photoresist are 70% and 40% for i-line (365nm) and g-line (436nm), respectively. More than 90% of x-ray energy penetrates through the resist film, while in optical lithography about half of the incident light can be utilized. The sensitivity is generally described by incident energy radiation instead of absorbed energy. The sensitivity of a diazonaphthoquinone-novolak photoresist is about 100mJ/cm². It means that the x-ray resist should be more sensitive as 5 times the conventional photoresist as far as absorbed energy in a resist film is concerned.

6.6.2 Positive X-ray Resists

Positive x-ray resists are provided in Table 6.2. Poly(methylmethacrylate) PMMA is a positive electron beam and x-ray resist with extremely high resolution. Patterns of 175A line width were demonstrated by Flanders using an

X-ray Lithography 183

EB bombardment C_K source [31]. One of the problems with PMMA is its low sensitivity. Workers at IBM incorporated Tl into PMMA [32] to enhance the sensitivity of PMMA. Incorporation of Tl as a salt of copolymer of methylmethacrylate and methacrylic acid results in enhancement of sensitivity by a factor of 25. However, it was reported that decrease in film thickness during development is large for unexposed regions to achieve this high sensitivity.

Table 6.1 Mass absorption coefficient of selected elements [28]

$\mu_m(cm^2/g)$

Element	Z	Pd 4.36 Å	Rh 4.60 Å	Mo 5.41 Å	Si 7.13 Å	Al 8.34 Å	Cu 13.36 Å	C 44.70 Å
C	6	100	116	184	402	627	2714	2373
N	7	155	180	285	622	970	4022	3903
O	8	227	264	416	908	1415	5601	6044
F	9	323	376	594	1298	2022	6941	8730
Si	14	1149	1337	2115	279	423	1959	36980
P	15	1400	1630	2579	371	564	2405	41280
S	16	1697	1975	232	483	733	3079	47940
Cl	17	2013	197	301	628	953	3596	50760
Br	35	1500	1730	2649	3680	1456	3101	32550
Fe	26	630	726	1118	2330	3536	10690	13300
Sn	50	675	772	1160	2318	3437	9623	6332
Tl	81	1083	1213	1032	1904	2697	6276	13030
Most Absorbing Elements		Cl S Br P Si Heavy Atoms	S Br P Si Heavy Atoms	Br P Si Heavy Atoms F	Br Heavy Atoms F O Cl	Heavy Atoms F F O N	Heavy Atoms F O N Cl	Cl S P Si Heavy Atoms
Least Absorbing Elements		C N O	C N Cl	C S N	Si P C	Si P C	Si P C	C N O

Polyfluorobutylmethacrylate (FBM) is a high sensitive electron beam resist and x-ray resist developed by Kakuchi et al. [33]. Submicrometer resolution was achieved and sensitivity of this resist for Si_K and Mo_L lines was reported to be 30 to 60 mJ/cm². Improvement in adhesion during the development was demonstrated using a copolymer of fluorobutylmethacrylate with a small amount of glycidylmethacrylate. This version was used for fabrication of prototype devices with Si_K x-ray step-and-repeat exposure at NTT [34]. The problem with these acrylate polymers is low resistance to dry etching. Due to the poor

Table 6.2 Positive x-ray resists

	structure	source	sensitivity (mJ/cm2)	
Main chain scission	PMMA $-\!\!+\!CH_2-\underset{COOCH_3}{\overset{CH_3}{\underset{\|}{C}}}\!\!\!+\!\!-$	Al	600	[32]
	PMMA $-\!\!+\!CH_2-\underset{COOCH_3}{\overset{CH_3}{\underset{\|}{C}}}\!\!\!+\!\!-\!\!+\!CH_2-\underset{COO^- Tl^+}{\overset{CH_3}{\underset{\|}{C}}}\!\!\!+\!\!-$ FBM	Al	24	[32]
	$-\!\!+\!CH_2-\underset{COOCH_2CF_2CFHCF_3}{\overset{CH_3}{\underset{\|}{C}}}\!\!\!+\!\!-$	Mo	52	[33]
Dissolution inhibition	NPR (PMPS +novolak resin) PMPS $-\!\!+\!CH_2-\underset{\underset{\underset{CH_3}{\|}}{\underset{CH_2}{\|}}}{\underset{CH_2}{\overset{CH_3}{\underset{\|}{C}}}}\!\!-\!SO_2\!+\!-$ novolak resin	Mo	75 (He) 450 (Air)	[35, 36]

resistance to dry etching, acrylate polymer resists cannot be used in a single layer resist system.

The problem of dry-etch resistance in the application of positive resist was circumvented by development of NPR at Bell Laboratories [35]. Described as an electron beam resist, NPR is comprised of an electron sensitive poly(2-methyl-penten-1-sulfone) and dry etch resistant novolak resin. When NPR is evaluated as an x-ray resist, its sensitivity is dependent on exposure atmosphere; the sensitivity is 450mJ/cm² under air environment and 75 mJ/cm² under inert gas [36]. Since non-swelling development is carried out by aqueous alkali solution, this resist has a high potential of resolution.

6.6.3 Negative X-ray Resists

Negative x-ray resists are provided in Table 6.3. These are classified into acrylate polymers and styrene polymers: the dry etch resistance of the former resists are poor and that of the latter is high. Resists except PSTTF listed in Table 6.3 undergoes radiation induced crosslinking which leads to insolubilization. The sensitivity of this type of resist strongly depends on the molecular weight of the polymer: the higher molecular weight polymers have higher sensitivity (see Table 6.3). However, swelling of the irradiated region during development deteriorates the pattern definition. Swelling is apparently decreased if one uses a low molecular weight polymer. However, the use of low molecular weight polymers results in low sensitivity. It is difficult to satisfy both requirements of sensitivity and resolution. This problem was recognized in the study on PCMS by Choong et al. [37] at Hewlet-Packard. Use of low dispersity polymers and a judicious choice of developer solvents are the keys minimizing the swelling.

A chlorinated acrylate polymer DCPA [30] was designed for x-ray lithography using an EB bombardment Pd source, as Cl atom strongly absorbs Pd x-rays. To improve adhesion, a negative EB resist poly(glycidyl methacrylate-co-ethyl acrylate), COP, was added to DCPA, designated as DCOPA [38]. High sensitivity of 10 mJ/cm^2 was achieved when a polymer with very high molecular weight of 2×10^6 was used. Allylacrylate polymer EK-88 developed by workers at Kodak [39] has high sensitivity irrespective of low molecular weight, though resolution of EK-88 seems to be insufficient due to swelling. Because of insufficient resolution of crosslinking type of resists due to swelling, Taylor et al. [40] proposed a dry developable resist system which is described below.

Styrene-type polymers, poly(4-chloromethylstyrene) (PCMS) [37], chloromethylated polystyrene (CMS) [41], and chlorinated poly(methylstyrenes) (CPMS) [42] developed as negative electron beam resists with dry etching resistance can be used as x-ray resists (Table 6.3). Here again, these polymers have difficulty in satisfying both requirements for sensitivity and resolution.

Hofer et al. [43] reported a new class of polymeric negative resist, poly(tetrathiafulvalene) (PSTTF), which shows no evidence of swelling during development. PSTTF, was originally developed for a conductive polymer. X-ray irradiation of PSTTF films doped with halocarbon acceptor such as CBr_4 results in ionic-crosslinks formation between polymer-bond TTF cations which leads to the generation of differential solubility behavior. Nonswelling negative patterns were obtained when a nonpolar solvent is used as a developer. However, the problems with this resist are difficulties in synthesis and its stability. The resist solution should be prepared just before the coating. Irradiation should be conducted immediately after coating without prebaking.

Table 6.3 Negative x-ray resists

		source	sensitivity (mJ/cm2)	Mw (x10-4)	
Acrylate Polymers	**DCPA** −(CH$_2$−CH)− | COOCH$_2$CHClCH$_2$Cl	Pd	5-13 (inert) 8.5 (N$_2$)	120-50	[30] [38]
	DCOPA CH$_3$ | −(CH$_2$−C)− −(CH$_2$−CH)− + DCPA | | COOCH$_2$—CH$_2$ COOC$_2$H$_5$ \\O/	Pd	11.5 (N2) 5.5 (N$_2$) 36 (air)	240 200 200	[38] [37] [32]
	EK-88 CH$_3$ | −(CH$_2$−C)− −(CH$_2$−CH)− | | COOCH$_2$CH=CH$_2$ COOCH$_2$CH$_2$OH	W Pd	9 (-) 170 (air)	4 4	[39] [36]
Styrene Polymers	**PCMS** −(CH$_2$−CH)− | C$_6$H$_4$−CH$_2$Cl	Pd	23-470 (air)	44-3	[37]
	CMS −(CH$_2$−CH)−(CH$_2$−CH)− | | C$_6$H$_4$−CH$_2$Cl C$_6$H$_5$	Pd Mo	150 (air) 8-29 (-)	20 110-18	[37] [41]
	PSTTF −(CH$_2$−CH)− | C$_6$H$_4$−CH$_2$OC$_6$H$_4$−(tetrathiafulvalene) + CBr$_4$	Al	44 (-)	30	[43]
	CPMS Cl Cl | | −(CH$_2$−CH)−(CH$_2$−C)−(CH−CH)− | | | C$_6$H$_4$−CH$_2$Cl C$_6$H$_5$ C$_6$H$_4$−CH$_3$	Pd	17 (-)	50	[42]

6.6.4 Approaches for Highly Sensitive X-ray Resists with High Resolution

The solvent induced swelling during the wet development limits the utility of highly sensitive x-ray resists such as DCPA and DCOPA. Taylor et al. [40] have studied plasma developable x-ray resists that do not suffer from problems with swelling. A scheme of this process is shown in Fig. 6.21.

These resists consist of metal containing monomer "rm" and absorbing host polymers "P" containing Cl or Br, where "rm" contains radiation sensitive functional group and inorganic atoms which form metal oxides during oxygen plasma development. These monomers "rm" are locked into the host polymer "P" by grafting and polymerization under the x-ray irradiation. Then the unlocked monomer is removed by heating in a vacuum. The resulting negative relief image is developed by O_2 reactive ion etching. Interaction of organometallic moiety with this plasma forms a metal oxide layer which acts as a practical etch mask to provide negative pattern. A mixture of DCPA as a base polymer and bis-acryloxybutyltetramethyldisiloxane (BSBTDS) has good sensitivity (1.5 ~ 4.5 mJ/cm^2) with Pd_L x-ray. Due to its poor resistance to dry-etching and round shape in pattern after development, a three-layer process is required. The problems with this process is low normalized-remaining resist thickness (30%), which results in difficulty in linewidth control.

High sensitivity of this dry developable resist is achieved by radiation induced chain polymerization. In this type of chain reaction an active species produced by photoelectron or Auger electron causes many chemical events. Since radiation energy is utilized efficiently, this type of chain reaction would provide a high sensitive resist.

In order for drastic improvement in sensitivity, a photo-induced chain reaction system as described in the "chemical amplification" [44] section of photoresists, have been applied to x-ray resists. The chemical amplification system utilized strong acids produced from acid generator photodecomposition to catalyze the reaction of acid sensitivity groups in either polymer backbone or on the side chain. X-ray irradiation easily induces the decomposition of an acid generator to produce an acid. Since the chemical amplification system utilizes the drastic change in solubility for an aqueous base developer, it shows high resolution capability. Recent reports on device fabrication using x-ray lithography have been demonstrated using chemical amplification resists [45-47]. Generally a deprotection reaction is used for positive resists and an acid-hardening reaction of melamine derivatives is used for negative resists. One of the disadvantages of this type of resist is acid diffusion into the unexposed area.

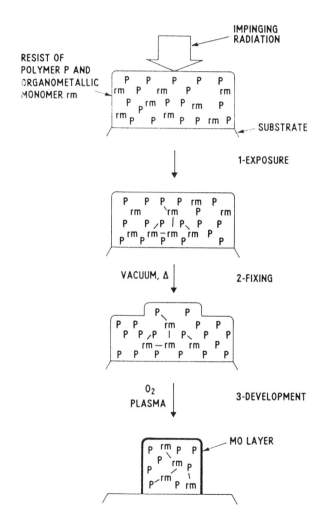

FIG. 6.21 Schematic representation of a plasma developed x-ray resist process. [Reprint with permission from *J. Vac. Sci. Technol.*, **19**, 872(1981).]

X-ray Lithography

The range of the diffusion depends on the process condition, especially post-exposure-baking temperature, which was investigated in detail [48].

As the aspect ratio (height/width) increases, associated with the demand for high resolution, one should pay attention to resist pattern collapse during the development as shown in Fig. 6.22. Tanaka et al. [49] observed the resist patterns before (in the liquid) and after development by an atomic force microscope. They concluded that the resist pattern collapses during the dry-off of liquid rinse. To avoid this problem, they proposed special processes during development such as heating during and flood exposure during rinsing [50]. These processes induce the crosslinking of the resists, leading to a hardening of the resist pattern before drying.

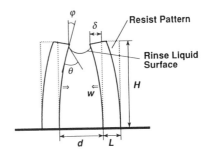

FIG. 6.22 Pattern collapse during the development. [Reprint with permission from *Jpn. J. Appl. Phys.*, **32**, 6059(1993).]

6.7 X-RAY REDUCTION PROJECTION SYSTEM

Some efforts for reduction projection x-ray lithography were reported. Workers of AT&T Bell Laboratories demonstrated an x-ray reduction projection exposure system based on the Schwaltzschild reflection type of 20:1 reduction in 1990 (Fig. 6.23) [51]. Since then, much attention has been focused on x-ray lithography again, though reduction projection systems had already been reported by workers of the Lawrence Livermore Laboratories [52] and NTT [53]. They demonstrated 0.1 μm line and space patterns which is impossible to obtain by photolithography.

The x-ray reduction projection system can be accepted as an extension of the

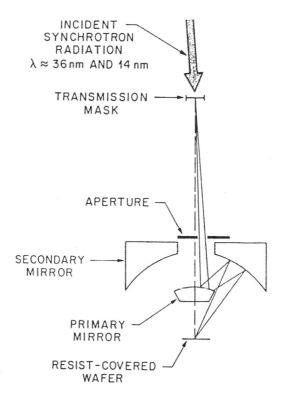

FIG. 6.23 Schematic diagram showing the Schwarzschild objective used with an eccentric aperture and off-axis illumination. [Reprint with permission from *J. Vac. Sci. Technol.*, **B8**, 1509(1990).]

projection system (steppers) used in the present photolithography as shown in Fig. 6.24 [54]. The wavelength used for x-ray reduction projection system is ~10 nm, which is much shorter than those for conventional photolithography such as i-line (365 nm) and even for ArF (193 nm). Therefore, x-ray optics with smaller NA has a resolution capability below 0.1 μm as shown in Fig. 6.24.

Patterning on the x-ray mask can be alleviated depending on the reduction magnitude, because, in reduction x-ray lithography, minimum feature size on the mask is much larger than that for proximity printing, depending on reduction factors. The recent x-ray optics proposed by NTT group is shown in Fig. 6.25 [55].

However, one of the most difficult issues in an x-ray reduction system is the fabrication of mirrors. The aspherical multilayer mirrors are necessary to reduce the number of mirrors. Since the wavefront error of these mirrors is required to

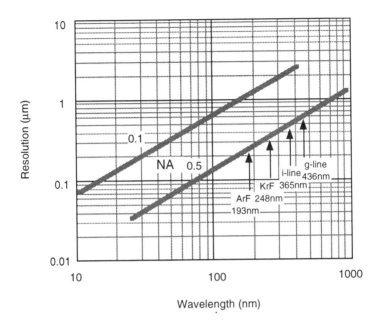

FIG. 6.24 Relation of resolution with wavelength in x-ray reduction projection lithography.

be less than $\lambda/14$, a higher accuracy of mirrors are needed for shorter wavelengths. Multilayer mirrors are used to get high transmittance for ~10 nm wavelength. The multilayer mirrors are prepared by periodical vapor deposition of dielectrics with different refractive indices as shown in Fig. 6.26 [53]. The pitch of dielectrics determines the wavelength of maximum reflectivity and the accuracy of the pitch determines the reflectivity. To get high transmittance, atomic level uniformity is required for periodical deposition. The use of wavelengths shorter than 10 nm is limited by the present status of mutilayer mirrors fabrication due to the difficulty in preparing smaller pitch multilayers.

The reduction projection system demands different types of resist materials from those for x-ray proximity printing. Most of the materials have a high absorption coefficient at these wavelength region. The extinction depth of x-rays along the film thickness is remarkably small to be less than 0.1 µm. The workers of AT&T Bell Laboratories demonstrated pattern formation with a three-layer resist process using PMMA as an imaging layer. Another type of resist process such as surface imaging is required for practical applications.

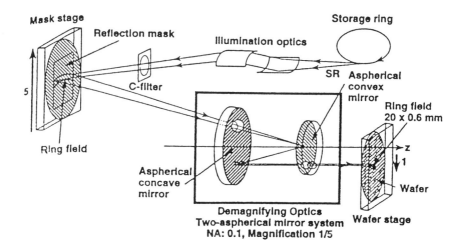

FIG. 6.25 Configuration of the two-aspherical-mirror system. It consists of an SOR source, illumination optics, carbon filter, a reflection mask, the demagnifying optics, and a wafer. The demagnifying optics consists of aspherical concave and convex mirrors. The numerical aperture (NA) is 0.1 and the magnification is 1/5. [Reprint with permission from *J. Vac. Sci. Technol.*, **B13**, 2914(1995).]

6.8 CHARACTERISTICS OF X-RAY LITHOGRAPHY

Advantages of 1:1 x-ray lithography are summarized as follows based on the reported articles, though some are controversial as described below.
(1) By using soft x-rays, wavelength-related diffraction problems which limit the resolution of optical lithography are effectively reduced.
(2) High-aspect-ratio pattern fabrication can be achieved due to the transparency of resist film to x-rays.
(3) Many defect-causing particles in the light optics regime are transparent to x-rays.
(4) There is a possibility of high throughput since large areas can be irradiated.
(5) Large depth-of-focus gives process latitude.
(6) There is practically no field size limitation.

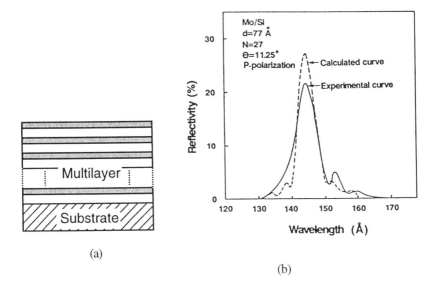

FIG. 6.26 The structure of x-ray multilayer and the measured P-polarization reflectivity of Mo/Si multilayer mirror. [Reprint with permission from *J. Vac. Sci. Technol.*, **B7**, 1648(1989).]

However, these advantages sometimes give rise to disadvantages. Since the refractive index of an x-ray is near one, optical elements for x-rays are limited to mirror and Fresnel zone plates. It is, therefore, difficult to collimate or focus the x-rays. Since the transmittance of the resist film is high, only several percents of x-ray energy is deposited in the resist film. This high transmittance requires extremely high resist sensitivity to avoid x-ray damage of the devices. The proximity printing requires a system to control the gap between mask and wafer and the alignment. This gap also derives such problems as penumbra and magnification. For the application of x-ray lithography to actual LSI fabrication, alignment technology should also be established which is not described here. Above all, x-ray mask fabrication is the most difficult issue for actual use of x-ray lithography.

It is always difficult to predict the limitation of minimum feature size that can be fabricated with optical lithography. It is no doubt, however, that the obstacles for x-ray lithography described above should be overcome before the end of the optical lithography era. At that time, the choice of lithography will be determined by the stage of development of either electron beam or x-ray lithography.

REFERENCES

1. E. Spears and H. I. Smith, High resolution pattern replication using soft x-rays, Electron Lett., 8, 102(1972); E. Spears and H. I. Smith, X-ray lithography - A high resolution replica processes, Solid State Technol., 15(7), 21(1972).
2. B. Fay, L. Tai, and D. Alexander, Recent printing and registration results with x-ray lithography, Proc. SPIE, **537**, 57(1985).
3. P. N. Dunn, X-ray's future: a cloudy picture, Solid State Technol., **37**(6), 49(1994).
4. H. I. Smith, Fabrication technique for surface-acoustic-wave and thin film optical devices, Proc. IEEE, **62**, 1361(1974).
5. T. Terasawa and S. Okazaki, Phase-shifting technology for ULSI patterning, IEICE Trans. Electron. **E76-C**, 19(1993).
6. R. Feder, E. Spiller, and J. Topalian, X-ray lithography, Polm. Eng. Sci., **17**, 385(1977).
7. N. Atoda, Resists for synchrotron radiation lithography, Hoshasenkagaku (Radiation Chemistry), **19**, 41(1984); N. Atoda, Proc. of Int'l Conf. on Adv. Microelectronic Devices and Processing, p. 109(1994).
8. T. Watanbe, Hoshasen to Genshi Bunshi (Japanese, Radiation Effect on Atoms and Molecules), S. Shida Ed. Kyoritsu Pub. p. 30 (1966).
9. K. Early, M. L. Schattenburg, and H. I. Smith, Microelectronic Eng., **11**, 317(1990).
10. K. Deguchi, K. Miyoshi, H. Ban, T. Matsuda, T. Ohno, and Y. Kado, Effect of photo- and Auger electron scattering on resolution and linewidth control in SOR lithography, Jpn. J. Appl. Phys., **29**, 2207(1990).
11. P. Tischerrand and E. Hundt, Profile Structure in PMMA by x-ray lithography, Proc. Symp. 8th Electron Ion Beam Sci. Technol., vol.**78-5**, p.444(1987).
12. T. Ogawa, K. Mochiji, Y. Soda, and T. Kimura, The effect of secondary electrons from a silicon substrate on SOR x-ray lithography, Jpn. J. Appl. Phys., **28**, 2070(1989).
13. J. R. Maldonado, G. A. Coquin, D. Maydan, and S. Somekh, Spurious effects caused by the continuous radiation and ejected electrons in x-ray lithography, J. Vac. Sci. Technol., **12**, 1329(1975).
14. N. Atoda, H. Kawamatsu, H. Tanino, S. Ichimura, M. Harita, and K. Hoh, Diffraction effect on pattern replication with synchrotron radiation, J. Vac. Sci. Technol., **B1**, 1267(1983).

15. S. D. Hector, M. L. Schattenburg, E. H. Anderson, W. Chu, V. V. Wong, and H.I. Smith, Modeling and experimental verification of illumination and diffraction effects on image quality in x-ray lithography, *J. Vac. Sci. Technol.*, **B10**, 3164(1992).
16. B. Leslite, A. Neukermans, T. Simon, and J. Foster, Enhanced Brightness x-ray Sources, *J. Vac. Sci. Technol.*, **B1, 1251**(1983).
17. J. L. Gaines and R. A. Hansen, An improved annular-shaped electron gun for x-ray generator, *Nucl. Instrum. Method*, **126**, 99(1975).
18. E. Spiller, D. E. Eastman, R. Feder, W. D. Grobman, W. Gudat, and J. Topalion, Application of synchrotron radiation to x-ray lithography, *J. Appl. Phys.*, **47**, 5450(1976).
19. J. B. Murphy, D. L. White, A. A. MacDowell, and O. R. Wood II, Synchrotron radiation sources and condensers for projection x-ray lithography, *Appl. Opt.*, **32**, 6920(1993).
20. N. P. Economou and D. C. Flander, Prospect for high bright x-ray sources for lithography, *J. Vac. Sci. Technol.*, **19**, 868(1981).
21. C. Stalling, K. Childers, I. Roth, and R. Schneider, Imploding argon Plasma experiments, *Appl. Phys. Lett.*, **35**, 524(1979).
22. A. L. Hoffman, G. F. Albrecht, and E. A. Crawford, High brightness laser/plasma source for high throughput submicron lithography, *J. Vac. Sci. Technol.*, **B3**, 258(1985).
23. N. M. Ceglio, A. M. Hawryluk, and G. E. Sommargren, Front-end design issues in soft-x-ray projection lithography, *Appl. Opt.*, **32**, 7050(1993).
24. S. Ohki, M. Kakuchi, T. Matsuda, A. Ozawa, T. Ohkubo, M. Oda, and H. Yoshihara, Ta/SiN-structure x-ray masks for sub-half-micron LSIs, *Jpn. J. Appl. Phys.*, **28**, 2074(1989).
25. Y. Somemura, K. Deguchi, K. Miyoshi, and T. Matsuda, X-ray phase-shifting mask for 0.1-μm pattern replication under a large proximity gap condition, *Jpn. J. Appl. Phys.*, **31**,4221(1992); Y. Somemura, and K. Deguchi, Effects of Fresnel diffraction on resolution and linewidth control in synchrotron radiation lithography, *Jpn. J. Appl. Phys.*, **31**,938(1992).
26. J. Xiao, M. Kahn, R. Nachman, J. Wallance, Z. Chen, and F. Cerrina, Modeling image formation: Application to mask optimization, *J. Vac. Sci. Technol.*, **B12**, 4038(1994).
27. Y. C. Ku, H. I. Smith, and I. Plotnik, Low stress tungsten absorber for x-ray masks, *Microelectronic Eng.*, **11**, 303(1990).
28. G. N. Taylor, X-ray resist materials, *Solid State Technol.*, **23**(5), 73(1980); G. N. Taylor, X-ray resist Trends, *Solid State Technol.*, **27**(6),

124(1984).
29. K. Deguchi, ULSI lithography no kakushin (Innovation of ULSI lithography), p.245, Science Forum (1994).
30. G. N. Taylor, G. A. Coquin, and S. Somek, Sensitive chlorine-containing resists for x-ray lithography, *Polym. Eng. Sci.*, **17**, 420(1977).
31. D. C. Flanders, Replication of 175A lines and spaces in polymethylethacrylate using x-ray lithography, *Appl. Phys. Lett.*, **36**, 93(1983).
32. I. Haller, R. Feder, M. Hatzakis, and E. Spiller, Copolymer of methylmethacrylate and methacrylic acid and their metal salts as radiation sensitive resists, *J. Electrochem. Soc.*, **126**, 154(1979).
33. M. Kakuchi, S. Sugawara, K. Murase, and K. Matsuyama, Poly(fluoromethacrylate) as highly sensitive, high contrast positive resists, *J. Electrochem. Soc.*, **124**, 1648(1977).
34. T. Hayasaka, S. Ishihara, and H. Kinoshita, A step-and-repeat x-ray exposure system for 0.5µm pattern replication, *Proc. Symp. Electron and Ion Beam Sci. Technol.*, **83-2**, 347(1983).
35. M. J. Bowden, L. F. Thompson, S. R. Fahrenholtz, and E. M. Doerries, A sensitive novolak-based positive electron resist, *J. Electrochem. Soc.*, **128**, 1304(1981).
36. K. Mochiji, Y. Soda, and T. Kimura, Sensitivity improvement of x-ray resist by overlaying oxygen blocking film, *J. Electrochem. Soc.*, **133**, 147(1986).
37. H. S. Choong and F. J. Kahn, Poly(chloromethylstyrene): A high performance x-ray resist, *J. Vac. Sci. Technol.*, **B1, 1066**(1983).
38. J. M. Moran and G. N. Taylor, Mixture of poly(2,3-dichloro-1-propylacrylate) and poly(glycidyl metahcrylate-co-ethyl acrylate) as an x-ray resist with improved adhesion, *J. Vac. Sci. Technol.*, **16**, 2014(1979).
39. Z. C. Tan, C. C. Petropoulos, and F. J. Rauner, High-sensitivity, high resolution, high thermal resistant negative electron x-ray resist, *J. Vac. Sci. Technol.*, **19**, 1348(1981).
40. G. N. Taylor, T. M. Wolf and J. M. Moran, Organosilicon monomers for plasma-developed x-ray resists, *J. Vac. Sci. Technol.*, **19**, 872(1981).
41. S. Imamura and S. Sugawara, Chloromethylated polystyrene as deep UV and x-ray resist, *Jpn. J. Appl. Phys.*, **21**, 776(1982).
42. N. Yoshioka, K. Suzuki, T. Yamazaki, A high performance negative x-ray resist: CPMS-X(Pd), *Proc. SPIE*, **537**, 51(1985).
43. D. C. Hofer, F. B. Kaufman, S. R. Kramer, and A. Aviam, New high resolution charge transfer x-ray and electron beam negative resist, *Appl.*

Phys. Lett., **37**, 314(1980).
44. H. Ito and C. G. Willson, Chemical amplification in the design of dry developing resist materials, *Polym. Eng. Sci.*, **23**, 1012(1983); H. Ito and C. G. Willson, Application of photoinitiators to the design of resists for semiconductor manufacturing, "Polymers in Electronics", *ACS Symp. Ser.*, **242**, 11(1984).
45. K. Deguchi, Synchrotron radiation x-ray lithography for the fabrication large-scale-integrated circuits, *J. Photopolym. Sci. Technol.*, **4**, 445(1993).
46. R. DellaGardia, C. Wasik, D. Puisto, R. Fair, L. Liebman, J. Rocque, S. Nash, A. Lamberti, G. Collini, R. French, B. Vampatella, G. Gifford, V. Nastasi, P. Sa, F. Volkringer, T. Zell, D. Seeger, and J. Warlaumont, Fabrication of 64 Megabit DRAM using x-ray lithography, *Proc. SPIE*, **2437**,112(1995).
47. K. Fujii, T. Yoshihara, Y. Tanaka, K. Suzuki, T. Nkajima, T. Miyatake, E. Miyatake, E. Orita, and K. Ito, Applicability test for synchrotron radiation x-ray lithography in 64 -Mb dynamic random access memory fabrication processes, *J. Vac. Sci. Technol.*, **B12**, 3949(1994).
48. J. Nakamura, H. Ban, K. Deguchi, and A. Tanaka, Effect of acid diffusion on resolution of a chemically amplified resist in x-ray lithography, *Jpn. J. Appl. Phys.*, **30**, 2619(1991).
49. T. Tanaka, M. Morinaga, and N. Atoda, Mechanism of resist pattern collapse during development process, *Jpn. J. Appl. Phys.*, **32**, 6059(1993).
50. T. Tanaka, M. Morigami, H. Oizumi, T. Soga, T. Ogawa, and F. Murai, Prevention of resist pattern collapse by resist heating during rinsing, *J. Electrochem. Soc.*, **141**, L169(1994); T. Tanaka, M. Morigami, H. Oizumi, T. Ogawa, and S. Uchino, Prevention of resist pattern collapse by flood exposure during rinse process, *Jpn. J. Appl. Phys.*, **33**, L1803(1994).
51. J. E. Bjorkholm, J. Borkor, L. Eicher, R. R. Freeman, J. Gregus, T. E. Jewell, W. M. Mansfield, A. A. MacDowell, E. L. Raab, W. T. Silftvast, L. H. Szeto, D. M. Dennant, W. K. Waskiewicz, D. L. White, D. L. Windt, O. R. Wood II, and J. H. Bruning, Reduction imaging at 14 nm using multilayer-coated optics: Printing of features smaller than 0.1 µm, *J. Vac. Sci. Technol.*, **B8**, 1509(1990).
52. A. M. Hawryluk and L. G. Seppala, Soft-x-ray projection lithography using an x-ray reduction camera, Soft x-ray projection lithography using an x-ray radiation camera, *J. Vac. Sci. Technol.*, **B6**, 2162(1988).
53. H. Kinoshita, K. Kurihara, Y. Ishii, and Y. Torii, Soft x-ray reduction lithography using multilayer mirrors, *J. Vac. Sci. Technol.*, **B7**,

1648(1989).
54. S. Okazaki, Lithographic technologies for future ULSI, *Appl. Surf. Sci.*, **70/71**, 603(1993).
55. T. Haga and H. Kinoshita, Illumination systems for extreme ultraviolet lithography, *J. Vac. Sci. Technol.*, **B13**, 2914(1995).
56. A. Heuberger, X-ray lithography, *Solid State Technol.*, **29**(2), 93(1986); M. N. Wilson, A. I. Smith , V. C. Kempson, A. L. Purvis, R. J. Anderson, M. C. Townsend, A. R. Jorden, D. E. Andrew, V. P. Suller, and M. W. Poole, *Jpn. J. Appl. Phys.*, **29**, 2620(1990).

7

Electron Beam Lithography

7.1 INTRODUCTION

The magnification of an optical microscope is at most 2000. On the other hand, an electron microscope using an accelerated electron beam can achieve several million magnification. Electrons can be focused into very fine points (0.1μm and less) and deflected by electrostatic and magnetic fields. They are scanned and controlled by a computer. It is natural to consider that one can obtain high resolution patterns using an accelerated electron beam instead of optical lithography. The experiment to confirm this idea, was started in the middle of the 1960s [1]. Another advantage of electron beam lithography over photolithography and x-ray lithography is its pattern generation capability. In both photolithography and x-ray lithography, masks for pattern transfer are fabricated using electron beam lithography. Even with these advantages, electron beam lithography suffers from long exposure time: low throughput.

The breakthrough of electron beam lithography was the announcement of EBES (Electron Beam Exposure System) by Bell Laboratories[2]. This system has been successfully used for mask fabrication for photolithography, which made the time required for writing a mask shorter by more than two orders of magnitude compared with the time required by an optical pattern generator. Before that, the photomask had been fabricated by complicated steps. The mask patterns of 10 times magnitude on a silver photosensitive plate prepared by a pattern generator were transferred to a mask substrate using a reduction projection exposure system. Electron beam lithography for mask fabrication has been widely accepted since the development of EBES. Almost all photomasks for LSI (Large Scale Integrated circuits) are fabricated by electron beam lithography.

Another application of electron beam lithography is "direct" writing patterns on a wafer. An electronic engineer designs patterns for VLSI and transfers this data to a computer, which can make the pattern on a wafer by a computer controlled electron-beam exposure. However, only a small part of LSI devices have been fabricated by "direct" writing due to its low writing speed (throughput). As the minimum feature size decreases, more exposure shot number increases, resulting in longer exposure time. However, due to the resolution limitation of optical lithography and large demand for ASIC (application specific integrated circuit) in future, electron beam lithography has received great attention as one of the candidates for future lithography.

7.2 ELECTRON BEAM EXPOSURE SYSTEM

Electron beam exposure tools have been developed based on the electron microscope. A focused electron beam is scanned, which causes a chemical reaction in resist film. An electron exposure tool consists of a source of electrons (electron gun), electrostatic and magnetic electron optics (condenser lens, objective lens, and deflector), blanking plates and a wafer coated a resist film as schematically shown in Fig. 7.1. The electrons emitted from a electron

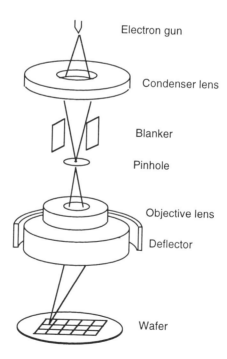

FIG. 7.1 Schematic representation of an electron beam exposure system.

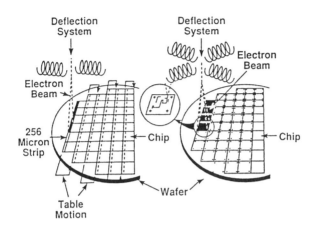

FIG. 7.2 Scanning modes for electron beam systems: (left) raster scan coupled with continuous table motion, and (right) vector scan, step and repeat. [Reprint with permission from ACS Professional Reference Book, American Chemical Society, Washington, DC (1994).]

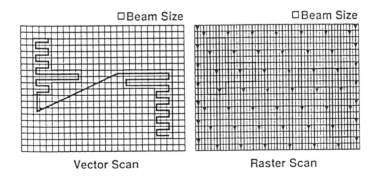

FIG. 7.3 Comparison of vector-scan and raster-scan writing schemes. [Reprint with permission from ACS Professional Reference Book, American Chemical Society, Washington, DC (1994).]

source are accelerated by potential difference and focused on a pinhole by a condenser lens. The electrons which pass through the pin hole are confined to a point on a resist film by a projection lens. The electron beam can be magnetically and electrostatically deflected and moved on a selected area of a resist film. The blanking plates control "on" and "off" of electron current. The control of the blanking plate combined with electrostatic and magnetic optics can make the pattern exposure on a resist film. The whole arrangement, including a mobile stage, which carries the wafer is enclosed in a carefully managed vacuum.

In an early stage of an electron exposure machine, an electron beam is scanned over the surface of the resist in a serial fashion (raster scan) as shown in Fig.7.2 [3]. Scanning is accomplished by moving the wafer in one direction and scanning the electron beam via the deflector plates in a direction perpendicular to the stage motion. Both movements have to be controlled to fine tolerance of the order of less than one-fourth of width of the smallest fine line feature in the pattern. This control is achieved by interferometrically matching two optical

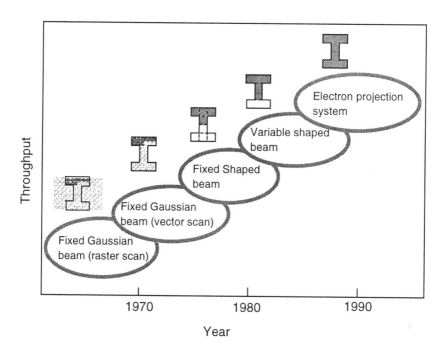

FIG. 7.4 Improvement in throughput by changing beam shape and shot number.

grids in the corner of the scan field. In this way overlay and butting errors have been reduced to 0.1 μm.

Although electron beam lithography has high resolution capability and pattern generation capability, it suffers from low throughput compare to optical lithography.

7.3 IMPROVEMENT IN THROUGHPUT OF ELECTRON BEAM LITHOGRAPHY

Depending on the scanning modes, electron beam exposure systems can be divided into two groups (Figs. 7.2 and 7.3): those using a raster scan mode and those using a vector scan. In the raster scan mode every point on the film surface is addressed. The stage moves in rows and the electron beam exposes the

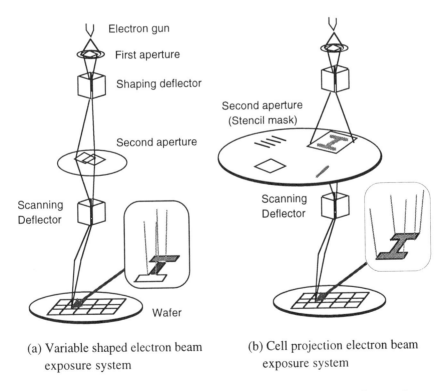

(a) Variable shaped electron beam exposure system

(b) Cell projection electron beam exposure system

FIG. 7.5 Comparison of variable shaped beam and projection electron beam exposure systems.

wafer in columns, which is used for EBES [2]. Since in many integrated circuit patterns only limited area of wafer are to be exposed, considerable time can be saved sequentially addressing only those areas that need to be exposed. This is the so-called "vector" scan method [4,5] as compared with the raster scan method in Fig. 7.2 and Fig.7.3 [3]. After the exposure of a certain area, the stage is moved rapidly with great precision in a step-and-repeat fashion.

These designs are using a fixed Gaussian beam to write the pattern. Fig. 7.4 shows development of various writing procedures including focused point beam, fixed shaped beam, variable shaped beam and pattern projection [6]. Exposure time could be saved by using a square-shaped beam instead of a Gaussian beam, which would expose a square area in a single flash [7]. A variable-shaped beam system can expose a certain local area with one shot by controlling the second

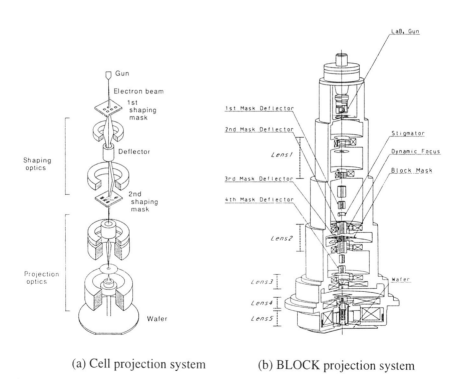

(a) Cell projection system   (b) BLOCK projection system

FIG. 7.6 Schematic view of an electron beam cell projection and block exposure column.[Reprint with permission from *J. Vac. Sci. Technol.*, **B8**, 1836(1990) and *J. Vac. Sci. Technol.*, **B11**, 2357(1993).]

aperture position as shown in Fig. 7.5(a). The shaped beam system can significantly reduce the exposure shot number [8,9].

The requirement for the minimum feature size of recent ULSI devices is as small as the focused beam size of a former point beam system. Large integration and reduction in minimum feature size require more shot numbers per chip. Therefore, much improvement in throughput of the variable shaped beam system is necessary. The ultimate in exposure time saving would be achieved with an electron projection system (parallel exposure) using a stencil mask as shown in Fig. 7.5(b). The exposure tool of a combination of projection and variable shape method, such as cell projection[10] and block exposure[11], have been reported. Schematic representaion of these exposure systems are shown in Fig. 7.6. As integration of LSI devices proceeds to shrink memory cell size, several memory cell patterns of recent advanced memory devices can be exposed with the maximum beam size of the conventional variable beam system. Since the memory patterns include a lot of repeated patterns, the aperture for part of memory cell patterns can be projected onto the wafer with one exposure as shown in Fig. 7.5(b). This system can significantly reduce the exposure shot number for repeated patterns, though stencil masks are needed.

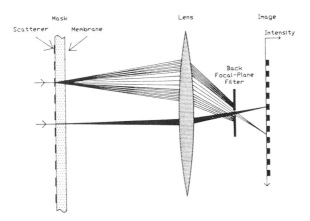

FIG. 7.7 Principle of operation of the SCALPEL (SCattering of Angular Limitation in Projection Electron-beam Lithography). [Reprint with permission from *Appl. Phys. Lett.*, **57**, 153(1990).]

A new approach to projection electron beam lithography was described by workers of AT&T Bell Laboratories [12]. This technique archives image contrast with a transparent mask comprised of two materials that differ in their scattering properties. This concept has been called SCALPEL (SCattering with Angular Limitation in Projection Electron-beam Lithography), which consists of a high energy electron beam and projection mask as described in Fig. 7.7. As electrons pass through a mask, which consists of a low atomic number membrane supporting a pattern of a high atomic number materials, scattering occurs resulting in different angular distributions of electrons from the patterned and unpatterned region. In the back-focal plane of the projecting lens system the electrons are distributed according to their angle of scatter. An aperture in this plane acts as an angularly limiting filter. The contrast obtained is governed by the size of the aperture.

Although the projection system is effective for repeated patterns in the memory cell region, it still takes time for random patterns in advanced ASIC LSI. Several approaches to improve the throughput for random patterns have been proposed such as multiple aperture [13], microcolumns [14], and blanking aperture array [15].

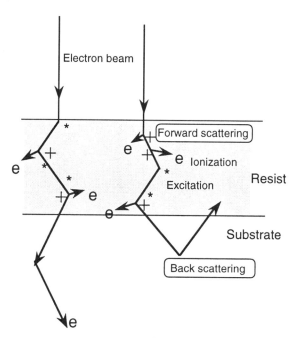

FIG. 7.8 Electron beam scattering in a resist and substrate.

Electron beam lithography is one of the candidates for future lithography as the next generation following KrF excimer laser lithography. Although several approaches to improve the inherent disadvantage of throughput have been reported as described above, further improvement is needed to realize the application of electron beam lithography to mass production.

7.4 FACTORS DETERMINING RESOLUTION

Since the acceleration voltage used in electron beam lithography ranges from 20 to 50 keV, the de Broglie wavelength of those electrons falls around 1 pm. It is actually difficult to confine an electron beam into 1 pm due to the electron repulsion and aberration of electron optics. It is possible to confine the beam into ~nm at the expense of throughput, as one can expect from the resolution of an electron microscope. Although the required resolution for electron beam lithography is about 0.1 µm at present, it is not easy to get this resolution with acceptable throughput for actual use.

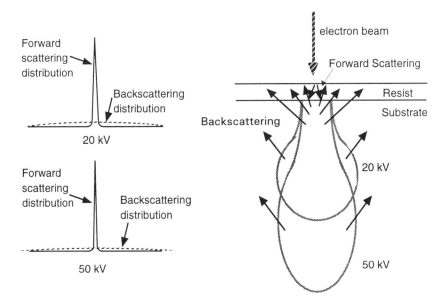

FIG. 7.9 Scattering of electrons in resist film. The curve at the left of the figure shows the exposure distributions due to the forward scattered and backscattered electrons.

The main factor which limits the resolution capability is the scattering of electrons in resist materials. When the accelerated electrons enter a solid material, the direction of electrons is changed continuously by elastic and inelastic scattering. Some of the electrons return to the resist film via scattering by atoms in the underlying substrate. The scattering in a resist film is called "forward scattering" and the scattering from the substrate is "backward scattering" as schematically described in Fig. 7.8. The scattering behavior depends on the electron beam energies, resist film thickness and substrate atomic number as shown in Fig. 7.9 [3]. The simulation of electron trajectories for various acceleration energies in a resist and a substrate is shown in Fig. 7.10 [16].

As the electron energy increases, the energy loss per unit path length and the scattering cross sections decrease. Therefore, the lateral spread of the forward-scattered electrons and energy dissipated per electron decreases, while the lateral extent of the backscattered electrons increases due to scattering from the deep

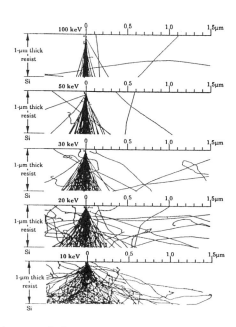

FIG. 7.10 Visual images of electron trajectories at various acceleration voltages. No blur in the incident beam is assumed. [Reprint with permission from *J. Vac. Sci. Technol.*, **B12**, 3874(1994).]

level of the substrate. As the resist film thickness increases, the cumulative effect of the small angle collisions by the forward scattered electrons increases. Thus, the area exposed by the scattered electrons at the resist-air surface is larger in thick films than in thin films. Since the energy of the electron is deposited via inelastic scattering, the trajectories of electrons result in a deposit energy profile in a resist film. Electron scattering broadens the incoming beam considerably, leading to a "deformed profile" compared to the initial beam profile in the vacuum.

As one can see from Figs.7.9 and 7.10, an electron beam with higher energy gives a better energy-deposited profile in a resist film. In addition backward scattering caused from the deep level of underlying substrate results in uniform and low level. As a result, the energy deposited profile for higher acceleration energy is improved.

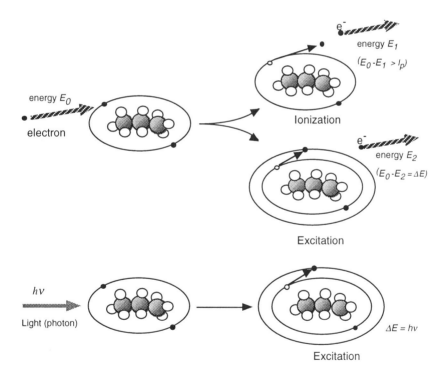

FIG. 7.11 Interaction of an electron with molecules and interaction of light with molecules.

7.5 INTERACTION OF ELECTRONS WITH ATOMS AND MOLECULES IN A RESIST

The primary interaction of incident electrons in the resist film involves elastic and inelastic scattering by atoms or molecules. The energy used for electron beam lithography, 20-50 keV, is much larger than the energy of the chemical bonds of molecules including polymers in resist materials and larger than the ionization potential of its constituent atoms [17].

The inelastic scattering leads to ionization and excitation (Fig. 7.11). The electron produced in the primary ionization step carries a large amount of kinetic energy that can ionize other atoms and molecules. This processes is repeated over and over again to the energy that cannot ionize them, resulting in a lot of electrons with a wide range of energy. These secondary electrons also repeat ionization and excitation of molecules in a resist film. The secondary electrons lose their energy to thermal energy. Some of the secondary electrons may recombine with their original partners of ions and produce molecules in higher excited states. Some electrons may attach themselves either to specific molecules (forming molecular aions) or to the solvent in the form of solvated electrons. The excited states produced by direct excitation and by recombination either emit radiation, or fragment into ions and radicals, decay to lower excited states by internal conversion, or finally return to the ground state. Often this sequence leads to final products similar to those obtained by photochemistry. The above sequences are schematically depicted in Fig. 7.12.

In photochemistry radiation quanta are absorbed by specific molecules and promote these into well-defined excited states corresponding to the photon energy. As the energy of quanta increases and starts to exceed some of the bond energies in the molecules, photochemistry leads to fragmentation. The distinguishing feature of electron-beam induced chemistry is that energy absorption is not any more associated with a particular molecule, but occurs at random in the material. Electron beams can ionize and excite to higher levels of molecules. In photolithography, absorption of photosensitive compounds is tuned to the exposure wavelength, while molecules and polymers with no absorption at the UV-visible region may be used in electron beam lithography. Another difference between electron induced chemistry and photochemistry is that photochemical excitation of molecules results in photon extinction, while the electron just loses its energy and does not disappear by excitation or ionization.

All these events take place in the vicinity of the primary high energy electron and various product species are formed along its trajectory or in spurs (groups of excitation and ionization molecules) and side branches induced by secondary electrons.

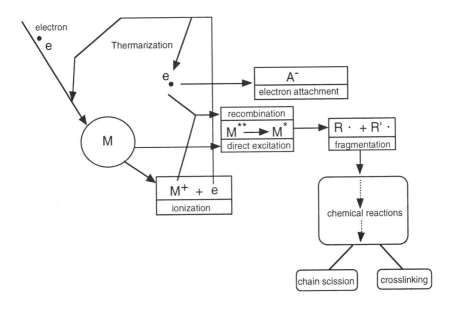

FIG. 7.12 Primary processes induced by electron beam exposure.

7.6 FACTORS DETERMINING ELECTRON BEAM RESIST SENSITIVITY

The trend to use a higher acceleration energy of electrons requires very sensitive resists. Higher acceleration energy leads to a shorter interaction of the electron with molecules, which results in low deposited energy in a resist film. The energy loss process of a high energy electron (deposition of electron energy in a resist film) can be described by the following Bethe energy formula [18]

$$\frac{dE}{dx} = \frac{-2\pi e^4 n_e}{E} \ln[(\frac{E}{I})(\frac{e}{2})^{1/2}] \qquad (7.1)$$

where E is the electron energy, x is the distance of electron trajectory, e is the electron charge, n_e is the density of atomic electrons and I the mean excitation energy. On the basis of Bethe's equation, one can easily understand that energy loss per unit length for a higher energy electron is lower than that of a lower energy electron.

Most electrons for initial energy of 30 to 50 keV pass through the resist film depositing only several percentage of its initial energy. The energy loss per unit length of electron trajectory in Bethe's equation is a function of electron energy, mean excitation energy, and the electron density of atoms comprising a resist film. As one can see from Bethe's equation, the deposited energy of an electron beam is a strong function of electron energy rather than the composition of resists. Therefore, "apparent" sensitivity described by irradiation dose such as C/cm^2 shows higher values with higher acceleration energy, which means lower sensitivity. The relation of resist sensitivity for iodinated polystyrene (IPS) and acceleration energy of electron beams is shown in Fig. 7.13 [19]. This figure shows that the required exposure dose increases with acceleration energy for the resist used.

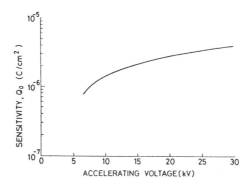

FIG. 7.13 Dependence of electron beam acceleration energy on resist sensitivity for iodinated polystyrene. [Reprint with permission from *J. Appl. Polym. Sci.*, **29**, 223(1984).]

The relation between electron sensitivity and photo-sensitivity is shown in Fig. 7.14. The apparent UV sensitivity described in mJ/cm^2 is obtained by calculating the deposited electron energy in 1 µm resist film using the Bethe equation [16]. As shown in Fig. 7.14, the sensitivity of 1µC/cm^2 for 30keV electron corresponds to 1mJ/cm^2 for UV light. The sensitivity of DNQ-novolak type resist currently used in LSI industry is 100~200mJ/cm^2, which means that the required sensitivity for electron beam lithography is about 100 times higher than that for optical lithography. It is difficult to obtain high sensitive resists in the conventional way as used for a DNQ type resist. The use of chain reactions such as an acid-catalyzed reaction in chemical amplification is needed to improve the sensitivity. High sensitivity is also one of the important parameters which govern the throughput of electron beam lithography.

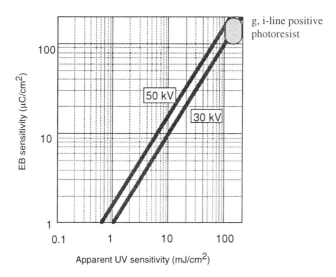

FIG. 7.14 Comparison of EB sensitivity with UV sensitivity. Apparent UV sensitivity was estimated by absorbed electron energy in resist film using Bethe equation.

7.7 ELECTRON BEAM RESISTS

Many chemical reactions following excitation or ionization are induced when a resist film is exposed to an electron beam. These reactions of polymers are classified into two types (Fig. 7.15, Table 7.1), though the induced reactions are complicated as descried in Fig. 7.12. One is chain scission and the other is crosslinking.

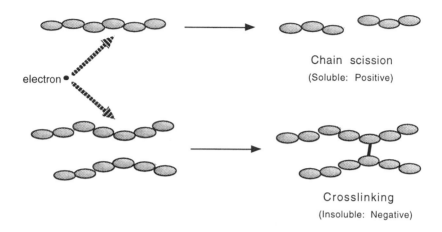

FIG. 7.15 Electron beam induced chain scission and crosslinking reaction of polymers.

A chain-scission type polymer degrades upon exposure to an electron beam, which makes the polymers soluble in a solvent. When this type of polymer film is performed with pattern-wise exposure with an electron beam and developed with an organic solvent, the exposed area of decomposed polymer is dissolved away, resulting in patterns of unexposed area. Therefore, this type of polymer can be used as a positive electron beam resist. A crosslinking type of polymer forms a three-dimensional network upon exposure to the electron beam, which makes the exposed area insoluble in a solvent. This type of polymer can be used as a negative electron beam resist.

7.7.1 Positive Electron Beam Resists

(a) Chain scission type polymers

Polymethylmethacrylate (PMMA) is a typical radiation degradation polymer. Workers of IBM showed that chain-scission type polymers can be used as a positive electron beam resist [20]. Since then PMMA is widely used as an electron beam resist with high resolution capability. The reason for the high resolution capability of PMMA is ascribed to its high contrast and non-swelling type dissolution in a developer. One of the disadvantages of PMMA is low

Table 7.1 Electron beam resists

Positive	Main chain scission	\multicolumn{2}{l	}{PMMA: $+CH_2-C(CH_3)(COOCH_3)+$; FBM: $+CH_2-C(CH_3)(COOCH_2CF_2CFHCF_3)+$; EBR-9: $+CH_2-C(Cl)(COOCH_2CF_3)+$; PBS: $+CH_2-CH(CH_2CH_3)-SO_2+$; ZEP: $+CH_2-C(Cl)(COOCH_3)+CH_2-C(CH_3)(C_6H_5)+$}
	Dissolution Inhibition	\multicolumn{2}{l	}{NPR (PMPS + novolak resin); PMPS: $+CH_2-C(CH_3)(CH_2CH_2CH_3)-SO_2+$; novolak resin}
Negative	Cross linking	Epoxy group	PGMA: $+CH_2-C(CH_3)(COOCH_2-CH\underset{O}{-}CH_2)+$; COP: $+CH_2-C(CH_3)(COOCH_2-CH\underset{O}{-}CH_2)+CH_2-CH(COOC_2H_5)+$
		polystyrene derivatives	halogenated styrenes: $+CH_2-CH(C_6H_4X)+$; CMS: $+CH_2-CH(C_6H_4CH_2Cl)+$
	alkali developable	\multicolumn{2}{l	}{polyhydroxystyrene: $+CH_2-CH(C_6H_4OH)+$; azide: $N_3-C_6H_4-SO_2-C_6H_4-N_3$}

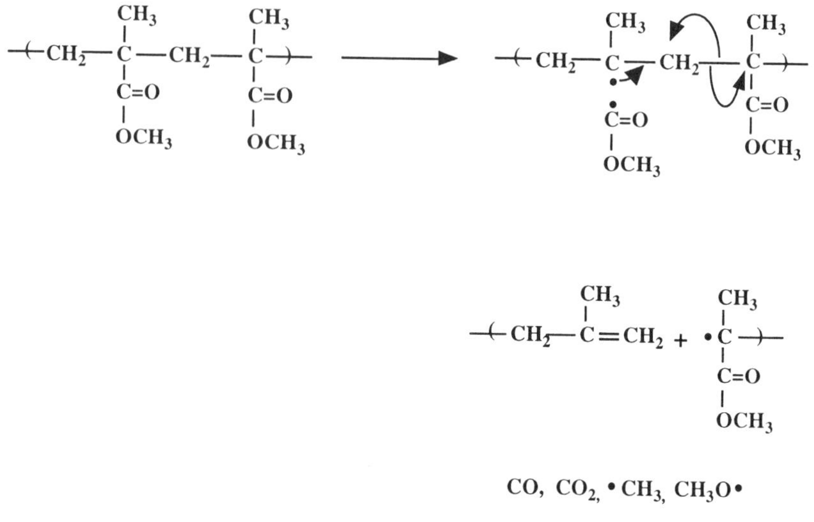

$$\begin{array}{c}\text{CH}_3\\|\\\text{+CH}_2-\text{C}-\text{CH}_2-\text{C}+\\|\\\text{C=O}\quad\text{C=O}\\|\quad\quad|\\\text{OCH}_3\quad\text{OCH}_3\end{array}\longrightarrow\begin{array}{c}\text{CH}_3\quad\quad\text{CH}_3\\|\quad\quad\quad|\\\text{+CH}_2-\text{C}-\text{CH}_2-\text{C}+\\\bullet\quad\quad\quad|\\\text{C=O}\quad\quad\text{C=O}\\|\quad\quad\quad|\\\text{OCH}_3\quad\quad\text{OCH}_3\end{array}$$

$$\begin{array}{c}\text{CH}_3\quad\quad\text{CH}_3\\|\quad\quad\quad|\\\text{+CH}_2-\text{C=CH}_2\;+\;\bullet\text{C}+\\\quad\quad\quad\quad|\\\quad\quad\quad\quad\text{C=O}\\\quad\quad\quad\quad|\\\quad\quad\quad\quad\text{OCH}_3\end{array}$$

CO, CO_2, •CH_3, CH_3O•

FIG. 7.16 Radiation chemistry of poly(methylmethacrylate)PMMA.

sensitivity. Therefore, PMMA is rarely used for industrial applications, such as photomask fabrication, though PMMA is widely used to check resolution capability of an electron beam exposure system [21].

The radiation induced reaction is shown in Fig. 7.16 [22]. Upon irradiation to PMMA, the scission of side chain induces the elimination of carbon monoxide, leading to a tertiary radical in the main chain. β scission causes main-chain scission, resulting in a tertiary radical stabilized by acyl group. These events make the molecular weight of PMMA lower.

Lower molecular weight PMMA in an exposed area is more soluble than the initial one. The dissolution rate of PMMA increases with a decrease of molecular weight. Ouano et al. [23] demonstrated that the increase in free volume of an exposed area due to the evolution of gaseous products enhanced the dissolution rate of exposed PMMA.

Based on the knowledge of radiation chemistry, polymers containing a tertial carbon in the main chain lead to chain scission upon exposure to radiation. A lot of attempts to improve the sensitivity of PMMA [24] using modified PMMA have been proposed. Among them, Poly(hexafluorobutylmethacrylate) (FBM) [25], poly(trifluoroethyl-α-chloroacrylate) (EBR-9) [26], and poly(αchloro-acrylate-co-α-methylstyrene) (ZEP) [27] as shown in Table 7.1 have been commercialized.

Poly(olefine sulfone)s are chain-scission polymers upon exposure to an electron beam. Bell Laboratories have proposed poly(butene-1-sulfone) (PBS) [28] as a positive electron beam resist. PBS is an electron beam resist designed for EBES which was developed by Bell Laboratories for photomask fabrication. The sensitivity of PBS is 0.7 µC/cm^2 which meets the required sensitivity, <1µC/cm^2, for MEBES (20 keV).

(b) Positive electron beam resists with high-dry etch resistance

Positive electron beam resists described above show low dry etch (plasma) resistance. The requirement for dry-etch resistance for photomask fabrication is not so important, as thin Cr film can be etched in wet process without line width change. However, dry-etch resistance is needed, when patterns are directly fabricated on a silicon wafer by electron beam lithography. It is always difficult to obtain a positive electron beam resist with dry-etch resistance, since ion bombardment in a plasma also induces chain scission.

A positive electron beam resist which meets the contradictory requirements of high sensitivity and dry etch resistance has been reported. It is a new positive electron beam resist (NPR) that is composed of poly(2-methylpentene-1-sulfone) (PMPS) and a novolak resin [29]. PMPS acts as a dissolution inhibitor and undergoes chain scission effectively upon exposure to an electron beam. This chain scission leads to loss of dissolution inhibition capability, which makes the exposure area soluble in an aqueous base. PMPS acts in a similar role as diazonaphthoquinone in a positive photoresist. Since a novolak resin is a aromatic polymer, it shows similar dry-etch resistance as a DNQ type positive photoresist. However, this resist has a problem of phase separation induced by incompatibility of a mixture of polymers of a novolak resin and PMPS. Some novolak resins were evaluated as a base polymer for NPR [29]. Isoamylacetate was found to avoid incompatibility of two polymers and NPR is commercially available [30] (Hitachi Chemical RE-5000P).

PMPS itself shows self-development, which indicates film thickness loss during electron beam exposure. Since the ceiling temperature of PMPS (-34°C) is lower than room temperature, PMPS degrades spontaneously to monomers once chain scission is induced by electron beam exposure. Monomer evaporation results in a positive relief image.

7.7.2 Negative Electron Beam Resists

(a) Epoxy polymers

Workers of Hitachi reported polymers with an epoxy group such as epoxydized polybutadiene (EPB) and poly(glycidylmethacrylate) (PGMA) give a highly sensitive negative electron beam resists [31,32]. Copolymer of glycidylmethacrylate and ethylacrylate (COP) was reported as a high resolution resist by Bell Laboratories [33], which was designed for EBES. Electron beam exposure induces the ring opening crosslinking reaction, resulting in a three-dimensional network insoluble to developer. This reaction is a chain reaction, which gives high sensitivity.

(b) Negative electron beam resists of polystyrene derivatives

The negative electron beam resists described above give low dry-etch resistance. Although dry etch resistance is not an important requirement for photomask fabrication, it is important for direct fabrication of LSI patterns on a silicon wafer. Since polystyrene has an aromatic ring, it has dry-etch resistance but low sensitivity for practical use. Consequently many attempts were made to sensitize polystyrenes by introducing electron sensitive substituents into the phenyl ring of polystyrenes [24]. It was said in the late 1970s that every resist researcher was concerned with polystyrene derivatives. Chloromethylated polystyrene (CMS) is especially widely used [34]. The improvement in sensitivity can be ascribed to the low energy of the carbon-halogen bond.

    The reaction mechanism for iodinated polystyrene as an example of halogenated polystyrene is shown in Fig. 7.17 [19]. The scission of the phenyl-iodine bond occurs upon exposure to an electron beam. The recombination reaction of an iodine atom and the phenyl radical compete the hydrogen abstraction reaction of iodine atoms. These reaction schemes can explain the saturation of sensitivity with an increase in iodination degree as shown in Fig. 7.18.

    One of the serious disadvantages for polystyrene derivatives is swelling during development, leading to deteriorated patterns, such as snake-walk and bridging. The swelling problem is related to the crosslinking density: higher crosslinking density retards the swelling. However, sensitivity is proportional to the molecular weight of polystyrene derivatives: higher molecular weight gives low crosslinking density. Sensitivity and resolution is trade-off for this type of resist.

According to Charlesby's theory [35], mono-disperse polymers gives high contrast. The polystyrene derivatives of narrow molecular weight distribution have been reported with high resolution [19, 34].

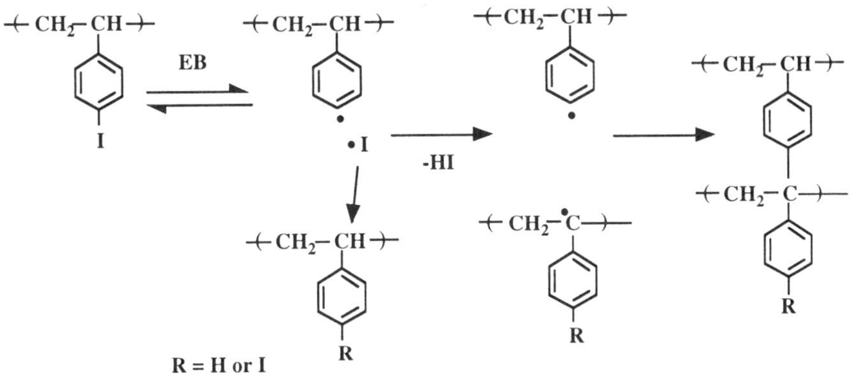

FIG. 7.17 Electron-beam induced reaction mechanism of iodinated polystyrene IPS.

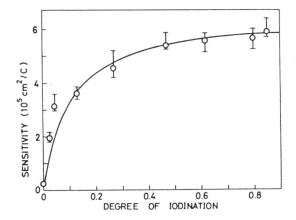

FIG. 7.18 Electron-beam sensitivity of iodinated polystyrene as a function of iodination degree. [Reprint with permission from *J. Appl. Polym. Sci.*, **29**, 223(1984).]

7.7.3 Alkali-developable High Resolution Negative Electron Beam Resists

One of the important advantages of electron beam lithography is high resolution. Negative electron beam resists described above do not meet the requirement for high resolution due to swelling during development. A non-swelling and high contrast resist is needed for high resolution. Workers at Hitachi reported that the resist based on poly(hydroxystyrene) (PHS) and a bisazide which was originally developed as a deep-UV resist can be used as an electron beam resist [36]. Electron beam exposure produces nitrene from a bisazide, which induces crosslinking of PHS.

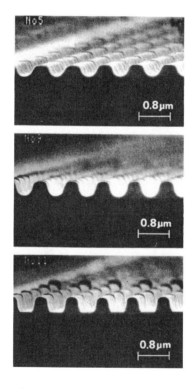

FIG. 7.19 Change in resist pattern with development time in an aqueous base for the resist composed of poly(hydroxystyrene) and a bisazide. [Reprint with permission from *ACS Symp. Ser.*, **242**, 167(1984).]

The crosslinking causes an increase in the molecular weight of PHS, leading to reduction in the dissolution rate for an alkali developer. They observed the dissolution behavior during the development using direct-writing patterned resist film with an electron-beam exposure machine. Fig. 7.19 shows patterns during development with the increase in development time. One can clearly recognize no evidence of swelling during the development. For polystyrene derivatives and cyclized rubber resists, this kind of development behavior can not be observed due to the swelling.

The relation between dissolution rate in alkaline developer and the molecular weight of PHS is shown in Fig. 7.20 [36]. The dissolution rate of PHS decreases rapidly with increasing molecular weight up to 10^4. This decrease also occurs in the exposed area where molecular weight is increased by crosslinking. Although there is no evidence of swelling during the development for molecular weight up to 10^4, the swelling problem was observed when a higher molecular weight ($M_w > 10^4$) PHS is used. Since high-molecular weight PHS cannot be used as a base polymer, this type of resist renders limitation in sensitivity of ~10 µC/cm^2.

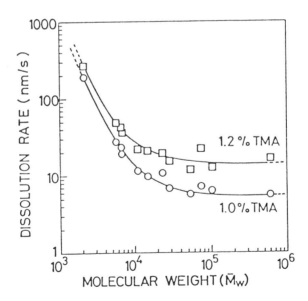

FIG. 7.20 Dissolution rate of poly(p-vinylphenol) as a function of molecular weight. ○: 1.0% tetramethylammonium hydroxide (TMAH) aqueous solution. □: 1.2% TMA solution. [Reprint with permission from *ACS Symp. Ser.*, **242**, 167(1984).]

7.7.3 Chemical Amplification Resists

Drastic improvement in sensitivity while maintaining high resolution can be achieved using chemically amplified resists using an acid-catalyzed reaction. The basic idea for a chemically amplified resist was already described in the former section. Most of the resists can also be applied to electron beam resists. Basically the deprotection reaction is used for positive resists [37] and the acid-hardening reaction of melamine derivatives is used for negative ones [38].

The acid-hardening type of resist has been evaluated extensively since this resist shows high contrast in alkali development. Liu et al. [39] evaluated the resist contrast by ratio of dissolution rate, R(0.4D)/R(D)=Srg(-), where R(0.4D) is the dissolution rate of 40% dose and R(D) is that of a dose D. This dissolution rate ratio is taken into account of the energy deposition by scattered electrons in an unexposed area caused by proximity effect such as forward and backward scattering. As shown in Fig. 7.21, Srg(-) shows difference of 200, though it depends on the post-exposure-baking temperature.

The announcement of this resist stimulates the application of chemically amplified resists to LSI fabrication due to its alkali developability, which is compatible with a production line.

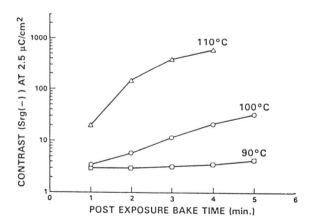

FIG. 7.21 Srg(-) at 2.5 $\mu C/cm^2$ as a function of both post-exposure temperature and time, where Srg(-) is the ratio of dissolution rate R(0.4D)/R(D) at D= 2.5 $\mu C/cm^2$ and D is the dose for patterning. [Reprint with permission from *J. Vac. Sci. Technol.*, **B6**, 379(1988).]

Electron Beam Lithography

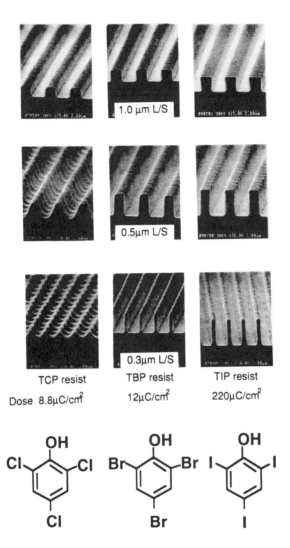

FIG. 7.22 Effect of acid diffusion of HCl, HBr, and HI on the pattern formation in chemical amplification resists. The resists composed of trihalophenols as acid generators, hexamethoxymethylmelamine as a crosslinker and a novolak resin were evaluated. Trichlorophenol, tribromophenol, and triiodophenol were used as acid generators. [Reprint with permission from Extended Abstract of 10th International Conference of Photopolymers, Ellenvile, N.Y., Poster Session, Paper 3, (1994).]

It is pointed out that acid diffusion in a resist may affect the resolution capability for CA resists. The effect of acid diffusion of HCl, HBr, and HI was investigated in acid-hardening type of resists [40]. HCl, HBr, and HI are generated from 2,4,6-trichlorophenol, 2,4,6-tribromophenol, and 2,4,6-triiodophenol, respectively. Fig. 7.22 clearly demonstrated that the large size acid, HI, renders the highest resolution among the acids investigated in spite of low sensitivity.

Workers of Hitachi have reported the contrast-enhancement EB resist using a base-catalyzed reaction during development [41]. They named the resists as contrasted boosted resists (CBR) as described in Chapter 4. The base-catalyzed reaction from water-repellant compounds to hydrophilic compounds during aqueous-base development enhances the change in dissolution rate of the unexposed area for a negative resist. As shown in Fig. 7.23, 0.225 µm line and space patterns were demonstrated with a high sensitivity of ~2.0 µC/cm².

FIG. 7.23 High resolution capability of advanced negative resist for electron beam lithography. The resist is composed of 1,3,5-tris(trichloromethyl)-triazine as an acid generator, hexamethoxymethylmelamine as a crosslinker, novolak resin, and 1,3,5-tris-(bromoacetyl)benzene as a contrast enhancer. [Reprint with permission from *Proc. SPIE*, **2724**, 438(1996).]

7.8 CONCLUSION

The advance of electron beam exposure has been focused on the improvement in throughput by beam shape and by decrease in shot number. The resolution capability has been improved by the increase in acceleration energy of the electron beam. Improvement in resolution and sensitivity of the electron beam resists has played an important role in the reduction of exposure time (improvement in throughput) in electron beam lithography. It is still controversial that electron beam lithography can be used in a mass production line [42].

REFERENCES

1. M. Hatzakis, Electron resists for microcircuit and mask production, *J. Electrochem. Soc.*, **116**, 1033(1969).
2. D. R. Herriott, R. J. Collier, D. S. Alles, and J. W. Stafford, EBES: A practical electron lithographic system, *IEEE Trans. Electron Devices*, **ED-22**, 385(1975).
3. M. Bowden, *Introduction to Microlithography, Second edition*, Chap. 2, ACS Professional Reference Book, American Chemical Society, Washington, DC (1994).
4. A. J. Speth, A. D. Wilson, A. Kern, and T. H. P. Chang, Electron-beam lithography using vector-scan techniques, *J. Vac. Sci. Technol.*, **12**, 1235(1975).
5. S. Asai, H. Inomata, A. Yanagisawa, E. Takeda, I. Miwa, and M. Fujinami, Distortion correction and deflection calibration by means of laser interferometry in an electron-beam exposure system, *J. Vac. Sci. Technol.*, **16**, 1710(1979).
6. S. Okazaki, Lithographic technologies for future ULSI, *Appl. Surf. Sci.*, **70/71**, 603(1993).
7. H. C. Pfeiffer, New imaging and deflection concept for probe-forming microfabrication systems, *J. Vac. Sci. Technol.*, **12**, 1170(1975)
8. G. J. Giuffre, J. F. Marquis, H. C. Pfeiffer, and W. Stickel, Practical results of EL2, *J. Vac. Sci. Technol.*, **16**, 1644(1979).
9. M. Fujinami, T. Matsuda, A. Takamoto, H. Yoda, T. Ishiga, N. Saitou and T. Komoda, Variably shaped electron beam lithography system, EB55:I. System design, *J. Vac. Sci. Technol.*, **19**, 941(1981).
10. Y. Nakayama, S. Okazaki, N. Saitou, H. Wakabayashi, Electron-beam cell projection lithography: A new high-throughput electron beam direct-writing technology using specially tailored Si aperture, *J. Vac. Sci. Technol.*, **B8**,

1836(1990); Y. Nakayama, H. Satoh, N. Saitou, S. Hirasawa, T. Yanagida, and H. Todokoro, Thermal characteristics of Si mask for EB cell projection lithography, *Jpn. J. Appl. Phys.*, **31**, 4268(1992).
11. K. Sakamoto, S. Fueki, S. Yamazaki, T. Abe, K. Kobayashi, H. NIshino, T. Satoh, A. Takemoto, A. Ookura, M. Oono, S. Sago, Y. Oae, A. Yamada, and H. Yasuda, Electron-beam block exposure system for 256 M dynamic random access memory, *J. Vac. Sci. Technol.*, **B11**, 2357(1993); A. Yamada, K. Sakamoto, S. Yamazaki, K. Kobayashi, S. Sago, M. Oono, H. Watanabe, and H. Yasuda, Deflector and correction coil calibrations in an electron beam block exposure system, *J. Vac. Sci. Technol.*, **B12**, 3404(1994).
12. S. D. Berger and J. M. Gibson, New approach to projection-electron lithography with demonstrated 0.1 µm line width, *Appl. Phys. Lett.*, **57**, 153(1990).
13. G. W. Jones, S. K. Jones, M. Walters, and B. Dudley, Microstructures for particle beam control, *J. Vac. Sci. Technol.*, **B6**, 2023(1988).
14. T.H.P.Chang, D. P. Kern, and L. P. Muray, Arrayed miniature electron beam columns for high throughput sub-100nm lithography, *J. Vac. Sci. Technol.*, **B10**, 2743(1920); L. P. Muray, U. Staufer, D. P. Kern, and T.H.P.Chang, Performance measurements of a 1-keV electron-beam microcolumn, *J. Vac. Sci. Technol.*, **B10**, 2749(1992).
15. H. Yasuda, S. Arai, J. Kai, Y. Ooae, T. Abe, Y. Takahashi, S. Fueki, S. Maruyama, S. Sago, and K. Betsui, Fast electron beam lithography system with 1024 beams individually controlled by blanking aperture array, *Jpn. J. Appl. Phys.*, **32**, 6012(1993).
16. F. Murai, J. Yamamoto, H. Yamaguchi, S. Okazaki, K. Sato, and H. Hayakawa, High-speed single-layer-resist process and energy- dependent aspect ratios for 0.2-µm electron beam lithography, *J. Vac. Sci. Technol.*, **B12**, 3874(1994).
17. A. Reiser, Photoreactive polymers, A Wiley-Interscience Pub., New York, p.302(1989).
18. H. A. Bethe and J. Ashkin, Experimental Nuclear Physics, E. Segre, Ed., Wiley, New York, 1953, vol. 11, p.1166.
19. T. Ueno, H. Shiraishi, and S. Nonogaki, Insolubilization mechanism and lithographic characteristics of a negative electron beam resist iodinated polystyrene, *J. Appl. Polym. Sci.*, **29**, 223(1984).
20. I. Haller, M. Hatzakis, and R. Srinivasan, High resolution positive resists for electron beam exposure, *IBM J. Res. Develop.*, **12**, 251(1968).

21. W. Chen and H. Ahmed, Fabrication of high aspect ratio silicon pillars of <10 nm diameter, *Appl. Phys. Lett.*, **63**, 1116(1993).
22. H. Hiraoka, Radiation chemistry of poly(methylmethacrylate), *IBM J. Res. Develop.*, **21**, 121(1977).
23. A. C. Ouano, A study of dissolution rate of irradiated poly(methylmethacrylate), *Polym. Eng. Sci.*, **18**, 306(1978); A. C. Ouano, Dependence of dissolution rate on processing and molecular parameters of resists, *ACS Symp. Ser.*, **242**, 79(1984).
24. T. Tamamura, S. Imamura, and S. Sugawara, Resists for electron beam lithography, *ACS Symp. Ser.*, **242**, 104(1984).
25. M. Kakuchi, S. Sugawara, K. Murase, and K. Matsuyama, Poly(fluoromethacrylate) as highly sensitive, high contrast positive resists, *J. Electrochem. Soc.*, **124**, 1648(1977).
26. T. Tada, Poly(trifluoroethyl-α-chloroacrylate) as a highly sensitive positive electron resist, *J. Electrochem. Soc.*, **126**, 1829(1979).
27. T. Nishida, M. Notomi, R. Iga, and T. Tamamura, Quantum wire fabrication by e-beam lithography using high-resolution and high sensitivity e-beam resist ZEP-520, *Jpn. J. Appl. Phys.*, **31**, 4508(1992).
28. M. J. Bowden, L. F. Thompson, and J. P. Ballantyne, Poly(butene-1-sulfone)-A highly sensitive positive resist, *J. Vac. Sci. Technol.*, **12**, 1294(1984).
29. M. J. Bowden, L. F. Thompson, S. R. Fahrenholtz, and E. M. Doerries, A sensitive novolak-based positive electron resist, *J. Electrochem. Soc.*, **128**, 1304(1981).
30. H. Shiraishi, A. Isobe, F. Murai, and S. Nonogaki, Novolak based positive electron beam resist containing a polymeric dissolution inhibitor, *ACS Symp. Ser.*, **242**, 167(1984).
31. T. Hirai, Y. Hatano, and S. Nonogaki, Epoxide- containing polymers as highly sensitive electron-beam resists, *J. Electrochem. Soc.*, **118**, 669(1971).
32. Y. Taniguchi, Y. Hatano, H. Shiraishi, S. Horigome, S. Nonogaki, and K. Naraoka, PGMA as a high resolution high sensitivity negative electron beam resist, *Jpn. J. Appl. Phys.*, **18**, 1143(1979).
33. L. F. Thompson, E. D. Feit, and R. D. Heindenreich, Lithography and radiation chemistry of epoxy containing negative electron resists, *Polym. Eng. Sci.*, **14**, 529(1974).
34. S. Imamura, Chloromethylated polystyrene as a dry etching-resistant negative resist for submicron technology, *J. Electrochem. Soc.*, **126**,

1628(1979).
35. A. Charlesby, *Atomic Radiation and Polymers*, Pergamon, Oxford, 1960.
36. H. Shiraishi, N. Hayashi, T. Ueno, O. Suga, F. Murai, and S. Nonogaki, Phenolic resin-based negative electron beam resists, *ACS Symp. Ser.*, **346**, 77(1987).
37. H. Shiraishi, N. Hayashi, T. Ueno, T. Sakamizu, and F. Murai, Novolak resin-based positive electron beam resist system utilizing acid-sensitive polymeric dissolution inhibitor with solubility reversal reactivity, *J. Vac. Sci. Technol.*, **B9**, 3343(1991).
38. W. E. Feely, J. C. Imhof, C. M. Stein, T. A. Fisher, and M. W. Legenza, The role of latent image in a new dual image aqueous developable thermally stable photoresist, *Polym. Eng. Sci.*, **26**, 1101(1986).
39. H.-Y. Liu, M. P. deGrandpre, and W. E. Feely, Characterization of a high resolution novolak based negative electron beam resist with $4\mu C/cm^2$ sensitivity, *J. Vac. Sci. Technol.*, **B6**, 379(1988).
40. S. Migitaka, S. Uchino, T. Ueno, F. Murai, and J. Yamamoto, Effect of acid size on sensitivity and resolution of chemically amplified negative resists, Extended Abstract of 10th International Conference of Photopolymers, Ellenvile, N.Y., Poster Session, Paper 3, (1994).
41. S. Uchino, T. Ueno, S. Migitaka, J. Yamamoto, T. Tanaka, F. Murai, H. Shiraishi, and M. Hashimoto, Contrast-boosted resist using a polarity-change reaction during aqueous base development, *Proc. SPIE*, **2724**, 438(1996).
42. R. DeJule, E-beam lithography: The debate continues, *Solid State Technol.*, **39**(9), 85(1996).

8
Variations in the Microlithographic Process

8.1 PROCESSES AND MATERIALS FOR RESOLUTION ENHANCEMENT

Continuous improvement of the operation speed and circuit density of LSIs requires enhancement of the resolution and process latitude of resist processes and materials. The practical resolution attainable depends on both the optical system used in the patterning step and the photoresist materials in which the pattern is initially formed. When attempting to use lithography above its resolution limit, it is necessary to determine a method for solving the resolution problem, in either the optical system or the resist process. In the case of the optical system the light source, the mask and the projection lens system are modified to enhance resolution, and this is called the "super resolution technique" and is based on Fourier's optics theory. Three methods have been studied so far, which correct the distribution function as follows: (1) intensity of light source illuminating mask, (2) complex amplitude transmittance through a mask and its Fourier transformation, and (3) pupil function of a projection lens. These are known as "annular illuminations" [1] (and their improved methods "modified illumination" [2]), "phase shifting mask" [1] and "pupil filtering" [3], respectively.

A variety of methods related to the resist process and a new class of lithographic material were developed to expand the resolution limit and process latitude. The contrast enhancement lithography (CEL) method developed by Griffing and West [4], a multilayer resist process, enhances resolution by depositing a photobleachable layer on a photoresist film. Photobleachable materials, known as contrast enhancement materials (CEMs), are needed for this new technique. The portable conformable mask (PCM) method developed by Lin [5] enhances resolution by means of a thin built-up mask layer on a deep-UV resist. A multilayer resist with dry development (typically O_2-RIE) [6] is efficient for avoiding swelling and other pattern deformation due to wet development, and micropatterns with a high aspect ratio can be easily obtained. Surface imaging [7] is a kind of thin film lithography. A very thin modified layer is formed by chemical treatment after patterning exposure, which acts as a high resolution built-in mask and determines the resolution of the following development step. Much effort has been made to remove any wet treatment from resist development in order to achieve fully-dry development, and several unique approaches have been proposed. Resolution enhancement is achieved in the methods described above by adding an extra treatment or component, *e.g.*, light exposure, chemical vapor exposure, overcoat materials and a newly designed imaging resist, to the conventional process. There are many varieties of microlithographic processes.

8.2 CONTRAST ENHACEMENT IN OPTICAL IMAGE TRANSFER

The concept of CEL (Contrast Enhancement Lithography) [4] is explained by Fig. 8.1. Very thin CEM (Contrast Enhancement Material) film is composed of a photobleaching compound and a binder resin. The process requires that the CEM layer first be applied to the photoresist surface. Following conventional exposure, the layer is removed and the resist is developed in the usual way. As a result of the presence of the photobleachable compound, the contrast of the illumination that reaches the photoresist is increased. The effect of CEM is to forestall exposure of the underlying photoresist until bleaching is complete, thus providing a controllable threshold dose independent of the resist properties. Although an additional dose is required to clear the CEM film, the photoresist layer receives the contrast optical image.

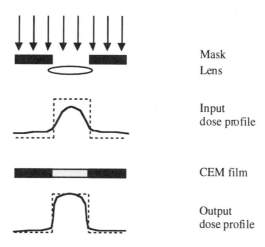

Fig. 8.1 Principle of CEL process.

Variations in Microlithographic Process

Bleaching of a diazonaphthoquinone photoresist is described by the Dill model [8]. The photochemical process of CEM is also analyzed practically using this model. The bleaching characteristics of CEL are measured, and parameters A, B and C can be calculated by

$$A = \frac{1}{d} \ln\left[\frac{T(\infty)}{T(0)}\right], \quad (8.1)$$

$$B = -\frac{1}{d} \ln[T(\infty)], \quad (8.2)$$

and

$$C = \frac{A+B}{AI_0 T(0)\{1 - T(0)\}} \frac{dT(0)}{dt}, \quad (8.3)$$

where $T(0)$, $T(\infty)$, I_0, d, and t are the transmittance of CEM before exposure, that after bleaching fully, intensity of incident light, film thickness, and exposure time, respectively. Each parameter in these equations means: A the absorption coefficient of the unexposed CEM layer, B that of the bleached film and C the bleaching speed of CEM. The three parameters play an important role in characterizing the dose transmission of the CEM film. Contrast enhancing factor α is introduced to evaluate the performance of the CEM [9]. This factor indicates the normalized value of the slope of the dose transmittance against the input dose exposed to the CEM as depicted in Fig. 8.2.

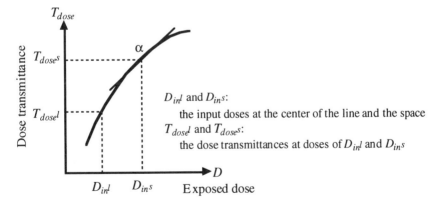

$D_{in}{}^l$ and $D_{in}{}^s$: the input doses at the center of the line and the space
$T_{dose}{}^l$ and $T_{dose}{}^s$: the dose transmittances at doses of $D_{in}{}^l$ and $D_{in}{}^s$

Fig. 8.2 Typical curves of the dose transmission with CEL film and contrast enhancing factor α.

Using α, the output dose contrast can be written as

$$C_{out} = C_{in}\left[(1+C_{in}) + \alpha(1-C_{in})\right] / \left[(1+C_{in}) - \alpha C_{in}(1-C_{in})\right] \quad (8.4)$$

where C_{in} and C_{out} are the contrast of input dose and that of output dose, respectively, for, *e.g.*, line and space patterns. Evidently, *B* should be small.

The parameter *A* reflects the density of the bleachable dye contained in the CEM, which will affect the transmission ratio between strongly and weakly exposed areas. The larger the *A* value, the higher the photoabsorption, enhancing the contrast effect. But too large a value of *A* will cause a decrease of the dose transmittance, increasing the required dose. It is important to mention the film thickness *d*. From Eq. 8.1, we can see that the light intensity transmittance ratio of $T(\infty)$ to $T(0)$ is related to the product of *A* and *d*. $T(\infty)$ can be considered to represent the approximate transmittance of a sufficiently exposed area, such as line patterns. A higher ratio than this is required by CEM characterization. Therefore, $A \cdot d$ can be considered to represent the contrast enhancing effect. Since the relationship between *A* and α are calculated as a function of *d*, optimum exposure conditions (for example, $A \cdot d$) can be determined when a is specified. The parameter *C* indicates the reaction speed of the dye. A large *C* value means a faster reaction of the dye and needs less exposure to attain the optimum exposure dose. Therefore, the output dose from CEM films decreases and will cause insufficient exposure to the bottom layer photoresist sensitivity. The optimum *C* value of the CEM film can be determined as long as the $A \cdot d$ and photoresist sensitivity are specified. When 0.5 µm line and space resolution is needed by the use of i-line exposure with NA=0.35, the required a value is 0.5. This contrast enhancement factor is give by a 0.35-µm-thick CEM film with an *A* value of 4.3 µm^{-1}. Also, the C value should be 0.016 mJ/cm^2 when a resist whose sensitivity is 80 mJ/cm^2 is used. Under these exposure conditions, about 50% of the exposed dose is lost through the CEM film.

Based on these considerations, Hirai *et al.* revised the "SAMPLE" [10] program for a multilayer structure and applied it to the simulation of g-line and i-line CEL using a water soluble CEM [11] developed by their group. The simulation results were almost the same as the resist pattern profiles. According to their conclusion, the CEL process will improve the maximum resolution limits by 20% - 30% even in critical cases, as compared with conventional cases. Babu *et al.* also performed numerically simultaneous bleaching of the contrast enhancing layer and the positive photoresist [12]. They calculated two cases, a GE's CEM-388/positive resist combination [13-17] and a polysilane/AZ-2400 resist combination [18], and confirm that their calculation results are in good agreement with the simulation result of the CEM-388 system reported by Mack [19, 20] using PLOLITH and the images reported by Hofer *et al.*

Various photobleachable dyes have been investigated and the CEL effect has been demonstrated. GE's CEM [13] was the first case of a contrast enhancement system for integrated circuit application. Arylnitrones were used as photobleaching dye, and the bleaching occurs by photoisomerization as:

$$\underset{^2R^\cdot}{\overset{O}{_{}}}N=\overset{R^1}{\underset{Ar}{}} \longrightarrow \underset{^2R}{\overset{\overset{O}{\cdot}}{_{}}}N-\overset{R^1}{\underset{Ar}{}} \longrightarrow \underset{Ar}{\overset{^2R_{\cdot}}{_{}}}N-\overset{O}{\underset{R^1}{}} \quad (8.5)$$

The wavelength of the absorption maxima λ_{max} can be shifted by the substituent effect (aryl group in the nitrone), and CEM-388 (λ_{max} 388 nm) and CEM-420 (λ_{max} 420 nm) were developed for i-line and g-line exposure, respectively. Especially when CEM-420 is applied, it was observed that the focus latitude was extended significantly [21]. The main chain of a polysilane degrades photolytically to give low molecular weight parts according to the mechanism:

$$-\left(\begin{array}{c}R^1\\Si\\R^2\end{array}\right)_n \longrightarrow -\left(\begin{array}{c}R^1\\Si\cdot\\R^2\end{array}\right)_m + \left(\begin{array}{c}R^1\\\cdot Si\\R^2\end{array}\right)_{n-m} \quad (8.6)$$

During this reaction λ_{max} shifts to a shorter wavelength, then the absorption coefficient at the initial λ_{max} becomes small. The high absorbance of substituted silane polymers coupled with their efficient photobleaching suggested potential applications as short wavelength contrast enhancing materials [18]. Hofer et al. have demonstrated CEL by imaging in the mid-UV spectral region, using a thin layer of poly(cyclohexyl methylsilane) as the contrast enhancing layer. Although polysilane improves resolution considerably, only low CEL gain is obtained when the exposure dose is low. More efficient bleaching characteristics are required for polysilane CEM. Diazonium salts, a well known dye, decomposes readily upon exposure according to the general reaction,

$$\underset{}{\text{Ar-N}_2^{\oplus}\text{N}\overset{\ominus}{\text{X}}} \longrightarrow \text{Ar}^{\oplus} + N_2 + \overset{\ominus}{X} \quad (8.7)$$

and the molar absorption coefficient decreases, for example, from 3×10^4 l/mol·cm to 10 l/mol·cm. Diazonium salts gives preferable A, B, C parameters due to their excellent bleaching ability. Diphenylamine-p-diazonium sulfate added to polyvinyl alcohol was examined [22]. In aqueous environment, the dye photoreacts to form a phenol, nitrogen, and sulfuric acid as

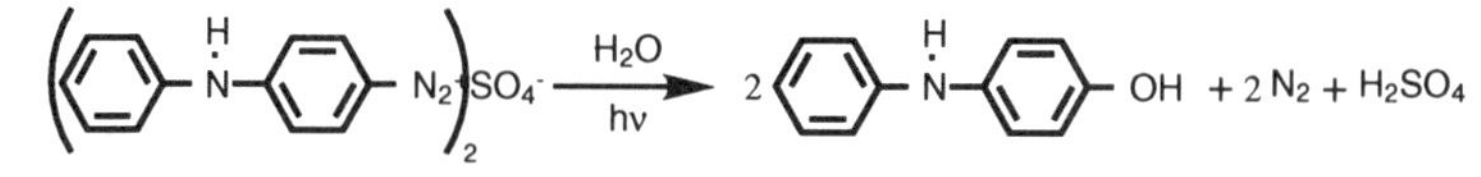

$$\left(\langle\underset{H}{\bigcirc}-\underset{|}{\overset{H}{N}}-\langle\bigcirc\rangle-N_2\right)_2 SO_4^- \xrightarrow[h\nu]{H_2O} 2 \langle\bigcirc\rangle-\underset{|}{\overset{H}{N}}-\langle\bigcirc\rangle-OH + 2N_2 + H_2SO_4$$

(8.8)

The merit of this CEM is that it is soluble in water, and is simpler than GE's CEM system. However, λ_{max} of the simple diazonium salts as shown in Eq. 8.8 was not suited to this wavelength, and an insufficient CEL effect was obtained. The diazonium salts adapted to g-line exposure gave submicron resist patterns with a steep profile [23]. The parameters of this CEM were: A 3.8 μm^{-1}, B 0.22 μm^{-1}, C 0.04 cm^2/mJ.

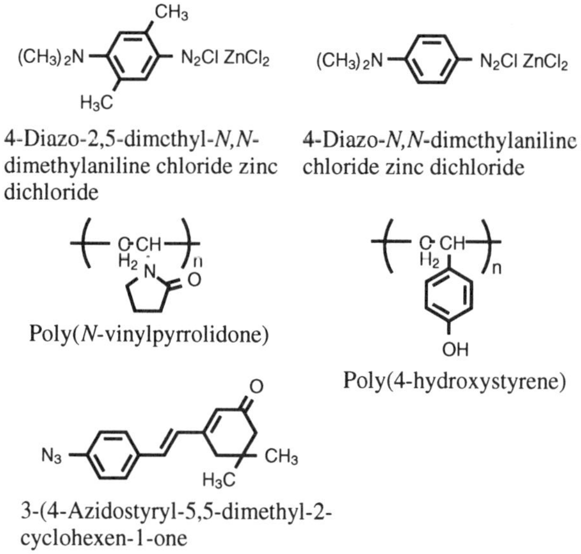

4-Diazo-2,5-dimethyl-N,N-dimethylaniline chloride zinc dichloride

4-Diazo-N,N-dimethylaniline chloride zinc dichloride

Poly(N-vinylpyrrolidone)

Poly(4-hydroxystyrene)

3-(4-Azidostyryl-5,5-dimethyl-2-cyclohexen-1-one

Fig. 8.3 Structural formulas for water soluble CEMs and resist components.

Uchino *et al.* used diazonium compounds which had dialkylamino groups on the aromatic ring for their negative resist system as depicted in Fig. 8.3 [24]. Zinc chloride double salt contributes to maintaining a stable CEM solution. The CEM compound was well suited to i-line lithography, and resolved 0.5 µm line and space patterns. The use of metal-containing diazonium salts would affect the reliability of Si-devices by the diffusion of residual metal ions into the substrates, therefore metal-free diazonium salts were developed [25]. An i-line CEM "D6" composed of 4-N,N-dimethylaminobenzenediazonium trifluoromethanesulfonate and poly(N-vinyl-2-pyrrolidone) did not form a mixing layer with the underlying resist (in this case RG3900B negative resist available from Hitachi Chem.) and was easily soluble in water [26]. The A, B and C parameters of D6 were 18 μm^{-1}, 0.12 μm^{-1} and 0.09 cm^2/mJ, respectively. These values, which were very close to that of the ideal CEL condition predicted by simulation, were $A=20$ μm^{-1}, $B=0$ μm^{-1}, $C=0.1$ cm^2/mJ. High resolution line and space patterns up to 0.35 μm were resolved using an i-line stepper (NA=0.42), and a focus latitude of ± 1 μm was obtained even at 0.4 μm lines and spaces.

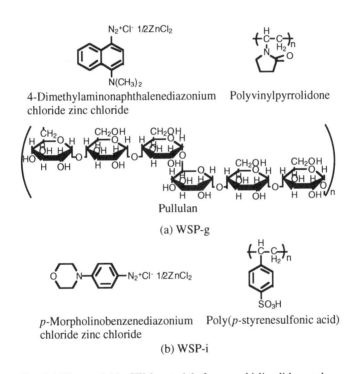

Fig. 8.4 Water soluble CEM materials for g- and i-line lithography.

Endo *et al.* have investigated CEL extending from the g-line to deep UV. The water-soluble photopolymer, WSP-g [27], is composed of 4-dimethylnaphthalene diazonium chloride zinc chloride salt, Pullulan and poly(N-vinyl-2-pyrrolidone) as depicted in Fig. 8.4a. WSP could be removed during alkaline development. The lack of nitrogen permeability of the Pullulan caused nitrogen bubbles from the underlying diazonaphthoquinone photoresist, which was prevented by the addition of poly(N-vinyl-2-pyrrolidone). The amount of addition should be as small as possible, otherwise, the alkaline dissolution rate of WSP is reduced. A, B and C parameters of WSP-g were 9.66 μm^{-1}, 0.25 μm^{-1} and 0.042 cm^2/mJ, respectively. In the case of line and space patterns of 0.6 μm formed by a g-line stepper with NA of 0.42, clear improvement in the pattern profiles were demonstrated by applying WSP. The effect of the contrast coefficient α (defined as Eq. 8.4) of WSP-g was 1.64.

WSP-i [28] is composed of 4-morpholinobenzendiazonium chloride zinc chloride salt and 4-styrenesulfonic acid as depicted in Fig. 8.4b. WSP-i could also be removed during alkaline development. Furthermore, diazonium salts are stable in an acidic aqueous solution, taking into account that the acidic polymer 4-styrenesulfonic acid has been used. A, B and C parameters of WSP-g were 10.01 μm^{-1}, 0.15 μm^{-1} and 0.078 cm^2/mJ, respectively [29]. Applying WSP-i, the contrast of the MPS-1400 photoresist increased from 2.06 to 3.02, and the sensitivity decreased by approximately 30%. Hole patterns of 0.6 μm and line and space patterns of 0.5 μm were clearly resolved using a stepper with NA of 0.42. WSP-EX [30] composed of 5-diazo-Meldrum's acid as a photobleachable compound, p-cresol novolk resin as a matrix polymer and diglyme as the solvent, was developed for KrF excimer laser lithography. Bleaching of 5-diazo-Meldrum's acid is caused by the reaction:

$$\underset{H_3C-O}{\overset{H_3C-O}{>}}\!\!\!\!\!\!\!\!\!\!\underset{O}{\overset{O}{\underset{\|}{\diagup}}}\!\!\!\!\!\!\!\!\!\!=N_2 \xrightarrow{\text{deep UV}} 3CO + \underset{H_3C}{\overset{O}{\underset{\|}{C}}}CH_3 \qquad (8.9)$$

A, B and C parameters of WSP-EX were 9.59 μm^{-1}, 0.51 μm^{-1} and 0.008 cm^2/mJ. The B parameter was not small enough, and the underlying resist, MP2400, is not suitable for deep UV exposure. The difference between the patterns with and without WSP was not large, but they were first studied by a combination of KrF excimer lithography and CEL. Endo *et al.* have improved their WSP-EX system [31]. Poly(styrene-co-maleic acid half isopropylate) was used instead of novolac in CEM, and its transmittance was superior to that of novolac (B=0.22). The resist was newly developed for deep UV lithography. The CEL effect was confirmed both by experiment and simulation using PROLITH.

Styrylpyridinium compounds are water soluble and more stable than diazonium salts. Both the heterocyclic partial structure and the quarternary salt structure in styrylpyridinium compounds affect the λ_{max} and ε_{max}, which enables design of a CEM suited to the specific wavelength [32]. Several CEM structures and their *A*, *B* and *C* parameters are listed in Table 8.1. The mechanism of bleaching has been proposed as depicted in Fig. 8.5. *Cis*-isomer, the equilibrium product of the initial *trans*-isomer, has less ε_{max} than that of the *trans*-isomer, and the dimer has no absorption at wavelengths longer than 300 nm, therefore a large CEL effect has been obtained in i-line lithography.

Table 8.1 CEL characteristics of styrylpyridinium compounds.

Photobleachable compounds	λ_{max} nm	ε_{max}	A, B, C parameters		
			A μm^{-1}	B μm^{-1}	C cm^2/mJ
for I-line					
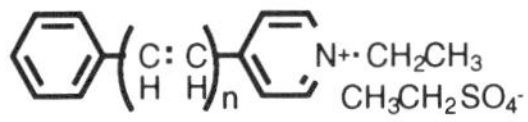 [structure]	380	44,500	11.54	0.49	0.26
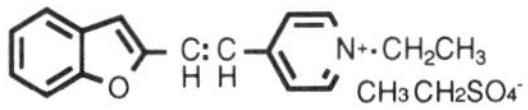 [structure]	390	38,600	8.28	0.38	0.27
for G-line					
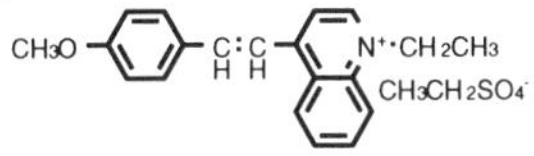 [structure]	413	27,400	5.75	0.22	0.04
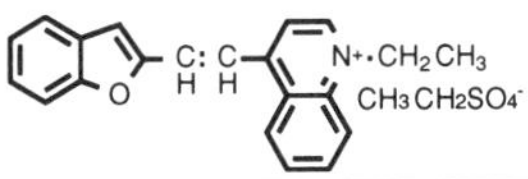 [structure]	425	33,900	6.00	0.20	0.06

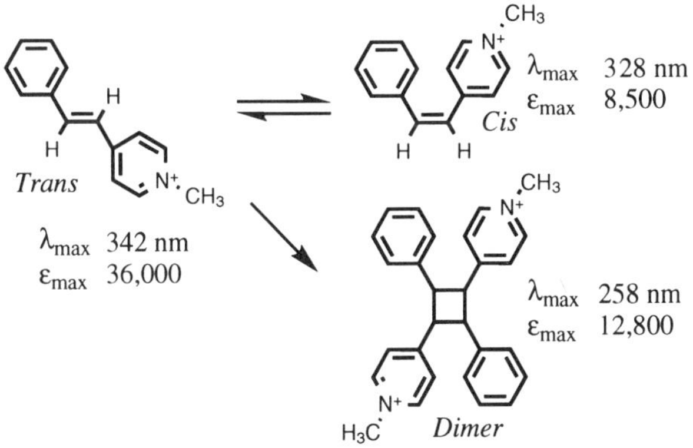

Fig. 8.5 Bleaching mechanism of styrylpyridinium salt.

Kaifu et al. reported two kinds of diazonaphthoquinone compounds and a diazonium salt as in Fig. 8.6 [33]. Si-LMR [34] is a silylated polymer derived from their LMR resist [35], and the cause of silylation is direct film formation on the bottom resist. NQND is analogous to positive photoresists. Bleaching ability similar to positive photoresists was expected. DZ-1 is a diazonium hexafluorophosphate, which is stable in the solution. Exposure parameters for these CEMs at 436 and 365 nm are listed in Table 8.2, and these have given good CEL effects.

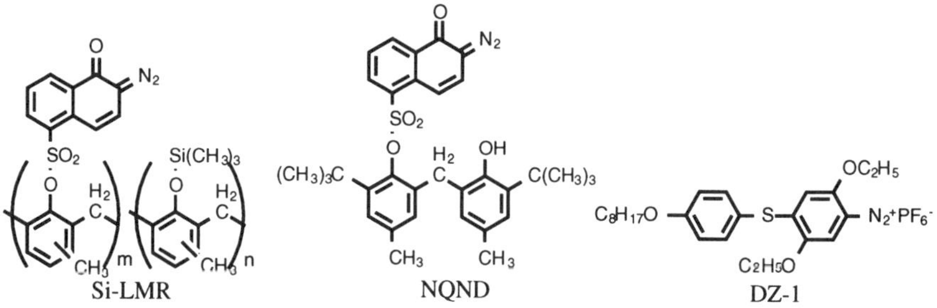

Fig. 8.6 Diazonaphthoquinoine compounds and diazonium salt as CEL.

Table 8.2 Exposure parameters for CEM depicted in Fig. 8.6 at 436 and 365 nm.

	Material	A μm^{-1}	B μm^{-1}	C cm^2/mJ
436 nm	Si-LMR	1.17	0.51	0.017
	NQND	1.40	0.83	0.035
	DZ-1	2.87	0.10	0.057
365 nm	Si-LMR	2.83	0.39	0.030
	NQND	3.39	0.47	0.060
	DZ-1	3.95	0.34	0.109

8.3 PORTABLE CONFORMABLE MASK

The portable conformable masking (PCM) technique, which is a kind of multilayer resist process, has been developed by Lin [36]. The concept of PCM is based on sequential exposure including high resolution imaging of a top layer resist followed by exposure of a thick underlying resist, and the variations of PCM are depicted in Fig. 8.7. The top resist exhibits large absorption in deep

UV regions even after UV exposure for imaging, therefore the top resist is developed to give the mask patterns for deep UV exposure of the bottom resist. PMMA was used for the bottom deep UV resist. An isolation layer is used to avoid intermixing if necessary, but may be omitted. Whether a capped or uncapped structure is formed depends on the developer used for the bottom resist. A capped structure would be suited to a lift-off process. Practically, a 0.2-μm-thick AZ1350J photoresist on a 1.9-μm-thick PMMA was exposed to near UV light, and developed by an AZ developer to form the patterns of AZ1350J. AZ1350J exhibits very large absorption in the deep UV region owing to the cresol moiety of the base resin and the diazonaphthoquionone component. Then, the PMMA film with AZ1350J patterns was exposed to deep UV (blanket exposure) with a dose of 500 mJ/cm^2, which was developed by MIBK to give an uncapped structure, or by chlorobenzene to give a capped structure [37]. Imaging exposure of AZ1350J could be achieved using an electron beam (e-beam or EB) because of the much lower e-beam sensitivity of PMMA than AZ1350J. Although the PCM method proposed by Lin is useful for high-aspect ratio patterning, the underlying resist PMMA has poor dry etching resistance compared with novolk-based photoresists.

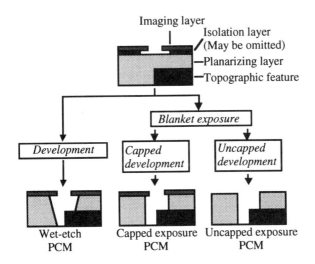

Fig. 8.7 Portable conformable mask system.

A unique approach has been developed by Yamashita et al. [38]. HR-PCM (High dry etching Resistivity PCM) consists of LMR-UV as an upper resist and AZ-5214 photoresist as a bottom resist. The merit of this PCM is the use of the novolak-based resist, which results in a higher dry etching compared with the PMMA bottom layer. The patterning process of HR-PCM is illustrated in Fig. 8.8. Figure 8.8a shows a normal PCM process. The light penetrating through the LMR-UV during flood UV exposure affects the bottom the AZ-5214 resist, then the bottom layer slightly dissolves in an alkaline developer to give thin patterns. Figure 8.8b shows the HR-PCM process. Image reversal of AZ-5214 resist is achieved by exposure using light penetrating through the LMR-UV layer followed by post-exposure baking. The image reversal process of AZ-5214 is based on the reaction [39,40]:

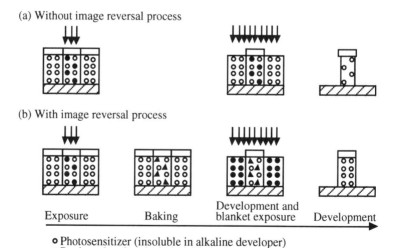

(8.10)

The resulting bottom patterns are hydrophobic and negative acting to an alkaline developer, which gives stable patterns. The dry etching durability of the HR-PCM process was improved to equal that of a novolak-based resist to PMMA.

Fig. 8.8 Schematic profiles of HR-PCM process.

Aqueous processible PCM has been studied. The alkaline developable resists depicted in Fig. 8.9 were employed. The sensitivity to deep UV light of the terpolymer (Fig. 8.9a) [41] was 600 mJ/cm^2, and 1.0 μm line and space patterns consisting of 0.75-μm-thick DNQ resist and 1.5-μm-thick terpolymer were resolved. PMGI (Fig. 8.9b) has been examined as PCM excimer laser lithography [42]. Bilayer resist patterns of 3.4-μm-height could be formed, but the wall profile of the bilayer patterns were not straight, due to the upper UV resist not optimizing for KrF excimer laser light. Then, WSP described in the previous section, was applied [42]. The combination process of PCM and CEL improved the wall profile significantly.

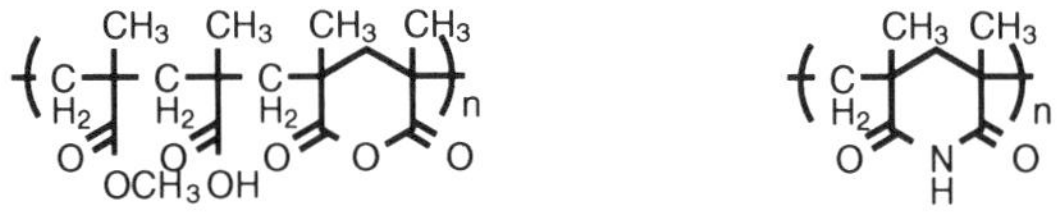

Fig. 8.9 Bottom resists for aqueous processible PCM.

(a) Poly(methyl methacrylate(MMA)-methacrylic acid anhydride(MAN)-methacrylic acid(MA))

(b) Poly(dimethylglutarimide)

PCM processes have been applied to device fabrication. Both the capped and uncapped processes of the AZ deep-UV PCM system have been tested using exploratory 1 μm double-poly-Si NMOS test chips. The poly-Si levels and contact holes were formed by uncapped PCM and the Al metalization was carried out by lift-off using the capped PCM [43,44].

8.4 LIFTOFF PROCESS

Hatzakis has reported a novel fabrication technique [45]. One of the main problems in the development of a submicron bipolar transistor is nonuniformity of wet etching in Al metallization. To address this problem a liftoff process has been developed. The process is explained by Fig. 8.10. A methacrylate mask was formed by e-beam exposure and Al was evaporated both on the substrate and the resist mask, then the resist mask was stripped off together with using Al by an organic solvent to result in an Al gate. The undercut profile of the methacrylate mask is important for separation of the Al gate line from the Al on the resist mask, which is caused by the proximity effect of e-beam exposure. It has been reported that devices have been fabricated with almost 100% yield. In addition, the process was applied to fabrication of a defect-reduced glass mask as shown in Fig. 8.11. The glass substrate was etched by wet etching through the resist mask with an undercutting profile, and Cr metal was evaporated. The metal-buried glass mask had the merit of decreasing defects and high edge resolution, which is effective for increasing chip yield by contact photolithography.

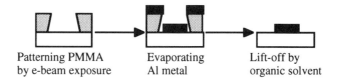

Patterning PMMA   Evaporating   Lift-off by
by e-beam exposure Al metal   organic solvent

Fig. 8.10 Typical lift-off process.

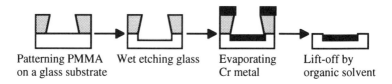

Patterning PMMA   Wet etching glass   Evaporating   Lift-off by
on a glass substrate   Cr metal   organic solvent

Fig. 8.11 Embedded chromium mask fabricated by lift-off process.

The liftoff process has been introduced to semiconductor manufacturing to improve in metal pitch for high density arrays, which has been accomplished by eliminating the etch bias component of subtractive etching. A chlorobenzene single-step liftoff process using a diazo-type resist has been developed [46], and was introduced to IBM 64 kb dynamic random access memory (DRAM) chip manufacturing [47]. The chlorobenzene soak took place after exposure of the mask pattern using projection printers, and before development. Soaking time was 10 to 20 minutes. During soaking, lower molecular weight novolak resin may be extracted into the chlorobenzene, in which case the surface of the resist film becomes less soluble in AZ developer. When such dissolution behavior occurs in the depth direction, overhung resist profiles such as those in Fig. 8.12 were obtained. Data concerning with the liftoff process using chlorobenzene soaking were summarized, and show the relationships among exposure, development, and penetration, and their effects on the overhang structure [48,49].

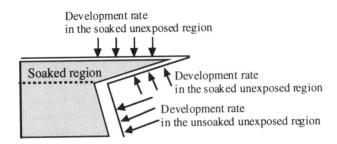

Fig. 8.12 Degradation in the overhang due to development of the top of the structure from both top and bottom surfaces.

Image reversal resist [39,40] can be applied to the liftoff process, and it is more practical than chlorobenzene soaking. Although a process using Genesis equipment is employed to generate vertical profiles for substrate-etching applications, the resist wall profiles are controlled by several parameters, *e.g.*, patterning exposure (E_1), amine vapor diffusion, and UV flood exposure (E_2) in the process flow depicted in Fig. 8.13. The profiles are easy to simulate using by PROSIM, and the patterning conditions are easily optimized. The process was applied to Al metallization and the submicron feature was demonstrated [50].

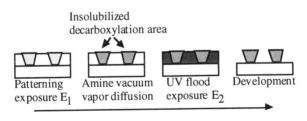

Fig. 8.13 Amine vapor diffusion image reversal process flow.

More recently, a very suitable resist to the liftoff process has been developed. LMR [38] and LMR-UV [51] are negative resists for deep UV and longer wavelength (typically g- and i-line) lithography, respectively. Their development is carried out by an organic developer and the insoluble mechanism is based on the slight increase of their molecular weight upon irradiation, not on the gelation [52]. Therefore, the overhang profiles can be obtained only by development, and they are controlled by the exposure dose and development time. The potential of the liftoff process using LMR resists has been demonstrated in the fabrication of GaAs devices [53,54].

8.5 MULTILAYER RESISTS COUPLED WITH DRY DEVELOPMENT

A new multilayer resist technology [6] has attracted much interest due to its capability of overcoming many of the difficulties encountered with conventional single layer resist pattern generation and is expected to become an essential technology for fabricating VLSI. The concept of multilayer resist schemes and their performance are well understood at present [36,55]. Pattern delineation in a multilayer resist is basically performed according to the following process steps (Fig. 8.14). First, an organic polymer layer coating which planarizes the surface topography is applied to the substrate. A thin top layer resist is then coated uniformly on the planarizing layer (bilayer system) or on an intermediate layer through its deposition process (trilayer system). After patterning the top resist by normal exposure and development, pattern transfer to the bottom planarizing layer is accomplished using a dry etching technique.

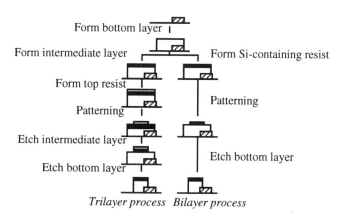

Fig. 8.14 Comparison between tri- and bilayer resist processes.

In the case of photolithography, a multilayer resist enables precise critical dimension control to be easily obtained because a thin top layer gives high resolution and a thick bottom planarizing layer can eliminate the reflected light from the substrate [56]. However, the aspect ratio is limited by pattern collapse and micro-loading in the subsequent etching process [57]. Then, the optical property and the thickness were optimized for each layer when a thinner bottom layer was applied [58]. By minimizing multiple interference in the trilayer system, a depth of focus of ±0.8 μm for 0.16 μm L&S patterns were obtained by KrF excimer laser lithography. In the case of e-beam lithography, a multilayer resist enables precise critical dimension control similar to that of photolithogaraphy. The resolution in e-beam direct writing strongly depends on the forward scattering and backscattering of electrons. A simulation study has been carried out for VLSI patterns containing some heavy metal silicide substrates [59].

An important issue in the multilayer resist technique is anisotropic and precise dry etching of a bottom layer. Etch selectivity between mask (intermediate inorganic layer) and planarizing layer is not a serious problem because a very high level can be attained by proper selection of masking materials. Reduction in gas pressure is effective for decreasing undercutting, in the same manner as conventional RIE. However, undercutting still remains at a pressure under 1 Pa, and cannot be disregarded for fine pattern formation of 1 μm or less. Reduction of substrate temperature has been found to be very effective in decreasing undercutting further [60]. The effect of low temperature etching with magnetically enhanced RIE has been considered [61]. Low temperature etching at -80 °C with O_2 gas reduced side etching and formed side wall protection by depositing a film, and provided a very steep profile even at the relatively high pressure of 20 mtorr. An additional gas such as hydrocarbon [62] or HBr [63] is also effective in decreasing undercutting. Using an O_2 + HBr mixed gas plasma conducting in an ECR plasma etching system, multilayer resist dry etching under a low substrate temperature has been applied to 0.25 μm design-rule 256 Mb dynamic random access memory fabrication.

Although conventional resists could be used in the trilayer process, the process would become more complicated. In contrast, a bilayer resist process [64] does not increase the number of process steps significantly, as shown in Fig. 8.14. A silicon-containing resist acts as both a patterning resist as a top layer and an inorganic middle layer in the trilayer process. The bilayer process does not demand high resolvability, but high resistance to O_2 plasma is required.

Poly(siloxane) is sensitive to electron beam irradiation and can be applied to patterning processes in semiconductor devices, which was demonstrated 20 years ago [65]. Table 8.3 shows the electron beam sensitivities of several poly(siloxane)s synthesized by hydrolysis-condensation of trichlorosilane derivatives. In particular, poly(siloxane)s that possess a polymerizable substituent such as the vinyl group exhibit high sensitivity. The poly(siloxane) resist patterns obtained by e-beam lithography were calcined, and used as ion implantation masks and passivation films for fabrication of bipolar transistors.

Table 8.3 Electron beam sensitivities of poly(cyclosiloxane)s obtained by hydrolyses of mono-substituted trichlorosilanes.

Mono-substituted trichlorosilane RSiCl$_3$		E-beam sensitivity of poly(cyclosiloxane) $\mu C/cm^2$ at 15 kV
vinyl	-CH=CH$_2$	4
methyl	-CH$_3$	130
ethyl	-C$_2$H$_5$	250
phenyl	-C$_6$H$_5$	800

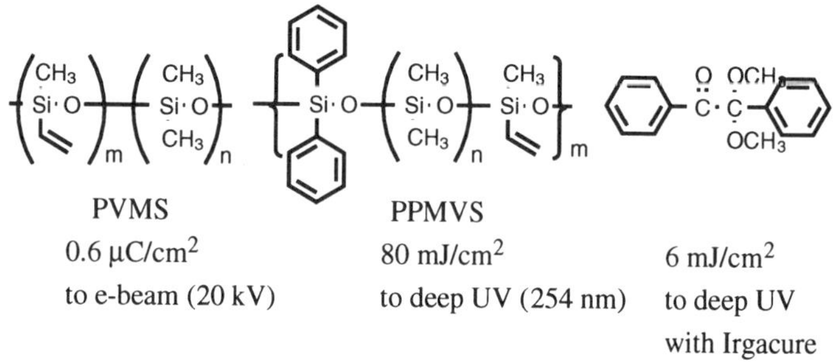

PVMS
0.6 $\mu C/cm^2$
to e-beam (20 kV)

PPMVS
80 mJ/cm^2
to deep UV (254 nm)

6 mJ/cm^2
to deep UV
with Irgacure

Fig. 8.15 Linear poly(siloxane)s for e-beam and deep UV lithography.

Thereafter, IBM's researchers [64] turned their attention to high resistance O$_2$-RIE, and proposed the bilayer resist process shown in Fig. 8.14. Figure 8.15 shows the linear poly(siloxane)s that they examined. The poly(siloxane) possessing vinyl groups on its side chain revealed a high sensitivity of 0.6 $\mu C/cm^2$ to e-beam irradiation with an accelerating voltage (V_{acc}) of 20 kV. Those possessing both phenyl and vinyl groups exhibited a sensitivity to deep UV irradiation (PE500 projection printer with UV2 filter). The O$_2$-RIE durability of these poly(siloxane)s is over 50 times that of a bottom resist, thus 0.4 μm bilayer resist patterns with an aspect ratio of 5 can be formed. Sensitivity to deep UV light can be enhanced by 2 to 5 times by the addition of Irgacure or dicumylperoxide which are radical polymerization initiators possessing deep UV sensitivity [66]. Amino-transformed poly(siloxane) with diazonaphthoquinone pendant groups can be used in g- and i-line lithography

[67]. This photosensitive poly(siloxane) acts as a negative resist based on the mechanism explained in Fig. 8.16.

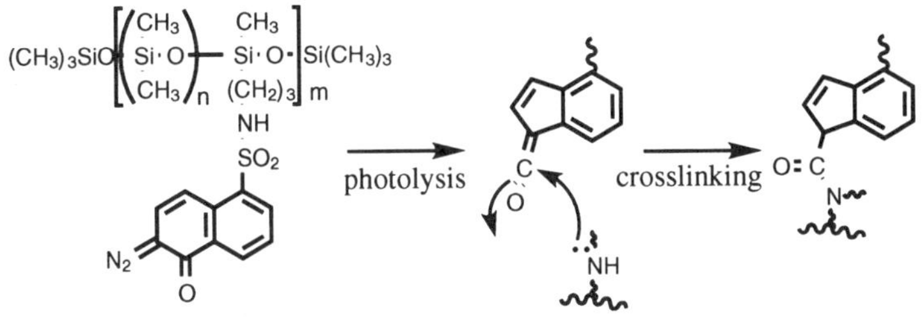

Fig. 8.16 Photocrosslinking mechanism of amino-transformed poly(siloxane).

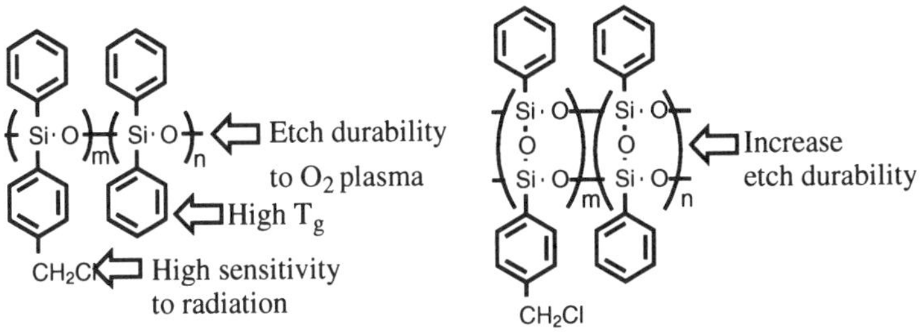

Fig. 8.17 Design principle of silicon-containing negative resist (SNR).

The poly(siloxane)s described above have been known for many years, therefore examinations were carried out early. Unfortunately the glass transition temperature (T_g) was less than room temperature (rt), which caused pattern swelling during development to degrade the resolution. Under these circumstances, new bilayer resists were developed when the usefulness of the bilayer process had been ascertained. NTT's researchers have developed an SNR resist [68] based on poly(diphenylsiloxane) as shown in Fig. 8.17. The design concept of SNR is analogous to that of CMS [69] developed in NTT, that is, the phenyl groups of SNR raise its T_g and the chloromethyl groups are sensitive to e-beam and deep UV irradiation. SNR is a negative resist and its insolubility is due to gelation which is a result of the radical cleavage of the chloromethyl groups. Therefore, the sensitivity of SNR depends on its molecular weight as listed in Table 8.4 [70]. SNR with a molecular weight of 38,000 exhibits a sensitivity of 5 μC/cm² to e-beam. The O₂-RIE durability to the bottom layer is over 30 and an aspect ratio of over 6 can be obtained in 0.2 μm bilayer patterns. Furthermore, high sensitivity e-beam resist, SNR-2 [71], has been developed by the same group. SNR (Mw 16,000) indicated the sensitivity of 8.5 μC/cm², and was slightly improved in O₂-RIE durability because of its ladder siloxane structure. SNR and SNR-2 are applicable to e-beam and deep UV lithography.

Table 8.4 Dependency of e-beam sensitivity of SNR on its molecular weight.

Mw	Sensitivity μC/cm², Vacc 20 kV
1.1×10^5	2.0
3.8×10^4	5.0
1.2×10^4	18

To apply SNR to near UV lithography, structural modification has been performed. The base resin of MSNR [72] was synthesized by the substitution reaction of the chloromethyl groups in SNR-2 by methacrylate ions, and 2,6-bis(azidebenzylidene)-4-methylcyclohexanone was added to the resin to formulate a photoresist, MSNR, as shown in Fig. 8.18. The sensitivity of MSNR to the i-line of a Hg lamp was 40 mJ/cm² and a resolution of 0.5 μm was demonstrated by the use of a stepper with NA=0.35. Naturally, MSNR was a highly sensitive e-beam resist that exhibited crosslinking reactivity of 0.002 C/cm² which is more sensitive than that of SNR-2 (0.072 C/cm²) [73]. However, such high e-beam sensitivity was due to the chain crosslinking reaction between the vinyl groups of MSNR, and the crosslinking was still proceeding even after irradiation was stopped. This phenomenon affected the resolution and MSNR is not suitable for an e-beam resist for microfabrication.

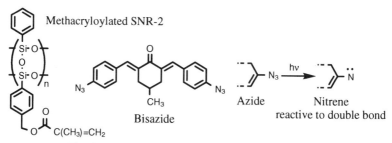

Fig. 8.18 MSNR resist system for mid-UV lithography.

From the point of view of coordination to the current resist process line, alkaline development is preferred since a diazonaphthoquinone-novolak resist system is used in production. Most poly(siloxane)s are hydrophobic. Therefore, a new alkaline soluble poly(siloxane) has been designed. Acetylated poly(phenylsilsesquioxane), APSQ, is derived from poly(phenylsilsesquioxane) by acetylation of aromatic rings, as shown in Fig. 8.19, and is soluble in alkaline solution. A silicone-based positive photoresist, SPP, composed of APSQ and a diazonaphthquinone photoactive compound has been developed [65]. SPP exhibited sensitivities to visible–deep UV light [75]. A resolution of 0.4 μm L&S was obtained and the developed pattern was well reproduced up to 0.35 μm L&S in g-line lithography (NA=0.6). A sensitivity of 100 mJ/cm^2 and a focus latitude of 0.75 μm in 0.3 μm L&S could be obtained by i-line lithography (NA=0.5). SPP for deep UV (typically, an KrF excimer laser) used diazomeldrum acid instead of diazonaphthoquinone [75]. Although SPP had a high sensitivity of 16 mJ/cm^2 to the KrF excimer laser, its resolution was only 0.48 μm and critical dimension controllability could not be maintained. SNP, a negative working resist, is composed of APSQ and 1-azidopyrene, as shown in Fig. 8.19 [76]. Although SNP had almost the same sensitivity and critical dimension controllability to SPP, increasing the content of 1-azidopyrene in the resist improved the pattern profile only slightly.

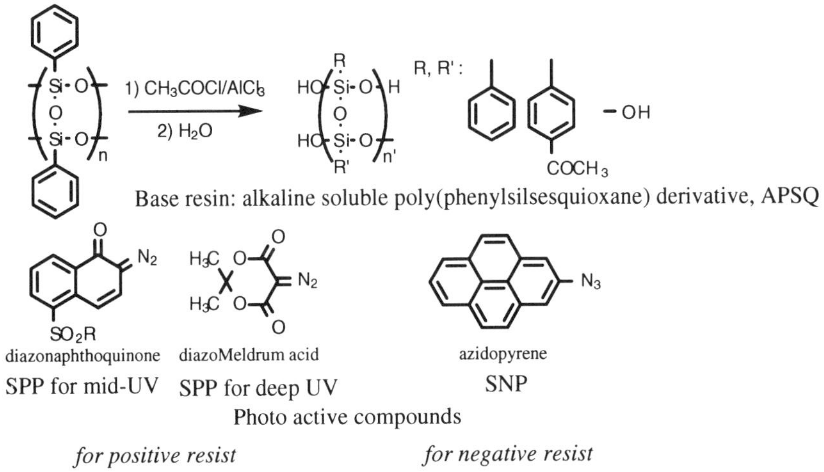

Fig. 8.19 Alkaline soluble photoresist based on design modification of SNR.

These works were carried out in the early stages of short wavelength lithography, therefore the resist systems were not optimized for deep UV lithography. However, new poly(siloxane) resists have been developed utilizing siloxane chemistry just as the "chemical amplified resist system" [77] has been proposed. CSNR (chemically amplified silicone-based negative resist) [78] is an example of a chemically amplified polysiloxane resist. CSNR is composed of alkaline-soluble silicone polymer, ASSP, and diphenyl(4-phenylthiophneyl)sulfonium hexafluoroantimonate. ASSP is synthesized by hydrolysis of phneyltrimethoxysilane and 2-(3,4-epoxycyclohexyl)ethyltrimethoxysilane. The silanol groups in ASSP condense easily to siloxanes under the Brønsted acid generated from the antimonate upon irradiation:

$$\equiv Si\text{-}OH + HO\cdot Si \equiv \xrightarrow{H^+} \equiv Si\cdot O\cdot Si \equiv \qquad (8.11)$$

The change from hydrophilic to hydrophobic property in the film causes insolubilization in an alkaline developer. The molecular weight of CSNR increases by post-exposure baking (PEB), which results in the condensation reaction described above. This was confirmed by IR spectroscopy and the activation energy of the condensation was estimated as 45.8 kcal/mol from the sensitivity dependence on PEB temperature. CSNR did not bleach upon KrF excimer laser exposure, however, the absorption coefficient was as low as 0.23 μm^{-1}, therefore a high resolution of 0.22 μm and good critical dimension controllability were obtained.

As mentioned above, poly(siloxane)s possessing phenyl groups in their side chains have been examined extensively for occasional exposure wavelengths. Since the phenyl moiety in a poly(siloxane) molecule is rather large, how much does it affect O_2-RIE durability? The etch rates and e-beam responses of various organosilicon polymers are summarized in Table 8.5 [79]. Etching studies were performed at 10 mtorr pressure, 0.15 W/cm^2 and a flow rate of O_2. The etch rate data quoted here refers to the average etch rate required to remove a 2-4000-Angstrom-thick film from the surface of a silicon wafer. The etch rates in Table 8.5 are plotted against the Si-content of each polymer, and Fig. 8.20 shows the enhancement of O_2-RIE durability with the increase of Si-content. The etch rate of silylated poly(*p*-hydroxystyrene) was found to vary with the amount of Si in the polymer. At concentrations below approximately 7% of Si (by weight) in the polymer, the etch rate was high compared to poly(dimethylsiloxane) PDMS. At concentrations higher than 10%, the etch rate approached that of PDMS, as can be seen in Fig. 8.20.

Table 8.5 Electron beam sensitivities and O_2 plasma etch rates of organosilicon polymers.

Polymer	Mw ×10⁻³	Sensitivity μC/cm²	% silicon by weight	Etch rate nm/min
[Si(CH₃)₂-O]ₙ	71.8	6.0	37.8	1.9
(CH₃)₃SiO[Si(CH₃)₂-O]ₙSi(CH₃)₃	260	7.0	37.8	1.7
[Si(C₆H₅)₂-O-Si(C₆H₅)₂-O-Si(CH₃)(C₆H₅)-O]ₙ	100	>150	17.9	1.8
[Si(CH₃)-O-Si(CH₃)-O]ₙ (ladder)	1.5	>80	37.3	1.8
[Si(C₆H₅)-O ladder]ₙ	1.0	21.6	21.6	1.7
[Si(CH₃)(CH₂CH₂CF₃)-O]ₙ	14	18	18.0	5.6
(CH₃)₃Si[O-Si(CH₃)₂]ₓ[phenyl-Si(CH₃)-O-Si(CH₃)-O]ᵧ	200	29.8	29.8	1.6
[Si(CH₃)₂-CH₂]ₙ	60	39	39.0	1.7
[Si(CH₃)₂-Si(CH₃)₂-O]ₙ	100	35	35.0	1.5
[Si(CH₃)₂-CH(CH₃)]ₙ	200	25	25.0	1.5
[Si(C₂H₅)₂]ₙ	200	22	22.0	2.4
[CH₂-CH(Si(CH₃)₃)]ₙ	300	28	28.0	1.8
[CH₂-CH(C₆H₅)(OH)]ₓ[CH₂-CH(C₆H₅)(OSi(CH₃)₃)]ᵧ	30	>1000	0 - 14.5	62 - 2.5
[N=Si(CH₃)₂-N=Si(CH₃)(H)]ₙ	10	>50	39.0	2.5
[Si(CH₃)(C₆H₅)]ₙ	200	>100	23.0	1.9

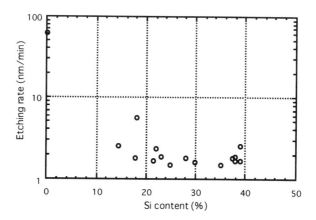

Fig. 8.20 O2-RIE rates of silicon-containing polymers.

Ladder type poly(siloxane)s other than the SNR series described above were examined since their Si-content was higher than that of other silicon polymers. Researchers at Hughes Research Lab. [80] evaluated poly(methylsilsequioxane-*co*-phenylsilsesquioxane), whose silanols were capped by trimethylsilyl groups as depicted in Fig. 8.21. The polymer was a negative working resist, and resolved 0.6 μm space patterns by e-beam lithography at a dose of 40 μC/cm^2 and 0.12 μm line patterns by Ga$^+$ beam lithography at a dose of 8×10^{12} Ga$^+$/cm^2. This poly(siloxane) copolymer was less sensitive to irradiation because there were no radiation sensitive groups and in contrast phenyl groups existed. It is well known that aryl groups themselves are inactive to energetic irradiation at least at the lithography use level due to the stabilization effect of aromatic resonance.

$$\begin{array}{c} Cl \\ R\cdot Si\cdot Cl \\ Cl \end{array} \xrightarrow{OH^-/H_2O}_{\text{hydrolysis condensation}} \begin{array}{c} R\ \ R\ \ R\ \ R \\ HO\ Si\cdot O\ Si\cdot O\ Si\cdot O\ Si\cdot O-- \\ O\ \ O\ \ O\ \ O \\ -Si\cdot O\ Si\cdot O\ Si\cdot O\ Si\ OH \\ R\ \ R\ \ R\ \ R \\ R=C_6H_5, CH_3 \end{array} \xrightarrow{(CH_3)_3SiCl}_{\text{protecting silanol}}$$

Fig. 8.21 Soluble ladder polysiloxane.

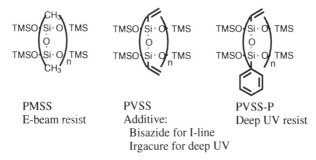

PMSS
E-beam resist

PVSS
Additive:
Bisazide for I-line
Irgacure for deep UV

PVSS-P
Deep UV resist

Fig. 8.22 Highly sensitive ladder type polysiloxane resists.

Figure 8.22 shows the ladder poly(siloxane) resists developed at Fujitsu Laboratory. Poly(methylsilsesquioxane) PMSS [81] has been developed as an e-beam resist, and showed a high sensitivity of 1-2 µC/cm^2 at a molecular weight of 140,000. The reactivity (Mw×D_g^i) of PMSS was 0.08, which was close to that of CMS (0.13). The Si-content is as high as 40%, therefore O_2-RIE durability is superior to that of the silicon-containing polymers up to that time. PMSS resolved 0.5 µm patterns at a dose of 1.4 µC/cm^2. PVSS [82] showed sensitivity 4 times higher than PMSS to e-beam exposure. Sensitivity of PMSS to i-line and deep UV could be realized by the addition of 2,6-bis(azidobenzyllidene)-4-methylcyclohexanone) and 2,2-dimethoxy-2-phenyl-acetophenone, respectively. Gelation of PMSS and PVSS upon e-beam irradiation was caused by the intermolecular coupling between the radicals produced by the C-H cleavage and double bond cleavage in the methyl and vinyl side chain, which was confirmed by IR spectroscopy [83]. Interestingly, vinyl-phenyl derivatives, PVSS-P [84], showed a relatively high sensitivity of 75 mJ/cm^2 to deep UV light. It was confirmed that the phenyl group was excited by deep UV light to cause radical polymerization of the vinyl groups.

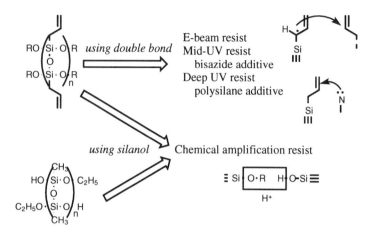

Fig. 8.23 Usable reactive sites in simple polysiloxane reists.

Poly(allylsilsesquioxane) PAS [85] (Fig. 8.23) was investigated. PAS exhibited a very high sensitivity of <1 µC/cm² to e-beam, however, the critical dimension was not well controlled due to the chain reaction of the allyl groups. Therefore, PAS was evaluated as a deep UV resist [86]. Various photo-degradable compounds were examined for radical polymerization of the allyl moiety in PAS. Some poly(silylene)s could be used. Hexakis(dimethylsilylene)/PAS showed a sensitivity of 140 mJ/cm² to KrF excimer laser light and a resolution of 0.26 µm L&S. Pentaphenylmethylsilylen enhanced the sensitivity up to 17 mJ/cm² since the absorption was larger than that of hexakis(dimethylsilylene) at 248 nm. The insolubilization mechanism has been considered as follows. The triplet state silylene is generated upon deep UV exposure, which subtracts the hydrogen attached to the allylic position of PAS to produce the allylic radical, then crosslinking occurs. PAS whose silanols were not capped could be applied to the chemically amplified resist based on silanol condensation [87]. The sensitivity to deep UV light was up to 0.5 mJ/cm², and it was confirmed that a tetrafunctional silane such as tetraphenoxysilane could be used as a crosslinker. PMSS has also been applied to the chemical amplification system. The alkyl groups are acid-labile protecting groups for silanols, therefore the alkyl protected PMSS shown in Fig. 8.23 is used as a negative resist [88]. This resist was quite transparent under UV light, therefore it has been used for direct formation of the phase shifter [1].

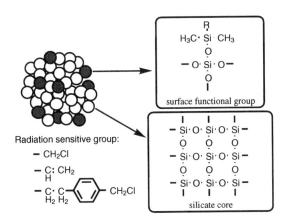

Fig. 8.24 Microresinoid siloxane resist.

As understood from the structural design of SNR, the introduction of rigid groups such as the phenyl group raises the glass transition temperature. Crosslinking toward three dimensions is another approach for improving the heat resistance, which restricts the movement of the polymer chain. Micro-resinoid siloxane developed by the NEC [89-91] group showed excellent heat resistance and high resolution. The chemical composition is shown in Fig. 8.24, and consists of an $SiO_{4/2}$ core with methyl and another functional group, R, around that core. The siloxane structure is similar to the silica-like structure shown in the figure, and is different from that of such siloxanes as the linear-chain $(R_2SiO_{2/2})_n$ or ladder-type $(RSiO_{3/2})_n$. The total concentration of Si and O is more than 70 wt%. Various functional groups, R, may be introduced into the resist. For e-beam lithography the silanols at the surface of the core were protected by chloromethyl groups, and for deep UV lithography these were protected by chloromethylphenethyl groups. Although the sensitivity to e-beam was not as high as 23 $\mu C/cm^2$, resolution of less than 0.1 μm was confirmed by point-beam exposure, and 0.2 μm L&S patterns were obtained using a variable-shaped e-beam exposure system. The pattern profiles of micro-resinoid siloxane did not change up to 400 °C, and bilayer resist patterning by O_2-RIE and undoped polysilicon patterning by Cl_2 RIE were demonstrated.

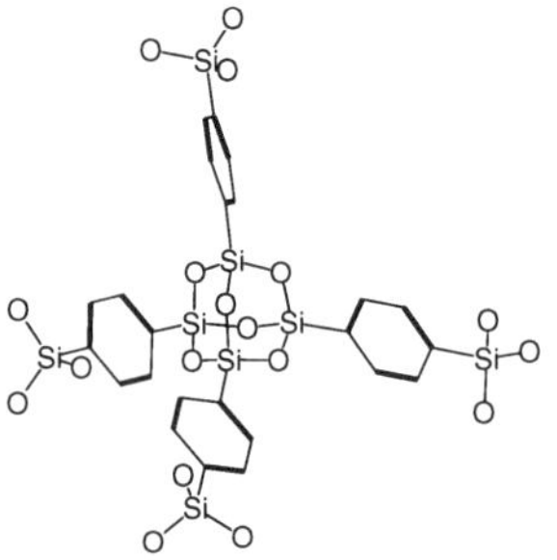

Functional group %	Mw ×10⁴	Sensitivity mJ/cm²	
		248 nm	193 nm
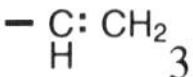 −C:CH₂ / H 31%	6.4	-	300
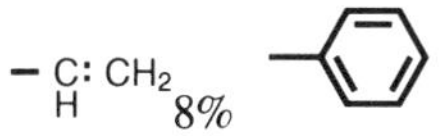 −C:CH₂ / H 8% —⟨⟩— 44%	1.9	1400	20
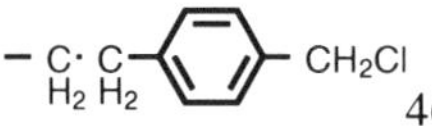 −C·C—⟨⟩—CH₂Cl / H₂ H₂ 46%	4.5	60	5

Fig. 8.25 Three-dimensional poly(silphenylene) resist, TSPS.

Three dimensional silphenylene siloxane TSPS [92] with a high softening temperature up to 400 °C has been developed by Fujitsu Laboratory. TSPS has been synthesized by hydrolysis of 1,4-bis(trialkoxysilyl)benzene followed by polymerization. The repeating unit of TSPS is shown in Fig. 8.25. Radiation sensitive groups were introduced to the silanol groups at the surface of the TSPS core. The TSPS core was, like SiO_2, similar to microresinoid siloxane, therefore the skeleton was rigid and swelling was suppressed during development. Sensitivity to deep UV was very high at 3 mJ/cm² (Mw was 480,000) and resolution was less than 0.35 μm, however, resolution could be improved by the use of a lower molecular weight TSPS since the sensitivity could be enhanced over 20 times by the addition of photoactive compounds. Quarter micron resolution was achieved when TPSS of Mw 13,000 and 2,2-dimethoxy-2-phenylacetophenone was used. Furthermore, TSPS protected by chloromethylphenyl groups show very high sensitivity to ArF excimer laser light [93]. High density bilayer resist patterns of 0.14 μm dimension were resolved at a dose of 7 mJ/cm². Another approach to the SiO_2-like resist described above has been reported [94].

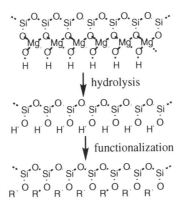

Fig. 8.26 Silicate-like resist using natural structure of Clay mineral.

Clay minerals are composed of silicate salts, and are hydrolyzed in hydrochloric acid followed by functionalization for solubility in a common organic solvent, as shown in Fig. 8.26. Silylated clay mineral resist SCMR with dimethylvinylsilyl groups as R showed e-beam sensitivity of 8 $\mu C/cm^2$ and resolution of 0.2 μm. As described, microresinoid siloxane, TSPS and SCMR are three-dimensional SiO_2-like resins, which are kinds of spin-on-glass (SOG) in the LSI process. Therefore a bilayer resist could be formulated from commercially available SOG, OCD-T17, and a photoacid generator. The result is a chemically amplified resist based on silanol condensation and with a sensitivity of 0.7 $\mu C/cm^2$ and a resolution of 0.3 μm [95].

Several silicone resin possessing phenol moieties as depicted in Fig. 8.27, have been developed to set the current resist line. Organosilicon positive photoresist OSPR [96,97] composed of poly(4-hydroxyphenyl-methyl-silsesuquioxane) and diazonaphthoquinone compound, and natural OSPR showed similar spectroscopic properties. Linear poly(siloxane)s contain phenol structures in their main chain or its side chain [98,99].

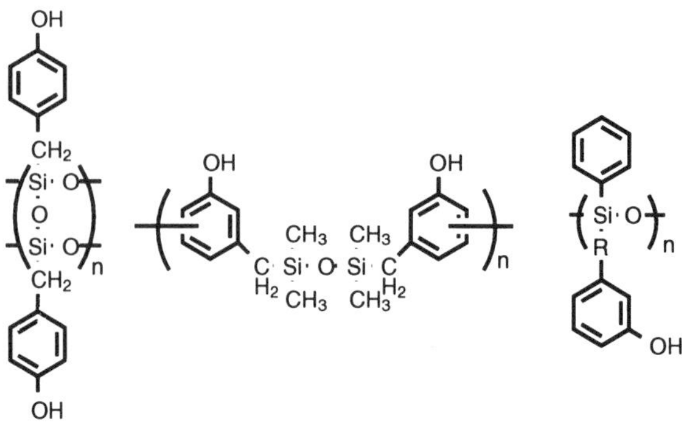

Fig. 8.27 Phenol-containing poly(siloxane)s for alkaline developable photoresist.

Silicon-containing resists have now been developed originally as both imaging resists (top layer) and SiO_2 (intermediate layer) in the trilayer resist method, therefore an important issue of resist design has been to increase the Si concentration in the resin. Consequently, SiO_2 with a patterning function is an ideal resist, and a new resist system which gives equivalent results has been developed. Figure 8.28 shows the retrosynthesis of a SiO_2 network. The main chain of the polymer is designed such that one-dimensional SiO_2 is cut out of three-dimensional SiO_2. A glass precursor resist (GPR) system is composed of a photoacid generator and poly(di-t-butoxysilane) [100]. The poly(siloxane) is equivalent to one-dimensional SiO_2 since t-butyl groups can be removed easily in the presence of Brønsted acid. GPR has been converted to silicate glass upon e-beam or deep UV exposure at very low temperatures (PEB temperature). The mechanism of the conversion is condensation of silanols through the silanol intermediate produced by deprotection of GPR as shown in Fig. 8.29, which has been confirmed by spectroscopic study and the dose-dependency of the etching rate in O_2 plasma [101,102]. Complete conversion to SiO_2 has been achieved by a lithographically useful dose and a PEB at 100 °C, and 0.1 μm L&S patterns have been resolved by e-beam lithography under these condition [103]. A SiO_2 mask for etching tungsten film can be prepared directly from GPR, the etching selectivity is over 5 times that of a conventional resist and is very close to that of a nondoped silicate glass obtained by chemical vapor deposition [104].

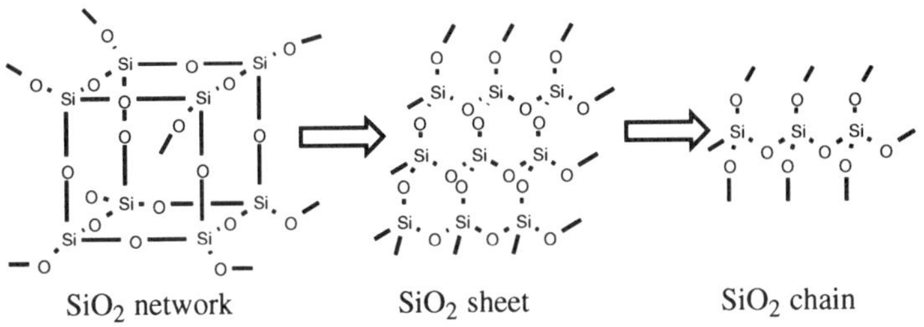

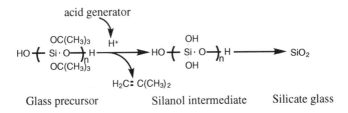

Fig. 8.28 Retrosynthesis of a SiO$_2$ network.

Fig. 8.29 Molecular and reaction design of glass precursor resist (GPR)

Polysilanes (poly(silylene)s as formal nomenclature) are a class of polymers with a silicon-silicon backbone that were first reported by Kipping [105]. However, the poly(dimethylsilane)s synthesized by this author were intractable. West and coworkers [106] were the first to synthesize a soluble polysilane, poly(phenylmethylsilane-*co*-dimethylsilane), which they termed silastyrene because of its structural analogy to poly(styrene). Characteristic of this polymer is strong UV absorption near 330 nm, with ε=8,000 per silicon atom when the polymer exhibits Mn=150,000 and PhMeSi/Me$_2$Si=0.59. They found that poly(phenylmethylsilane-*co*-dimethylsilane) appeared to become semiconducting upon contact with strong electron acceptors such as SbF$_5$. The polymer crosslinked upon deep UV (254 nm) irradiation, and the conductivity increased slowly but eventually reached a much higher level of ~0.5 Ω^{-1}cm^{-1}. The photochemistry of polysilanes has been studied extensively and it has been elucidated that simple polysilanes degrade to an oligomer or monomer upon UV irradiation as shown in Fig. 8.30 [107].

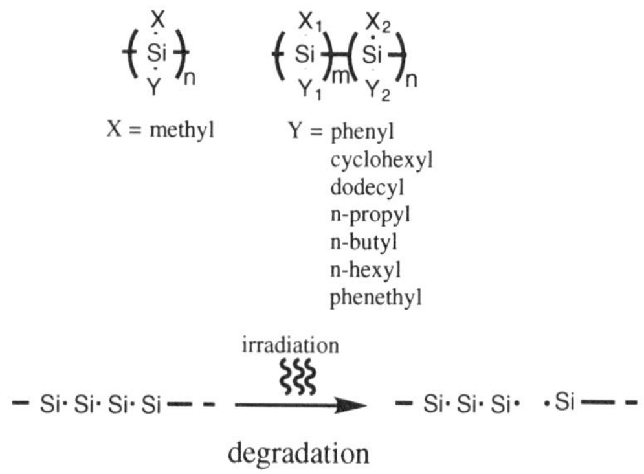

$\{ \underset{Y}{\overset{X}{Si}} \}_n$ $\{ \underset{Y_1}{\overset{X_1}{Si}} \}_m \{ \underset{Y_2}{\overset{X_2}{Si}} \}_n$

X = methyl   Y = phenyl
              cyclohexyl
              dodecyl
              n-propyl
              n-butyl
              n-hexyl
              phenethyl

— Si· Si· Si· Si — - $\xrightarrow{\text{irradiation}}$ — Si· Si· Si· ·Si — -
                         degradation

Fig. 8.30 Polysilane resist system.

Hofer and coworkers [108] changed the substituents of polysilanes systematically and studied their spectroscopic properties and photochemical reactions in depth. The polysilanes move their absorption maxima toward shorter wavelengths upon irradiation when they have nonaromatic substituents as shown in Fig. 8.30. This behavior is based on reducing the degree of polymerization from 260 to 8, then the polysilanes can be applied to CEL, lift-off processes and bilayer resists for mid-UV lithography [109]. Aliphatic polysilanes resolved a 0.75 μm feature at a dose of 100 mJ/cm^2 using a 0.167 NA mid-UV projection tool. Although poly(methylphenylsilane)s crosslink upon deep UV exposure as reported by West *et al.*, when t-butyl groups are introduced to the *p*-positions of the phenyl groups, crosslinking is suppressed and degradation becomes predominant. Poly(t-butylphenyl methylsilane) resolved a 0.75 μm feature by KrF excimer laser ablation at a dose of 550mJ/cm^2. Lithographic, photochemical and O$_2$-RIE properties of some polysilanes were also reported by Taylor and coworkers [110]. Polysilanes function as positive photoresists, but their sensitivities are rather low for commercial deep UV imaging tools. Therefore, photo-sensitization was studied by Wallraff *et al.*[111]. They addressed this problem using two approaches: (1) the syntheses of new polysilanes which are intrinsically more sensitive to photodegradation in the solid state and (2) the incorporation of additives which enhance the photosensitivity. Regarding the former, aryl substituted polysilane homopolymers, as shown in Fig. 8.31a, which are significantly more photolabile than standard poly(methylphenylsilane), as assayed by the rate of spectral photobleaching upon exposure to deep UV radiation, have been prepared and tested. When the phenyl groups attached directly to the Si of polysilane backbone, they bleached rapidly. In particular the *p*-position was blocked by

Variations in Microlithographic Process 261

alkyl groups, and bleaching was clearly observed since competitive reactions did not exist. Figure 8.31b shows the structures of the additives examined and their relative efficiencies based on poly(methylphenylsilane) bleaching rates. The reason for sensitization has not yet been clarified, but the paper pointed out the possibility that the polysilane and the sensitizer formed a redox system and electron transfer occurred efficiently from the excited polysilane to the photoacid generator as an electron acceptor. For example, the excited state oxidation potential for the poly(methylphenylsilane) (-2.79 eV) and the reduction potential for the quencher (e.g. phthalimide triflate, -0.91 eV) indicate that the electron transfer reaction is quite exothermic (~43 kcal/mol) and thus thermodynamically feasible.

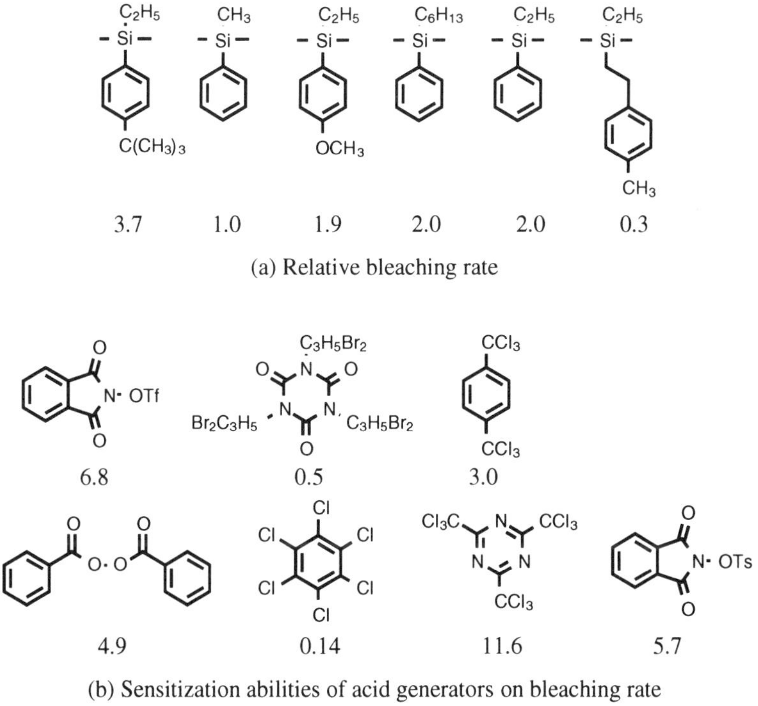

(a) Relative bleaching rate

(b) Sensitization abilities of acid generators on bleaching rate

Fig. 8.31 Relative bleaching rates of polysilanes.

Poly(silyne) is another class of polymer based on silicon-silicon bonds. This polymer has one alkyl substituent per silicon atom and its structural feature is not like that of poly(acetylene) but is a network consisting of σ–σ bonds of Si atoms. Synthesized poly(silyne) [112] had a molecular weight of 8,000 - 100,000 and was soluble in xylene, which showed a semiconductor-like property with a band gap of 2.5 - 3.0 eV. The sensitivity to ArF excimer laser light was 20 mJ/cm^2 in poly(n-butylsilyne). It was confirmed by Auger spectroscopy that the siloxane bonds were generated in the exposed area. Therefore, the exposed area did not dissolve in organic developer and could be used as a negative resist. Furthermore, the unexposed area was structurally similar to polysilicon, therefore it could be etched by HBr plasma. ArF projection lithography using a poly(silyne) resist showed a potential of 0.2 μm and 0.15 μm resolution by wet development and HBr plasma development, respectively.

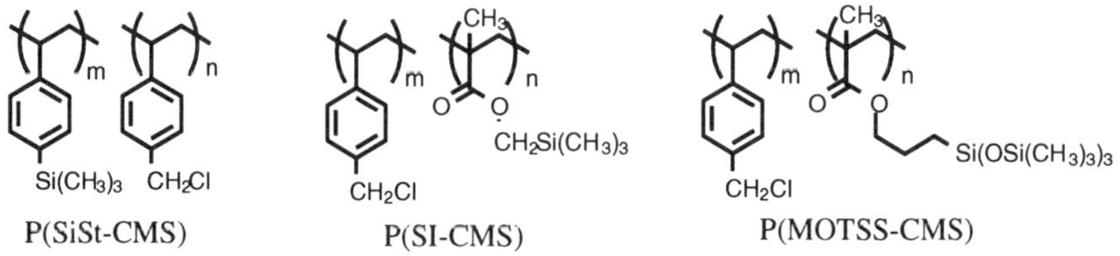

Fig. 8.32 Bilayer copolymer resists of Si-containing polymer and CMS.

The resists described previously are classes of siloxanes and polysilanes which have been studied widely and their silicon atom concentrations were relatively high since they have silicon-rich backbones, Si-O and Si-Si. Another class of silicon-containing resist will be described. Negative e-beam resist, P(SiSt-CMS), [113] incorporates silyl groups in the side chain as shown in Fig. 8.32, and consists of the repeated units of styrene derivatives. Although increasing the content of the CMS unit in the polymer decreased sensitivity, a content of 10 mol% in the CMS unit was enough to obtain practical sensitivity, the value (D_g^i) of which was 1~2 μC/cm^2. Oxygen ion beam etching was performed, and P(SiSt-CMS) showed a selectivity of 7~8 to the bottom layer and successful patterning of the bilayer at a dose of 4.5 μC/cm^2 [114]. Other radiation-active chlorinated polymers have been reported as shown in Fig. 8.32. The random copolymer, P(SI-CMS), comprised of trimethylsilylmethyl methacrylate (SI) and chloromethylstyrene (CMS) was shown to function as a negative acting e-beam and deep UV resist [115]. The resist exhibited e-beam and deep UV sensitivity ($D_g^{0.5}$) equal to 1.95 μC/cm^2 and 18 mJ/cm^2, respectively.

The ratio of etching rates of the planarizing layer HPR-206 to this material was 12 to 1 in O_2-plasma. Very recently, P(SI-CMS) has been evaluated as an ArF excimer laser resist [116]. By controlling the mole ratio of the monomers in the P(SI-CMS), absorbance values have been optimized for a film thickness of 0.2-0.3 µm for 90:10 SI:CMS, 0.35-0.45 µm for 95:5 SI:CMS and 0.55-0.65 µm for 98:2 SI:CMS. The resist showed sensitivities on the order of 4 to 20 mJ/cm^2 at Mw near 40,000. Poly(γ-methacryloyloxypropyltris(trimethylsiloxy)silane), PMOTSS, was an intrinsically negative resist with a sensitivity comparable to that of the polyhalostyrene. Although the homopolymer was rubbery, copolymers of MOTSS and halostyrenes are hard solids and could be used as e-beam resists for microfabrication [117]. Images were written in a 0.3 µm top layer of 1.3:1 P(CMS-MOTSS) by the AEBLE 150. A block copolymer of chlorinated polystyrene and poly(dimethylsiloxane) has been examined as a bilevel resist [118]. The electron lithographic performance of a block copolymer containing 15.5 wt.% silicon and 0.58 chlorines per methylstyrene unit (Mw=8.7×10^4) was determined to exhibit a sensitivity of $(D_g^{0.5})$=0.9 µC/cm^2, and contrast of 1.3. Polymers are generally not compatible with each others, and a microdomain has been observed in films in which 10% dimethylsiloxane units were incorporated in the block copolymer. When the quantity of dimethylsiloxane units was increased to 50%, a lamella structure appeared. These domains are 5-10 nm in size. Such a property of the film will affect the pattern quality in 0.1 µm or less microfabrication.

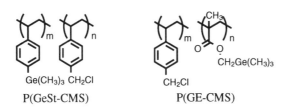

Fig. 8.33 Germanium-containing resists.

Figure 8.33 shows germanium-containing resists. These are copolymers of trimethylgermylmethyl methacrylate and chloromethylstyrene, and trimethylgermylstyrene and chloromethylstyrene, and their structures are very similar to those of P(SI-CMS) [119] and P(SiSt-CMS)[120], respectively, which are shown in Fig. 8.32. After patterning by dry development, germanium-containing resists could be removed easily by fuming nitric acid and a mixture of sulfuric acid/hydrogen peroxide. Such a chemical property, which is different to that of silicon containing resists, may be advantageous for introducing a bilevel resist method to common fabrication processes. The sensitivities of these resists are similar to their silicon analogs.

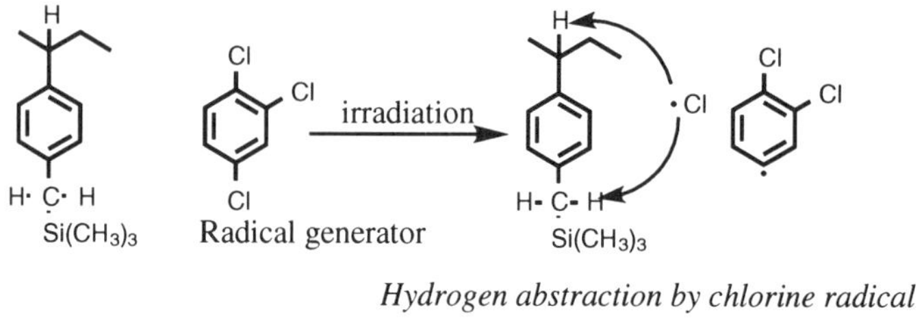

Hydrogen abstraction by chlorine radical

Fig. 8.34 Two component system by poly(trimethylsilylstyrene) and halides.

The origin of the sensitivities of the resists described above (Fig. 8.32 and 8.33) is the CMS units incorporated in the polymer chains. Figure 8.34 shows a two component approach based on the same mechanism as that of a CMS resist [121]. Trichlorobenzene generates a chlorine radical upon e-beam or deep UV irradiation and the radical subtracts active protons of poly(trimethylsilylmethylstyrene), then the resulting polymer radicals crosslink. This paper dealt with the insolubilization mechanism. Bisazide was not effective for crosslinking poly(trimethylsilylmethylstyrene), in contrast to poly(vinylphenol) in an MRS resist [122]. This result may be understood as the difference in the inductive effect to α-positions between the trimethylsilylmethyl group and hydroxyl group, and also that in the electronic state between chlorine and nitrene.

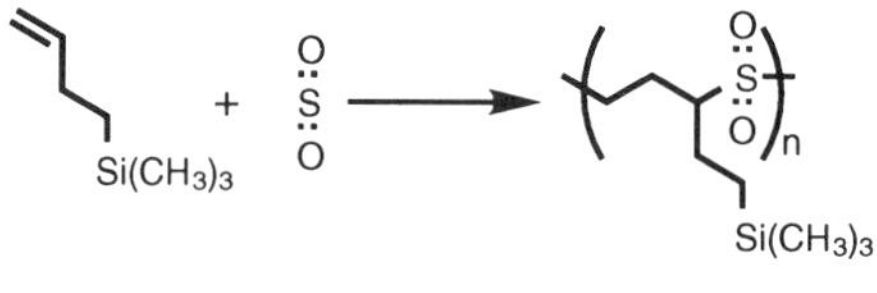

Fig. 8.35 Silicon-containing positive e-beam resist based on PBS resist.

Silicon-containing positive resists have been reported. The approach used for a PBS resist [123] has been applied to a bilayer resist. The PBTMSS shown in Fig. 8.35 is synthesized by copolymerization of 3-butenyltrimethylsilane and sulfur dioxide, in a similar manner to the process of PBS [124]. E-beam sensitivity as high as 1.0-1.5 µC/cm^2 was shown and T_g was 120 °C. Positive heat resistivity of the unexposed area is important since the property of a resist polymer directly determines the latitude in the following etching process. This series of studies has shown that a resist contrast of over 10 maintains an E_{th} sensitivity of under 10 µC/cm^2 [125]. The required doses for nanometer patterning are listed in Table 8.6. Using a bilayer of PBTMSS and a diamond-like carbon film, 60-nm-pitch laser grating has been fabricated on a GaInAsP substrate [126].

Table 8.6 Required e-beam dose for nm patterning.

Pitch nm	For grating µC/cm^2	For dot by weight
60	17	10.5
80	12	6
100	10	5
150	7	2.8
200	51.6	1.6

Although PBTMSS has found several unique applications in nanolithography, its development and plasma processing windows were narrow. Soluble 1:1 alternating copolymers of *p*-trimethylsilylstyrene and *p*-pentamethyldisilylstyrene with sulfur dioxide have been synthesized by free-radical copolymerization, and are illustrated in Fig. 8.36 [127]. Their thermal stability was improved compared with PBTMSS, and that in nitrogen (5% weight loss) was about 210-230 °C. The e-beam sensitivity was 3 µC/cm^2.

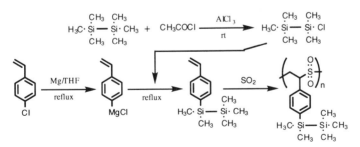

Fig. 8.36 Synthetic route for poly(*p*-trimethylsilylstyrene-co-*p*-pentamethyldisilylstyrene sulfone).

Poly(trimethylsilylpropyne) [128], as shown in Fig. 8.37, is a derivative of poly(acetylene), so its heat resistivity is high. The polymer is degraded by γ-ray irradiation in the presence of oxygen, therefore its use as a positive e-beam or deep UV resist was expected, however, the required dose of deep UV light is as high as 1 J/cm^2. Consequently, bromination has been performed for improving sensitivity [129]. Brominated poly(1-trimethylsilylpropyne) was also stable in air up to ~200 °C. The sensitivity of the brominated polymer was dependent on the degree of bromination and the time and temperature of PEB. When the degree of bromination was 0.1-0.2, the sensitivity and the contrast to deep UV light were 20-25 mJ/cm^2 and 3.5-4.0, respectively, at the PEB condition of 140 °C and 1 h. The etching rate ratio of the polymer *versus* a hard-baked novolak photoresist was better than 1:25.

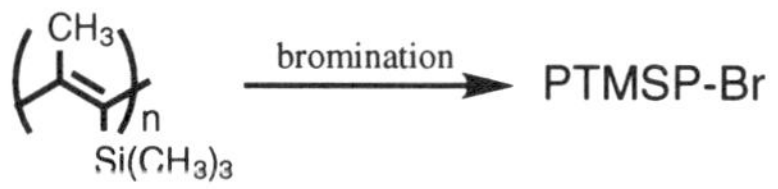

Fig.8.37 Brominated poly(trimethylsilylpropyne) as a positive resist.

Several silmethylene type polymers have been reported, and their structures are depicted in Fig. 8.38. TAS is composed of poly(triallylphenylsilane) and bisazide [130]. The polymer included five- and six-membered rings in its main chain, which contributed to better thermal stability than that of a linear polymer. The softening point ranged from 107-130 °C depending on its molecular weight. Substituted silacyclobutanes have been synthesized and their property as bilayer e-beam resists has been examined[131]. The mechanism of gelation is based on cross-linking at the carbon of the methyl group attached to Si atoms.

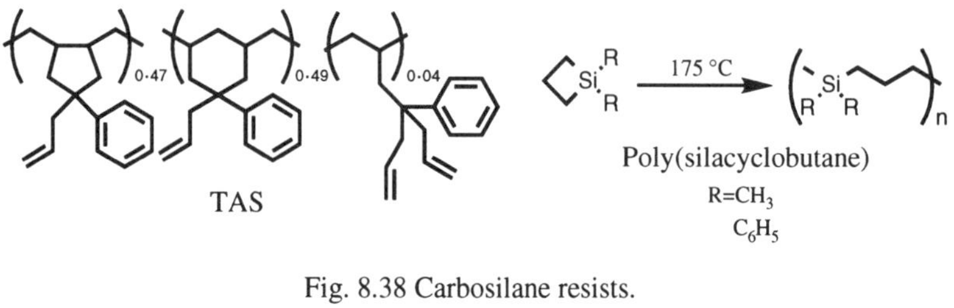

Fig. 8.38 Carbosilane resists.

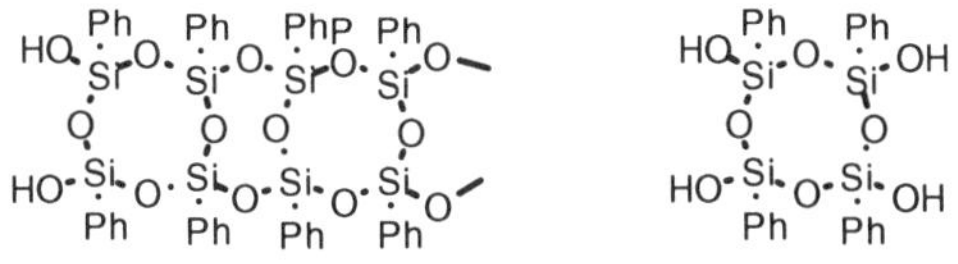

Variations in Microlithographic Process 267

$$\text{HO}\underset{\underset{\text{HO-Si-O-Si-O-Si-O-Si-O}}{\overset{\text{O}}{|}}}{\overset{\text{Ph}}{\underset{|}{\text{Si}}}}\text{O-}\underset{|}{\overset{\text{Ph}}{\text{Si}}}\text{-O-}\underset{|}{\overset{\text{PhP}}{\text{Si}}}\text{-O-}\underset{|}{\overset{\text{Ph}}{\text{Si}}}\text{O}\underset{\text{Ph}\ \text{Ph}\ \text{Ph}\ \text{Ph}}{}$$

Polyphenylsilsequioxane(PSQ)

$$\text{HO}\underset{\underset{\text{HO-Si-O-Si-OH}}{\overset{\text{O}}{|}}}{\overset{\text{Ph}}{\underset{|}{\text{Si}}}}\text{-O-}\underset{|}{\overset{\text{Ph}}{\text{Si}}}\text{-OH}\underset{\text{Ph}\ \text{Ph}}{}$$

Phenyl-T4

Fig. 8.39 Dissolution inhibitors used in a positive photoresist, ASTRO.

An alkaline developable bilayer resist can be formulated by adding a silicone compound to a commercially available positive or alkaline soluble resin. Poly(phenylsilsesquioxane) PSQ is compatible with OFPR-800 (diazonaphthoquinone positive photoresist, Tokyo Ohka Kogyo Co.) and its minimum unit, cis-1,3,5,7-tetrahydoxy-1,3,5,7-tetraphenylcyclotetrasiloxane, phenyl-T$_4$, is soluble in alkaline developer (Fig. 8.39). Optimum concentration of these silicone additives has given a bilayer resist, ASTRO [132], whose property was very close to that of OFPR-800. In the course of the study of ASTRO, it was found that some silanol compounds such as phenyl-T$_4$ act as dissolution promoters in the alkaline development of phenolic resins, while siloxanes work as dissolution inhibitors. Phenyl-T$_4$, PSQ and other several silanol compounds were examined for dissolution promotors in a negative chemical amplification resist sytem [133]. Insolubilization is due to the condensation reaction according to Eq. 8.11. The reaction does not cause gelation of silanol compounds but a slight increase of molecular weight indicating dimerization or origomerization, and it is well that known the resulting siloxane is hydrophobic. Diphenylsilanediol gave the best sensitivity and resolved 0.3 μm L&S patterns at a dose of 3 mJ/cm^2 by KrF excimer laser stepper. Silicones containing novolak or other phenolic resins are not soluble in an alkaline solution when the silicon concentration is high, however, several resins have been reported [134-136].

Most bilayer resists employ silicon or a related atom such as germanium for obtaining O$_2$-plasma resistance. However, a few bilayer resists containing other particular atoms have been known. The etching rates of oxygen plasma removal for halogenated poly(styrene) have been examined [137]. Among the halogenated polymers, iodinated poly(styrene), IPS, showed a higher plasma resistance as listed in Table 8.7. After plasma treatment of the IPS film, diiodo pentaoxide was assigned by ESCA study. Manifestation of O$_2$-RIE resistance occurred because a nonsublimed solid, I$_2$O$_5$ was formed at the surface of the IPS film during etching. Poly(diphenoxyphosphazene) has been demonstrated to exhibit good O$_2$-RIE resistance (Table 8.8) but low etch resistance in CF$_4$-RIE [138].

This polymer could be used as a negative bilayer resist in the presence of a sensitizer such as CBr_4. An organotin containing resist, TMAR [139], has been examined by X-ray lithography. The resist could be applied to a bilayer resist method since its O_2-RIE durability was improved 50 fold in comparison to that of PMMA. Since tin(IV) oxide is soluble in a strong acid, it is possible that the TMAR bilayer formed by O_2-RIE could be removed by an ordinary wet process.

Table 8.7 Relative rates (k_r) of oxgen plasma removal for polymers and resists.

Polymer		Mw (10^5)	k_r
Polystyrene		2.7	1.0
Iodinated polystyrene	2% I	2.7	0.89
	4% I	2.8	0.86
	13% I	3.1	0.72
	27% I	3.6	0.51
	61% I	4.5	0.10
	85% I	5.5	0.04
Clorinated polystyrene	10% Cl	9.4	1.25
	100% Cl	11	1.15
Brominated polystyrene	10% Br	10	1.12
	100% Br	20	0.98
Novolak resin		7	1.0
AZ1350J			0.84

Halogenated polystyrene
─ C-CH ─
  H₂ |
      ⌬   R=OH
      | H
      R   Cl
              Br
              I

Table 8.8 RIE rates of poly(diphenoxyphosphazene) and other resists.

Polymer	Rate (nm/min)
O_2-RIE: 40sccm, 60mTorr, 0.35W/cm²	
Poly(diphenoxyphosphazene)	3
AZ1350J	130
Silylated AZ1350J	3
Poly(chloromethylstyrene)	140
CF_4-O_2(12%) RIE: 40sccm, 60mTorr, 0.35W/cm²	
Poly(diphenoxyphosphazene)	170
AZ1350J	40
Silylated AZ1350J	35
Poly(chloromethylstyrene)	40

8.6 INORGANIC RESISTS

Inorganic materials have desirable properties in that they hardly swell due to their low molecular weight and are highly resistant to O_2-plasma. These properties enable inorganic materials to be applied to different classes of bilayer resist to those based on silicon chemistry. Flexibility in resist design is not as great as that for organic resists based on well established carbon chemistry, and therefore inorganic resists have not been studied well. Regardless of this difficulty in design, inorganic resists are a promising lithography technology for fabrication of nanoscale devices since it is expected that material science will develop further.

Chalcogenide glasses are amorphous materials containing VI family elements: *e.g.*, S, Se and Te, which offer unique capabilities for exploring critical issues and developing new technologies for microlithography. A photodarkening effect was found in Se-Ge glass films [140], and their etching characteristics in alkaline solutions were examined [141]. Se-Ge thin films were formed on various substrates by vacuum evaporation or rf sputtering in Ar atmosphere, and patterning of Se-Ge films could be performed by a conventional resist process since Se-Ge glasses were soluble in alkaline solutions. A Se_{75}-Ge_{25} film was exposed to increase its dissolution rate in 6 mol% ammonia, such that it could function as a positive working resist, however, the selectivity was very small. The selective etching effect occurred only in alkaline solutions, not in acid solutions. Se-Ge glasses are being etch masked for SiO_2, Si_3N_4 film [141] and polyimide films [142]. Although these initial works demonstrated the potential for developing a new fabrication technique, Se-Ge still has several disadvantages: *e. g.*, low sensitivity to light as 0.1-1 J/cm^2, an inherently small dissolution rate difference between exposed and unexposed areas and poor adhesion to some substrates.

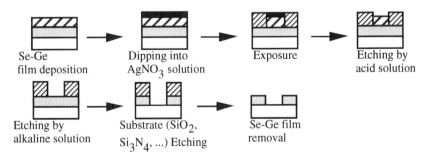

Fig. 8.40 Negative Se-Ge resist process.

The dissolution behavior of Se-Ge glasses was similar to that of a positive photoresist. In contrast, Se-Ge films were hardly soluble in alkaline when Ag was doped [143]. Figure 8.40 explains the negative process (Ag-doped Se-Ge film) and the previously described positive process (Se-Ge film). An Ag layer was formed by dipping in a $AgNO_3$ solution, and a thickness of 10 nm was sufficient for fabrication. Ag was doped into a Se_{75}-Ge_{25} film in the exposed area. Ag remaining on the unexposed areas was removed in a solution of HNO_3-HCl-H_2O, then the unexposed regions of the Se-Ge film were etched off by treatment in an alkaline aqueous solution. The negative process of Se_{75}-Ge_{25} improved the wet etching selectivity between the exposed and unexposed areas. Submicron feature patterns were successfully resolved using an Ag-doped Se_{75}-Ge_{25} resist and e-beam lithography [144]. Although a high resist contrast of ~8 was obtained, the sensitivity was still as low as 100-300 $\mu C/cm^2$ at an accelerating voltage of 20 kV. Photo-doping of Ag into Se_{75}-Ge_{25} resulted in diffusion along its grain boundary, and was confirmed by a SIMS study [145].

Heteropolytungstic acid has been investigated as a spin-coatable inorganic resist material. Peroxopolytungstic acid (HPA) with carbon as the heteroatom was an amorphous solid and was synthesized by dissolving metallic tungsten or tungsten carbide into an H_2O_2 solution. A typical experimental formula was $CO_2 \cdot 12WO_3 \cdot 7H_2O_2 \cdot nH_2O$, where n is about 25. The film became insoluble in water by deep UV, x-ray and e-beam irradiation, however, their sensitivities were low as a negative resist [146]. Substitution of 8-15 atm.% of W by Nb in HPA improved its sensitivity by one order to KrF or XeCl excimer laser [147]. Half-micron L&S bilayer resist patterns were fabricated at a dose of 300 mJ/cm^2 by an KrF excimer laser. The dependence on sensitivity was small in energy regions below 16 mJ/cm^2-pulse. At a power of over 56 mJ/cm^2-pulse reciprocity failure was observed, which was explained as the resist heating effect since the absorption at 248 nm was ~5 μm^{-1}.

Cadmium chloride is an inorganic resists which has the potential of an e-beam direct write resist [148]. The e-beam stimulated reaction :

$$CdCl_2 \xrightarrow{\text{e-beam}} Cd + Cl_2 \quad (8.12)$$

has been studied in $CdCl_2$ thin films between 303 and 483 K using 2 keV electrons. The authors claimed that decomposition itself resulted from radiolysis by e-beam and was not thermally induced. Irradiation at a low temperature of 30-60 °C left metallic Cd in the exposed area, but at a high temperature resulted in completely volatile products. The resolution was not accurately determined, however, it was speculated as about 100 nm since the grain size of the $CdCl_2$ used was the same size. Another cadmium halide also decomposed by e-beam

Variations in Microlithographic Process 271

irradiation according to Eq. 8.12. Cadmium iodide was preferable as a resist since it improved sensitivity (by approximately a factor of four). The trilayer scheme shown in Fig. 8.41 demonstrates contraction of a T-bar structure [149]. Furthermore, a pattern reversal process using CdF_2 was accomplished as shown in Fig. 8.42. Inorganic halides do not react chemically in halogen gas plasma, therefore the process was applied to fabrication of SiO_2 and a 0.1 µm structure was formed. Barium fluoride, strontium fluoride or their complex salts worked as negative resists by water development and showed a sensitivity of 25-100 µC/cm^2 which is less than that of cadmium halide positive resists [150]. The resistances of strontium and barium fluoride to reactive ion etching conditions were compared to those of vapor deposits of Ni, Cr, NiCr, Au and AuPd, as well as AZ-1350J photoresist. A summary of the results is shown in Table 8.9. Multiple quantum wells of GaAs have been fabricated using SrF_2 as a lift-off mask.

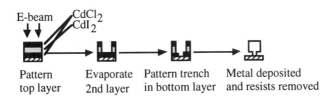

Fig. 8.41 Trilayer scheme for T-bar fabrication.

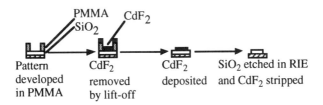

Fig. 8.42 Pattern reversal using CdF_2.

Table 8.9 Etch rates of prospective etch masks.

Etched material	Etch rate (nm/h)
Gallium arsenide	3000
AZ1350J	175
Au-Pd	150
Au	100-110
Glass	40-45
Cr	~75
Ni	~30
Ni-Cr	~50
Strontium fluoride	<10
Barium fluoride	<10

RIE etch conditions: 15 mTorr 9:1 Ar/BCl$_3$; 50W, 300V; 2 h etch time

It is thought that inorganic resists have high resolution capability of sub-10 nm feature. A very high resolution up to 1-2 nm has been demonstrated in AlF$_3$ resist, and it has been shown that AlF$_3$ resist can be used as a mask to replicate nanometer scale patterning onto Si$_3$N$_4$ using reactive ion etching. A direct metal structure could be formed for ultra small scale devices and processes. For example, continuous metal wires (1 μm long with a cross section of 8×10 nm) could be formed in this manner [151]. Matsui et. al., from NEC have examined an AlF$_3$ doped LiF resist. The resolution of this resist system was not greatly affected by secondary electrons since the activation energies of decomposition were small (LiF: 15 eV, AlF$_3$: 45 eV). The depth of self-development of the LiF(AlF$_3$) resist was dependent on the amount of doping of AlF$_3$, and the maximum etched depth was given by 10% AlF$_3$. This was because the grain boundary diffusion of Li and Al, which were the decomposition products of the resist, competed with the carbon-like film deposition originating from the remaining impurities in the exposure chamber. Al doping levels of 0% and 10% gave grain sizes of 100 nm and less than 10 nm, respectively. Taking into account this phenomena, it could be said that a smaller nominal pattern width is preferable in the LiF(AlF$_3$) resist system and is better suited to nanometer lithography. In practice, 5 nm line patterns have been successfully formed in a 60 nm pitch by an e-beam exposure machine at a dose of 0.4 μC/cm using a 1.5 nm beam diameter at 30 kV of acceleration voltage [152].

8.7 SURFACE SILYLATION OF SINGLE LAYER RESISTS

A multilayer resist scheme satisfies the requirements of microlithography: *e.g.,* resolution, aspect ratio of resist patterns and linewidth control. There are, unfortunately, several drawbacks to multilayer systems, the most important of which is their complexity. Although these systems prove adequate for building test design at small geometries, their introduction into the production line is unlikely. Then, a novel approach that integrated the performance of multilayer systems in a single layer technology was developed in the 1980s. The basic principle was proposed by Taylor, *et al.,* of AT&T [153]. They envisioned gas phase functionalized plasma developed resists that behave according to the two routes outlined in Fig. 8.43.

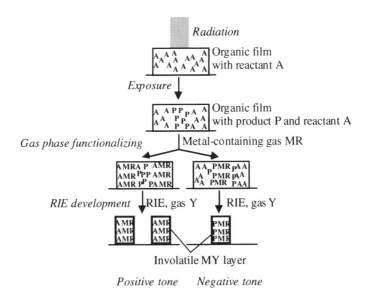

Fig. 8.43 Process flow of gas-phase-functionalized plasma developed resists.

A resist with reactive group A is irradiated to form product group P at the expense of A, which is depleted in the exposed regions. Next, the exposed film is treated with a reactive gas, MR, containing reactive groups R and inorganic atoms M capable of forming nonvolatile compounds, MY, which are resistant to removal under reactive ion etching conditions using gas Y. If gas diffusion occurs and reaction with either A or P occurs, selective functionalization results.

Subsequent plasma development can then produce either negative tone images when an inorganic compound bonds selectively to P or positive tone images when it selectively bonds to A. Tone is thus dependent on both the actinically generated species and selective functionalization.

The concentration of gaseous species C of a reactive gas diffusing into a polymer film at a depth x into the film from the film surface was estimated from the following equation:

$$C = C_0 erf\left(\frac{x}{2\sqrt{Dt}}\right). \qquad (8.12)$$

Their results are summarized in Table 8.10. The diffusion coefficient of poly(isoprene) is 500 times larger than that of poly(ethyl methacrylate), which obviously indicates that the depth of functionalization depends on T_g of the resist base resin.

Table 8.10 Dependence of C/C_0 on polymer and gas diffusion properties and on processing properties at 25 °C.

Gas	D (cm²/s×10⁶)	t (s)	x (μm)	C/C_0
A. Poly(isoprene), T_g=-73 °C				
O_2	1.73	1.0	1.0	1.45
Ar	1.36	1.0	1.0	1.42
SF_6	0.115	1.0	1.0	1.03
B. Poly(ethyl methacrylate), T_g=65 °C				
O_2	0.103	1.0	1.0	1.00
Ar	0.020	1.0	1.0	0.65
SF_6	0.00022	1.0	1.0	0.00
SF_6	0.00022	1.0	0.5	0.00
SF_6	0.00022	1.0	0.1	0.67
SF_6	0.00022	10.0	1.0	0.00
SF_6	0.00022	10.0	0.5	0.44
SF_6	0.00022	10.0	0.25	0.77
SF_6	0.00022	10.0	0.1	1.15

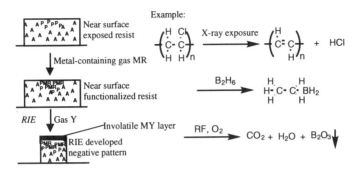

Fig. 8.44 Process flow of surface-functionalized plasma developed resists.

Then, the negative plasma developed resist scheme shown in Fig. 8.43 can be formed employing a low T_g resist, and Waycoat IC-43 which is composed of bis-azide and partially cyclized poly(isoprene), and several metal halides: *e.g.* $SiCl_4$, $SnCl_4$ and $(CH_3)_2SiCl_2$ have been examined [154]. Using a contact printer operated at 366 nm, 0.6 µm resolution was achieved by $SiCl_4$ treatment followed by O_2-RIE. Reaction in the near-surface 200 nm region is clearly favored by low diffusion coefficients associated with a higher T_g resist. The surface functionalization enables the formation of another negative plasma developed resist scheme shown in Fig. 8.44 [153]. Although B_2H_6 treatment of some poly(chloroalkyl acrylate) resists [155,156] did not give high selective functionalization, a 0.5 µm pattern was obtained by e-beam exposure.

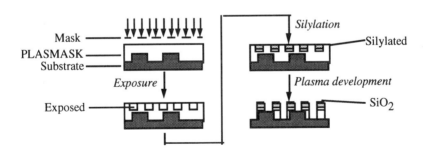

Fig. 8.45 The DESIRE process.

After AT&T's previous investigation, DESIRE (Diffusion Enhanced Silylation Resist) was proposed by Coopmans and Roland [157]. The DESIRE process was designed to be sensitive to near and mid UV and was therefore directly applicable to g-line and i-line exposure tools. The basic DESIRE process flow is shown in Fig. 8.45. First the PLASMASK® is coated onto the wafer. A standard UV exposure is performed, followed by selective silylation where silicon is built into the exposed area. During the development, the silicon-containing parts rapidly form a silicon-dioxide-rich layer preventing further erosion while the other regions can be etched away by means of an anisotropic RIE process. The overall process results in a negative tone relief image with nearly vertical sidewalls. All steps are compatible with normal device fabrication procedures and have been demonstrated using production-type equipment. PLASMASK® is a relatively high heat-resistant resist, and its chemical structure has not been revealed; however, it is thought to be a kind of DNQ-type photoresist. The sensitivity of the resist could be varied by modifying the subsequent silylation treatment or by choosing different development conditions. The light did not need to penetrate down to the substrate because activation was needed only in the top part of the resist where the silylation would take place. Therefore the depth of focus (DOF) requirement was about three times less than that required for the conventional process, which meant that the total resist thickness did not influence the DOF criterion. Lithographic performance obtained from early works is summarized in the literature [158]. For example, 0.7 μm L&S patterns were resolved on a highly reflective substrate (aluminum) using a low NA (0.3) g-line stepper and PLASMASK 150-G (for g-line exposure). A half-micron feature could be formed using an i-line stepper and PLASMASK 200-I (for i-line exposure). Selective silylation was conducted by a gas phase treatment with vaporized hexamethyldisilazane (HMDS). Silicon was incorporated into the resin during the process to a depth ranging from 100 to 250 nm depending on the silylation condition. In the unexposed parts the diffusion was very limited, typically less than 5 to 10 nm. Silylation conditions were optimized to control resist performance at exposure wavelengths of 436, 365 and 248 nm (sub-half to quarter micron rule) [159,160]. The profile of the silylated parts is very important because of they affect the final pattern profiles by dry development. A weaker silylation condition (lower silylation temperature) resulted in a more vertical sidewall and a thinner silylated profile. However, the lateral diffusion or silylation was not ignored, and therefore lateral swelling profiles and undercutting etched profiles similar to those in Fig. 8.46 were observed. Under KrF excimer laser exposure, silylated profiles were improved because of the high absorption of PLASMASK301-U (for KrF exposure).

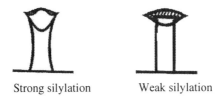

Fig. 8.46 Dry-developed profiles of a silylated resist.

Silylation is the most critical step in the DESIRE process, and alternative mono- and polyfunctional silylating agents have been examined and the kinetics of silylation have been studied [161]. Figure 8.47 shows the examined silylating agents and Table 8.11 summarizes the lateral swelling, etch selectivities and other results. Two silylating agents were studied in more detail: HMDS and TMDS, typifying low- and high-temperature silylation and swelling and nonswelling results, respectively. From RBS and IR measurements it has been defined that the diffusion of HMDS proceeded with the square root of the silylation time, indicating normal Fickian diffusion. In contrast, it has been observed that the silylation initially proceeded linearly with time, and for longer silylation times (or higher silylation temperatures), the silylation rate decreased and the silylation again followed a square root time dependence. Such behavior has been referred to as 'Case II' diffusion by Alfrey [162]. The chemistry and kinetics of gas phase silyaltion have also been studied by other researchers [163,164].

Liquid-phase silylation is promising for simplifying hardware requirements as well as improving silylation properties. Comparative studies between gas- and liquid-phase silylation processes have been carried out by Baik (IMEC) and Roland (UCB-JSR) [165,166]. The silylation agents shown in Fig. 8.47 were used. NMP (N-methyl-2-pyrrolidone) was used as the diffusion promoter and xylene as the solvent. It has been reported that B[DMA]DS has been identified as a promising candidate for liquid-phase silylation. This process exhibited several important advantages over a gas-phase silylation: (i) an improved silylation selectivity, resulting in better definition of the silylated areas, (ii) extremely high Si incorporation (up to ~25 wt.%), resulting in largely improved dry development selectivities, and (iii) improved thermal stability of the Si incorporation. Other mechanistic studies have been also carried out and the usefulness of the liquid-phase silylation technique has been demonstrated [167-170].

Mono-functional silylating agents

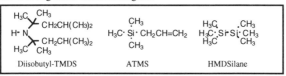

None-reacting mono-functional agents

Poly-functional agents

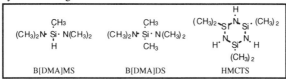

Fig. 8.47 Structures of the various silylating agents.

Table 8.11 Properties and performance of silylating agents.

Silylating agents	B.p. (°C)	Optimum silylation temp. (°C)	Surface roughness	Lateral swelling (%)	Residue formation	Etch selectivity
Mono-functional agents						
DMSDMA	103	80-85	0	0	yes	
TMDS	133	110-120	0	0	no	15
TMSDMA	117	115-125	1	3	no	13
TMSDEA	145	135-145	3	6	no	14
HMDS	161	155-165	4	6	no	11
HeptaMDS	175	155-165	5	4	yes	6
Diisobutyl-TMDS	175					
ATMS	114					
HMDSilane	146					
Poly-functional agents						
B[DMA]MS	132	165-175	9	11	yes	very poor
B[DMA]DS	146	165-175	10	12	yes	very poor
HMCTS	220	185-195	9.5	11	yes	very poor

Figure 8.48 shows the positive resist scheme of the silylation process. This has become possible due to the chemical crosslinking system similar to the chemical amplified resist using the melamine crosslinker [171]. In the unexposed area phenolic resin was silylated, while, in the crosslinked area (exposed area) silylation was prevented. This was suitable for deep UV lithography. After dry development, a positive image of the mask was obtained. This positive scheme has been applied to short wavelength optical lithography [172-176], e-beam lithography [177] and ion beam lithography [178].

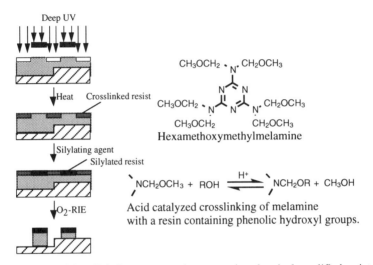

Fig. 8.48 Positive silylation process using a negative chemical amplified resist.

Different dry developable resist systems operating in the positive bilayer and negative top surface imaging modes were investigated for application in mid UV and deep UV lithography. They were called Si-CARL (Chemical Amplification of Resist Lines) and Top-CARL, respectively [179-181]. Their process flows are represented in Fig. 8.49, and it is understood that Top-CARL is more compatible with a normal single layer process. The chemical structures of binder polymer, photoactive compound and silylation agent are shown in Fig. 8.50a. The diazonaphthoquinone system showed relatively low sensitivity, and a t-BOC system (chemical amplification) shown in Fig. 8.50b has been proposed for KrF excimer laser lithography. A resolution of 0.35 µm was obtained at a dose of 10 mJ/cm^2. A new resist characterized as an iminosulfonate polymer has been applied to the surface imaging process [182,183]. Exposing the resist to UV light, the sulfonic acid groups was generated and the resulting strong acidic polymer catalyzed hydrolysis and the subsequent polymerization on the resist surface of alkoysilane as the silylating agent. The rate of surface silylation increased in the following order: $CH_3Si(OCH_3)_3$ > $CH_3Si(OC_2H_5)_3$ > $CH_3Si(OCH_3)_4$ > > $C_2H_5Si(OC_2H_5)_3$ ~ $C_6H_5CH_2Si(OC_2H_5)_3$ ~ $Si(OC_2H_5)_4$ ~ $Si(OC_3H_7)_4$. Some other applications have also been reported: *e.g.* surface titanation [184] and nanometer patterning [185].

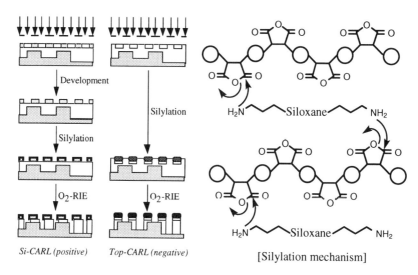

Fig. 8.49 CARL processes and their silylation mechanism.

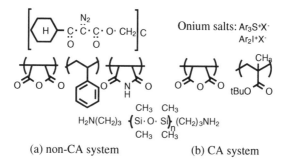

Fig. 8.50 The structures of the components used in CARL system.

8.8 OTHER SURFACE IMAGING

This section describes resolution enhancement by other surface imaging or surface modification. The mechanism of overhang formation in a diazonaphthoquinone/novolak photoresist film by the chlorobenzene soak process for lift-off stencil fabrication has been investigated [186]. A resist film which has undergone the chlorobenzene soaking process is generally composed of four regions as depicted in Fig. 8.51. The modified regions (B and D) denote the thin layer having a low PAC concentration formed on a resist film surface by this process. PAC is extracted by chlorobenzene from region B, which becomes an overhang layer after development. Such surface modification by extraction of part of the resist component was applied to improving the resist profile in the e-beam resist process [187]. A positive e-beam resist RE-5000P (NPR) was supplied by Hitachi Chemical Limited, and consisted of cresol novolak resin and PMPS (poly(2-methylpentene-1-sulfone)). The development characteristics shown in Fig. 8.51 (unsoaked case) result in film thickness loss in the unexposed area during development, thus the pattern edge at the top of the resist layer becomes somewhat rounded. Chlorobenzene treatment reduces film thickness loss in the unexposed area, and nonlinear development as shown in Fig. 8.51 (soaked case) is achieved. This is because lower molecular weight novolak resin is extracted by chlorobenzene, thus the development rate of the resist film surface increases. The soaking process was applied to an alkaline developable positive photoresist system [188]. Patterning exposure was performed after alkaline treatment of Shipley's photoresist S1400 to improve the resist edge profiles without changing the sensitivity. Chlorobenzene soaking treatment introduced a modified layer on top of the resist whose dissolution rate was very low in the unexposed area. The introduction of this layer suppressed the resist film thickness loss during development and improved the resolution. The modified layer in this case, however, was introduced only on the initial surface, so the layer could not protect the sidewalls which form during development. By repeating the interruptions at short intervals the dissolution in lateral directions is more efficiently suppressed and a more accurate resist profile is realized. The process flow and development interruption scheme are shown in Fig. 8.52 [189]. It was assumed that the surface insoluble layer formed because the residual ionized novolak resin was extracted by water and the concentration of dissolution inhibitor PMPS increased in the surface of the top and sidewall. This process was applied to the RE-5000P e-beam resist and a 0.4 μm hole was formed in a 1.5-μm-thick resist film.

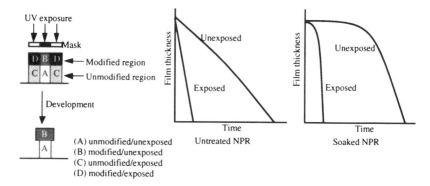

(A) unmodified/unexposed
(B) modified/unexposed
(C) unmodified/exposed
(D) modified/exposed

Fig. 8.51 Contrast enhancement of positive e-beam resist with a surface-modified layer (Mechanism and development characteristics).

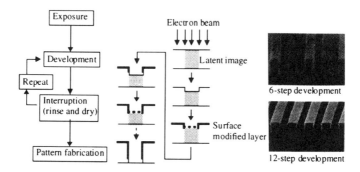

Fig. 8.52 Process for resist profile improvement with surface modification.

DESIRE and soaking processes improve the resolution or resist pattern profiles by wet treatment, while the resolution enhancement by the photoassisted process has also occurs. REL (Resolution Enhanced Lithography) [190-192] adopts deep UV flood exposure with simultaneous heating to improve the resolution as shown in Fig. 8.53. Deep UV flood exposure with heating formed an alkaline-insoluble layer on the resist surface, which resulted in the development of rate gradient in the depth direction. Surface insolubilization was caused by esterification between the ketene and phenolic compound (novolak resin and polyhydroxybenzophenone) and produced a partial crosslinked structure as shown in Fig. 8.53. REL was effective not only for resolution enhancement but also for accurate patterning on a highly reflective substrate. A similar approach to REL has been reported [193]. Bisazide was added to a diazonaphthoquinone positive photoresist effectively formed a crosslinked surface layer by the diffusion-controlled chemical reaction, and provided a method of controlling the wall profiles of resist images.

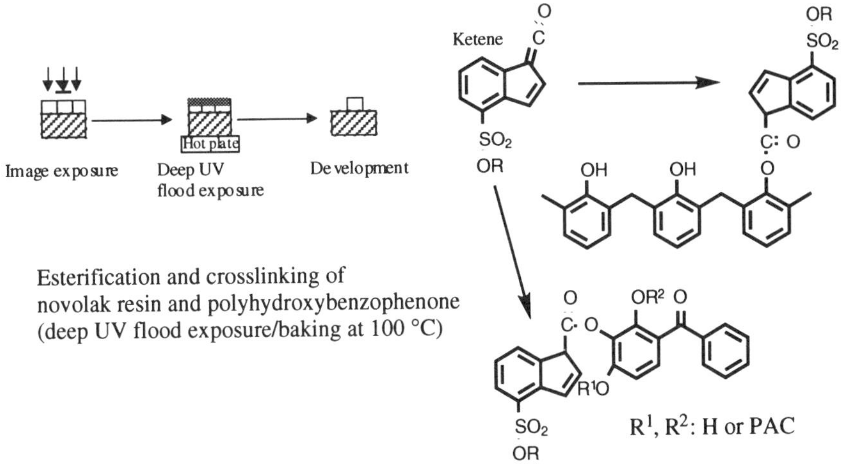

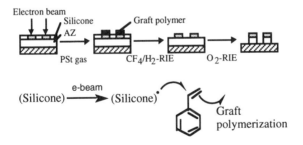

Fig. 8.53 REL process and its surface insolubilization mechanism.

Fig. 8.54 Schematic of the processing sequence and grafting mechanism for direct pattern formation by electron beam-induced vapor phase polymerization.

A unique patterning method has been reported. The method involved direct pattern fabrication by electron beam-induced vapor phase graft polymerization [194]. Selective pattern formation in irradiated areas was performed using PMMA or silicone resin as a base film and styrene as a grafting monomer. The base film patterns could be fabricated by dry etching with the mask of graft patterns. The process is shown in Fig. 8.54. This method has high resolution capability due to the well-focused beam induced surface reaction, however, because of the isotropic behavior in the growth of grafted polymer, the attainable resist thickness was relatively thin. The method had to adopt a multilayer scheme. A 0.1-µm-thick graft polymer on a silicone film could be obtained by irradiation at a dose of 80 µC/cm^2 followed by grafting for 1 h.

Figure 8.55 shows a schematic of the SIEL (Superficial Image Emphasis Lithography) process in which a high resolution surface image is transferred to the resist film [195-197]. The spin-coated single layer of a positive e-beam resist on a substrate is underexposed and develops to form concave patterns at the superficial layer of the resist. A uniform concave depth is achieved by dose control for every feature size. For superficial imaging, the interproximity effect caused by scattered electrons is nearly absent for a thick resist. These concave patterns are filled and planarized with a mask material exhibiting high resistance to the resist etchant. In order to facilitate planarization, low-viscous mask materials, such as silicone resin, are used. Mask patterns are formed over the resist by etching mask material uniformly. The desired line width is defined by stopping the etchback at the halfway point of the concave patterns. As a result, the difficulty of linewidth control due to degradation of pattern contrast caused by the spread in the incident beam edge slope can be avoided. The mask patterns are then transferred to the resist etching. The SIEL thus provides negative patterns on the substrate. SIEL process has been applied to 0.2 µm rule VLSI patterns and has given good line width control at the half micron feature on a Mo layer superior to that using the bilayer process. If the superficial layer was etched back near the concave bottom, very small patterns could be fabricated. Nanometer patterns down to 25 nm with an aspect ratio of 60 have been demonstrated. Although the SIEL process is an excellent patterning technique as described above, detection of the endpoint of etchback may be a problem.

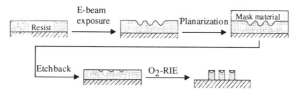

Fig. 8.55 SIEL process.

8.9 DRY DEVELOPMENT

Investigation of "dry lithography" has been underway for nearly 20 years and its purpose is thought to be resolution improvement by avoiding pattern swelling in wet development and simplifying and automatically controlling the whole resist process. The category is very wide and the inorganic resist process in Section 8.6 and surface silylation in Section 8.7 are examples of "dry lithography" or "dry development". This section will describe dry development using common organic resist materials and some organosilicon polymers.

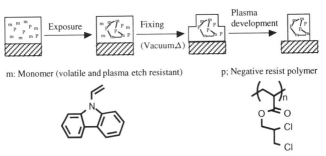

Fig. 8.56 Schematic diagram of plasma-developed resist.

Taylor *et al.* of AT&T Bell Labs. have designed a dry development system for x-ray lithography (characteristic x-ray source was used at that time), which consists of acrylate-type negative resists and aromatic compounds reactive to radical species [198]. The concept of dry development is explained in Fig. 8.56. The host polymer was 2,3-dichloro-1-propylacrylate (DCPA) as an example of an acrylate resist and the guest compound was *N*-vinylcarbazol which has excellent plasma etch resistance due to aromatization energy. X-ray exposure (Pd Lα) of the resist film caused crosslinking between DCPA and *N*-vinylcarbazol and the subsequent heating *in vacuo* removed the guest compound in the unexposed area. Thus, this treatment gave rise to a difference in plasma resistance between the exposed area and the unexposed area. The resist system consisting of DCPA and *N*-vinylcarbazol with an 8:2 ratio resolved 0.3 μm lines, however, the etching selectivity of the exposed part to the unexposed part and the pattern profiles were not good. Several vinyl silanes have been applied to the dry development system [199,200]. In the exposed area, the silicon compound was fixed and the film was treated with oxygen plasma or fluorine plasma to give negative patterns or positive patterns, respectively. A similar approach has been reported.

Poly(isopropenyl ketone) (PMIPK) reacts with the nitrene produced from bisazide to yield the β-amino ketone crosslinked polymer shown in Fig. 8.57 [201]. Etching selectivity was obtained because the aromatic structure was incorporated in the exposed area. The resulting patterns could act as an etching mask for a silicon substrate.

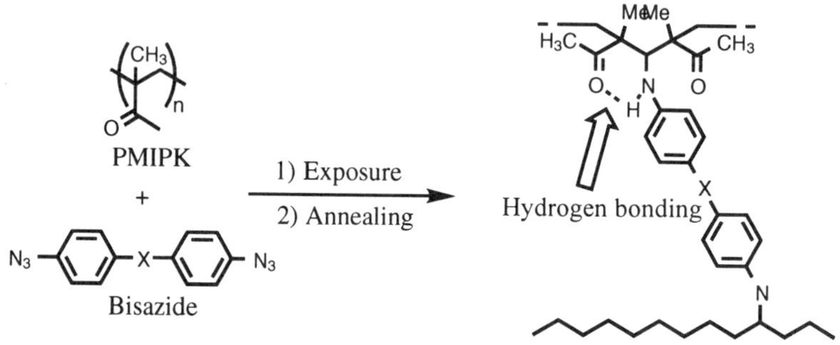

Fig. 8.57 Dry-development resist using crosslinking by a bisazide.

Photoablation can be used for microfabrication. Nitrated cellulose easily decomposes by exposure to Ar^+ beam and ArF excimer laser, then it is used as a positive self-development resist system [202,203]. The products shown in Fig. 8.58a caused by photoablation are confirmed by mass spectroscopy. Grating patterns of 320-nm-pitch have been formed by Ar^+ irradiation through a silicon nitride stencil mask. Photoablative organic polymers have markedly poor etch resistance. Some polysilanes are also photoablative, however, they have high etch resistance to oxygen plasma and are used for the bilayer resist process described in Section 8.5. A bilayer resist process including self-development can be constructed, which is pointing to all dry resist process. Self-development characteristics by KrF excimer laser exposure have been investigated for several poly(alkylsilane)s [204]. The volatile products caused by ablation were identified by mass spectroscopy and other analytical methods, according to which self-development originated from photodegradation as shown in Fig. 8.58b. Although full development of a 1-µm-thick poly(alkylsilane) film required a dose of 500-1000 mJ/cm^2, the dose for development could be reduced when used in a bilayer resist process.

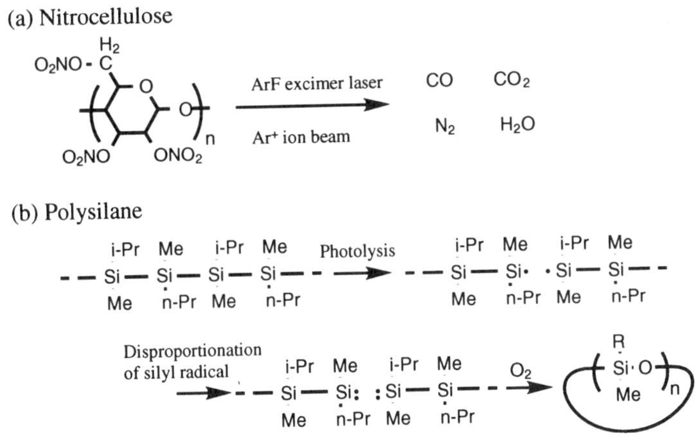

Variations in Microlithographic Process

(a) Nitrocellulose

$$O_2NO-CH_2 \quad \overset{O_2NO \quad ONO_2}{\underset{n}{\bigcirc}}-O \xrightarrow[\text{Ar}^+ \text{ ion beam}]{\text{ArF excimer laser}} CO \quad CO_2 \quad N_2 \quad H_2O$$

(b) Polysilane

$$--\underset{\text{Me}}{\text{Si}}-\underset{\text{n-Pr}}{\overset{\text{i-Pr}}{\text{Si}}}-\underset{\text{Me}}{\overset{\text{Me}}{\text{Si}}}-\underset{\text{n-Pr}}{\overset{\text{i-Pr}}{\text{Si}}}-- \xrightarrow{\text{Photolysis}} --\underset{\text{Me}}{\text{Si}}-\underset{\text{n-Pr}}{\overset{\text{i-Pr}}{\text{Si}}}\cdot \ \cdot\underset{\text{Me}}{\overset{\text{Me}}{\text{Si}}}-\underset{\text{n-Pr}}{\overset{\text{i-Pr}}{\text{Si}}}--$$

$$\xrightarrow[\text{of silyl radical}]{\text{Disproportionation}} --\underset{\text{Me}}{\text{Si}}-\underset{\text{n-Pr}}{\overset{\text{i-Pr}}{\text{Si}}}: \ :\underset{\text{Me}}{\overset{\text{Me}}{\text{Si}}}-\underset{\text{n-Pr}}{\overset{\text{i-Pr}}{\text{Si}}}-- \xrightarrow{O_2} \left(\underset{\text{Me}}{\overset{R}{\text{Si}\cdot O}}\right)_n$$

Fig. 8.58 Self-development resists.

A high resolution all dry process is ideal for resist processes. It was found that a polymerized organic film was deposited on the substrate in an exposure chamber. This was caused by electron beam induced polymerization of a residual contaminant such as a carbonic compound. Metal structures 10 nm high with sharply defined linewidths of 8 nm, called "contamination resists" have been produced using an electron-beam fabrication process [205]. The whole process was carried out in a dry medium and was significant in nm structure direct fabrication, but was determined by an uncontrollable factor. Weidman et al. have proposed an all dry lithography compatible with the current LSI process [206]. The process flow is illustrated in Fig. 8.59. A monosubstituted silane was deposited on plasma polymerization to achieve a poly(silyne) film on a substrate. Image-wise exposure to a KrF or ArF excimer laser was performed in air, then poly(silyne) was oxidized to poly(siloxane) in the exposed area. The unexposed area (poly(silyne)) could be etched by chlorine plasma to produce negative patterns. The etching selectivity of the unexposed parts to the exposed parts was 5:1. These patterns could act as etching masks for, e.g., the organic planarizing layer (bilayer resist process) and polySi. Bilayer resist patterns of 0.35 µm L&S were fabricated by KrF excimer laser lithography at a dose of 70 mJ/cm^2. In the exposed area the most organic moiety still remained and the subsequent oxygen plasma ashing removed this organic component. Then, the resulting porous film was densified by thermal annealing and could be applied to device structures [207,208]. This all dry lithography had potential resolution up to a 0.2 µm feature at a high sensitivity of 20 mJ/cm^2 using a Micrascan exposure system equipped with an ArF excimer laser source. This dry lithography is thought to be more practical than the other dry developments described in this section. The system can be constructed to cluster tools and will be introduced to mass-production in the near future.

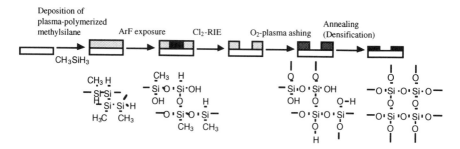

Fig. 8.59 All dry lithography using plasma polymerized methylsilane.

8.10 DIRECT FABRICATION OF SUBSTRATE TOWARD ULTRA SMALL SCALE DEVICES

Mesoscopic physical phenomena will be easily observed in small scale devices possessing sub-100 nm structures, and such phenomena are known as the Coulomb blockade effect, subband effect, resonance tunneling effect, etc. Several microfabrication techniques for preparation of ultra-small structures are available. The contamination resist described in the previous section has been demonstrated below 10 nm resolution. A poly(p-butoxycarbonyloxystyrene) resist shows a an high resolution of 18 nm on a Si_3N_4 membrane by use of STEM with accelerating voltage of 50 kV [209]. In nanometer scale patterning size fluctuation cannot be ignored when organic polymers are used for resist materials. To better understand nanometer scale pattern fluctuation in lithography a new model for electron beam exposed poly(methyl methacrylate) resist development has been developed [210]. The results indicate that the molecular scale resist structure plays an important role in PMMA edge roughness. This has also been confirmed qualitatively as a correlation of polydispersity and nano edge roughness in novolak based resist patterns [211]. The size fluctuation of nano structures on a SIMOX substrate was evaluated quantitatively in the course of preparing a single electron transistor [212]. It was evident that the fluctuation in linewidth was due to grain-like structures measuring 20-30 nm in diameter.

has no sensitivity to e-beam for practical use, however, SiO_2 acts as a positive resist to high energy e-beam. Silicon dioxide was written by a high energy focused e-beam (accelerating voltage=300 keV, beam diameter=3 nm) at a dose of 7.5 µC/cm and line patterns with a 15 nm period were formed [213]. Development could be achieved by several etchants for SiO_2, *e.g.*, p-etch solution ($HF:HNO_3:H_2O$=15:10:300), hydrofluoric acid and hot sodium hydroxide solution. In this case SiO_2 is a positive resist and the cause of dissolution of the exposed area is thought to be *via* induction of defects in the SiO_2 film by e-beam irradiation. Selective etching of SiO_2 in hydrofluoric acid was reported [214]. The etching proceeded under the contamination resist patterns and the mechanism has been explained as follows. A contamination resist is present as a kind of polymer radical in an exposure chamber, and purging the chamber with air causes the polymer radical to react with oxygen molecules to produce hydrophilic polymers. On the hydrophilic surface bis(hydrofluoric acid) ion, $(HF_2)^-$, is easily generated. Trench patterns with a minimum size of 3-5 nm have been delineated in SiO_2 film under conditions of accelerating voltage: 40 keV, line dose: 0.28-11 nC/cm and HF treatment at 110 °C. Figure 8.60 shows the direct fabrication of GaAs and AlGaAs epi-film by e-beam induced etching [215,216]. First, the surface of the GaAs film was oxidized in oxygen assisted by photoirradiation to produce a nm-thick oxide layer. The oxide layer was written by e-beam in chlorine gas ambient, and as a result the irradiated area was etched off as chlorides. The etch rates of GaAs and AlGaAs depended on the substrate temperature. This *in situ* e-beam lithography resulted in a structure of approximately 100 nm. The resolution of these direct fabrication processes can be enhanced further when a highly focused e-beam is used, and such exposure systems have recently become available [217,218].

Recently, scanning probe microscopies have been applied to nanometer lithography. Figure 8.61 shows the principle of the Ti oxidation process using a STM (scanning tunnel microscopy) tip as the anodization electrode. Titanium is oxidized in moisture when the STM tip is set close to the Ti substrate such that the Faraday current begins to flow. A planar type diode has been prepared using TiO_x as a tunnel oxide film [219,220]. Furthermore, a room temperature operated single electron transistor has been fabricated by by STM/AFM nano-oxidation process [221,222], and a 2DEG channel has also been formed in a delta-doped GaAs HEMT device [223].

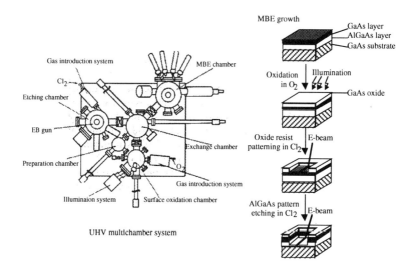

Fig. 8.60 Electron-beam-induced pattern etching of AlGaAs using an ultrathin GaAs oxide as a resist.

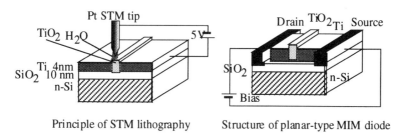

Principle of STM lithography   Structure of planar-type MIM diode

Fig. 8.61 STM lithography and its application to a nanoscale device.

References

1. M. D. Levenson, N. S. Viswanathan and R. A. Simpson: *IEEE Trans. Electron Devices* **ED-29**, 1828 (1982).
2. K. Kamon, T. Miyamoto, Y. Myoi, H. Nagata and M. Tanaka: *Jpn. J. Appl. Phys.* **30**, 3021 (1991).
3. H. Fukuda, T. Terasawa and S. Okazaki: *J. Vac. Sci. Technol.* **B9**, 3113 (1991).
4. B. F. Griffing and P. R. West: *IEEE Electron Device Lett.* **EDL-4**, 14 (1983).
5. B. J. Lin: *Proc.SPIE* **174**, 114 (1979).
6. J. M. Moran and D. Maydan: *J. Vac. Sci. Technol.* **16**, 1620 (1979).
7. F. Coopmans and B. Roland: *Proc. SPIE* **631**, 34 (1986).
8. F. H. Dill, W. P. Hornberger, P. S. Hauge and J. M. Shaw: *IEEE Trns. Electron Devices* **ED-22**, 717 (1975).
9. Y. Hirai, M. Sasago, M. Endo, K. Ogawa, Y. Mano and T. Ishihara: *J. Vac. Sci. Technol.* **B5**,434 (1987).
10. W. G. Oldham, S. M. Nandgaonkar, A. R. Neureuther and M. M. O'Toole: *IEEE Trans. Electron Devices* **26**, 717 (1979).
11. M. Sasago, M. Endo, Y. Hirai, K. Ogawa and T. Ishihara: *Proc. SPIE* **631**, 321 (1986).
12. S. V. Babu, E. Barouch and B. Bradie: *J. Vac. Sci. Technol.* **B6**, 564 (1988).
13. B. F. Griffing and W. E. Lorensen: *Digest of Technical Papres - 1983 Symposium on VLSI Technology*, pp. 76-77 (1983)
14. B. F. Griffing, P. R. West and B. A. Heath: *Electron Device Letters* **EDL-4**, 317 (1983).
15. P. R. West and B. F. Griffing: *Proc. SPIE* **394**, 33 (1983).
16. B. F. Griffing, P. R. West and E. W. Balch: *Proc. SPIE* **469**, 94 (1984).
17. B. F. Griffing and P. R. West: *Extended Abstracts of the 16th Conference on Solid State Devices and Materials*, 1984, pp. 7-10.
18. D. C. Hofer, R. D. Miller, C. G. Willson and A. R. Neureuther: *Proc. SPIE* **469**, 108 (1984).
19. C. Mack: *Proc. SPIE* **631**, 276 (1986).
20. C. Mack: *J. Vac. Sci. Technol.* **A5**, 1428 (1987).
21. B. F. Griffing and P. R. West: Solid State Technol. **May**, 152 (1985).
22. L. F. Halle: *J. Vac. Sci. Technol.* **B3**, 323 (1985).
23. H. Niki, M. Nakase, A. Kumagae, T. Sato and K. Ikari: *Extended Abstracts of the 17th Conference on Solid State Devices and Materials*, 1985, pp.361-364.

24. S. Uchino, T. Iwayanagi, T. Ueno, M. Hashimoto and S. Nonogaki: *Proc. SPIE* **771**, 11 (1987).
25. S. Uchino, M. Hashimoto and T. Iwayanagi: *Proc. ACS Division of Polymeric Materials: Science and Engineering* **60** (American Chemical Society, Washington D.C., 1989) 255.
26. T. Tanaka, S. Uchino, M. Hashimoto, N. Hasegawa, H. Fukuda and S. Okazaki: *Jpn. J. Appl. Phys.* **29**, 1860 (1990).
27. M. Endo, M. Sasago, Y. Hirai, K. Ogawa and T. Ishihara: *J. Vac. Sci. Technol.* **B6**, 1600 (1988).
28. M. Endo, M. Sasago, Y. Hirai, K. Ogawa and T. Ishihara: *J. Electrochem. Soc.* **136**, 508 (1989).
29. M. Endo, M. Sasago, A. Ueno and N. Nomura: *J. Vac. Sci. Technol.* **B7**, 565 (1989).
30. M. Endo, M. Sasago, H. Nakagawa, Y. Hirai, K. Ogawa and T. Ishihara: *J. Vac. Sci. Technol.* **B6**, 559 (1988).
31. M. Endo, Y. Tani, M. Sasago and N. Nomura: *J. Vac. Sci. Technol.* **B7**, 1072 (1989).
32. Y. Yonezawa, H. Kikuchi, K. Hayashi, T. Tochizawa, N. Endo, S. Fukuzawa, S. Sugito and K. Ichimura: *J. Photopolym. Sci. Technol.* **1**, 37 (1988).
33. K. Kaifu, T. Ito, M. Kosuge, Y. Yamashita, S. Ohno, T. Asano, K. Kobayashi and G. Nagamatsu: *J. Vac. Sci. Technol.* **B5**, 439 (1987).
34. R. Kawazu, Y. Yamashita, T. Ito, K. Kawamura, S. Ohno, T. Asano, K. Kobayashi and G. Nagamatsu: *J. Vac. Sci. Technol.* **B4**, 409 (1986).
35. Y. Yamashita, R. Kawazu, K. Kawamura, S. Ohno, T. Asano, K. Kobayashi and G. Nagamatsu: *J. Vac. Sci. Technol.* **B3**, 314 (1985).
36. B. J. Lin: *Solid State Technol.* **26**, 105 (1983).
37. B. J. Lin and T. H. P. Chang: *J. Vac. Sci. Technol.* **16**, 1669 (1980).
38. Y. Yamashita, H. Jinbo, R. Kawazu, S. Ohno, T. Asano, K. Kobayashi and G. Nagamatsu: *Proc. SPIE.* **771**, 273 (1987).
39. H. Moritz and G. Paal: *U. S. Patent* 4104070 (1978); *Chem. Abst.* **88** 14344u (1978).
40. S. A. MacDonald, R. D. Miller, C. G. Willson, G. M. Feinberg, R. T. Gleason, R. M. Halverson, N. W. MacIntyre and W. T. Motsiff: *Kodak Microelectronics Seminar, Interface '82,* 114 (1982).
41. C. Lyons and W. Moreau: *J. Electrochem. Soc.* **135**, 193 (1988).
42. M. Endo, M. Sasago, Y. Hirai, K. Ogawa and T. Ishihara: *J. Vac. Sci. Technol.* **B6**, 87 (1988).
43. B. J. Lin, V. W. Chao and K. E. Petrillo: *J. Vac. Sci. Technol.* **19**, 1313 (1981).

44. K. E. Petrillo, V. W. Chao and B. J. Lin: *J. Vac. Sci. Technol.* **B1**, 1219 (1983).
45. M. Hatzakis: *J. Electrochem. Soc.* **116**, 1033 (1969).
46. B. J. Canavello, M.Hatzakis and J. M. Shaw: *IBM J. Res. Develop.* **24**, 452 (1980).
47. R. A. Larsen: *IBM J. Res. Develop.* **24**, 268 (1980).
48. R. M. Halverson, M. W. MacIntyre and W. T. Motsiff: *IBM J. Res. Develop.* **26**, 590 (1982).
49. G. G. Collins and C. W. Halsted: *IBM J. Res. Develop.* **26**, 596 (1982).
50. S. K. Jones, R. C. Chapman and E. K. Pavelchek: *Proc. SPIE* **771**, 231 (1987).
51. H. Jinbo, Y. Yamashita, A. Endo, S. Nishibu, H. Umehara and T. Asano: *Jpn. J. Appl. Phys.* **28**, 2053 (1989).
52. T. Ito, Y. Yamashita, R. Kawazu, K. Kawamura, S. Ohno, T. Asano, K. Kobayashi and G. Nagamatsu: *Polym. Eng. Sci.* **26**, 1105 (1986).
53. K. Inokuchi, T. Saito, H. Jinbo, Y. Yamashita and Y. Sano: *Jpn. J. Appl. Phys.* **30**, 3818 (1991).
54. H. Jinbo, T. Ito and Y. Yamashita: *J. Vac. Sci. Technol.* **B13**, 2954 (1995).
55. M. J. Bowden: *Solid State Technol.* **24**, 73 (1981).
56. T. Tanaka, N. Hasegawa, S. Okazaki and Y. Azuma: *J. Vac. Sci. Technol.* **B10**, 723 (1992).
57. Y. Gotoh, T. Kure and S. Tachi: *Jpn. J. Appl. Phys.* **32**, 3035 (1993).
58. N. Asai, A. Imai, T. Ueno, Y. Azuma, T. Miyazaki, T. Tanaka and S. Okazaki: *J. Photopolymer Sci. Technol.* **7**, 23 (1994).
59. T. Hosoya, T. Matsuda, K. Iwadate and K. Harada: *Technical Report of IEICE (The Institute of Electronics, Information and Communication Engineers)* **SSD84-142**, 19 (1984).
60. T. Kure, H. Kawanami, S. Tachi and H. Enami: *Proc. Dry Process Symposium*, 117 (1990).
61. T. Sato, T. Ishida, M. Yoneda and K. Nakamoto: *IEICE Trans. Electron.* **E76-C**, 607 (1993).
62. H. Namatsu, Y. Oazaki and K. Hirata: *J. Electrochem. Soc.* **130**, 523 (1983).
63. K. Tokashiki, K. Sato, A. Nahori and E. Ikawa: *J. Vac. Sci. Technol.* **B11**, 2284 (1993).
64. J. M. Shaw, M. Hatzakis, J. Paraszczak, J. Liutkus and E. Babich: *Polym. Eng. Sci.* **23**, 1054 (1983).
65. E. D. Roberts: *J. Electrochem. Soc.* **120**, 1716 (1973).
66. J. M. Shaw, M. Hatzakis, J. Paraszczak and E. Babich: *Microelectronic Eng..* **3**, 293 (1985).

67. J. M. Shaw, E. Babich, M. Hatzakis and J. Paraszczak: *Solid State Technol.* 83 (1987).
68. M. Morita, A. Tanaka, S. Imamura, T. Tamamura and O. Kogure: *Jpn. J. Appl. Phys.* **22**, L659 (1983).
69. S. Imamura: *J. Electrochem. Soc.* **126**, 1628 (1979).
70. A. Tanaka and M. Morita: *Technical Report of IEICE (The Institute of Electronics, Information and Communication Engineers)* **SSD84-143**, 27 (1984).
71. S. Imamura, M. Morita, T. Tamamura and O. Kogure: *Macromolecules* **17**, 1412 (1984).
72. M. Morita, A. Tanaka and K. Onose: *J. Vac. Sci. Technol.* **B4**, 414 (1986).
73. A. Tanaka, M. Morita and K. Onose: *Proc. of the 25th Symposium on Semiconductors and Integrated Circuits Technology,* 48 (1984).
74. A. Tanaka, H. Ban, S. Imamura and K. Onose: *J. Vac. Sci. Twechnol.* **B7**, 572 (1989).
75. Y. Kawai, A. Tanaka and T. Matsuda: *Proc. of the 39th Symposium on Semiconductors and Integrated Circuits Technology,* 37 (1989).
76. Y. Kawai, A. Tanaka, Y. Ozaki, K. Takamoto and A. Yoshikawa: *Proc. SPIE* **1086**, 173 (1989).
77. H. Ito and C. G. Willson: *Polymer Eng. Sci.* **23**, 1012 (1983).
78. Y. Kawai, A. Tanaka, H. Ban, J. Nakamura and T. Matsuda: *Proc. SPIE* **1672**, 56 (1992).
79. E. Babich, J. Paraszczak, M. Hatzakis, J. M. Shaw and B. J. Grenon: *Microelectronic Eng.* **3**, 279 (1985).
80. R. G. Broult, R. L. Kubena and R. A. Metzger: *Proc. SPIE* **539**, 70 (1985).
81. K. Saito, Y. Yoneda, S. Fukuyama, S. Shiba, Y. Kawasaki and K. Watanabe: *Proc. of the 32th Symposium on Semiconductors and Integrated Circuits Technology,* 1 (1987).
82. K. Saito, S. Shiba, Y. Kawasaki, K. Watanabe and Y. Yoneda: *Proc. SPIE* **920**, 198 (1988)
83. K. Watanabe, K. Saito, S. Shiba, Y. Kawasaki and Y. Yoneda: *J. Photopolymer Sci. Technol.* **1**, 71 (1988).
84. K. Watanabe, S. Shiba, K. Saito and Y. Yoneda: *J. Photopolymer Sci. Technol.* **2**, 103 (1989).
85. M. Sakata, T. Ito and Y. Yamashita: *J. Photopolymer Sci. Technol.* **2**, 109 (1989).
86. M. Sakata, T. Ito and Y. Yamashita: *J. Photopolymer Sci. Technol.* **3**, 173 (1990).
87. M. Sakata, T. Ito and Y. Yamashita: *Jpn. J. Appl. Phys.* **30**, 3116 (1991).

88. H. Watanabe, Y. Todokoro and M. Inoue: *J. Vac. Sci. Technol.* **B9**, 3436 (1991).
89. S. Sugito, S. Ishida, Y. Iida, K. Mine and H. Muramoto: *Extended Abstracts of the 20th Conference on Solid State Devices and Materials*, 1988, pp. 561-564.
90. S. Sugito, S. Ishida and Y. Iida: *Microelectronic Eng.* **9**, 533 (1989).
91. S. Yamazaki, S. Ishida, H. Matsumoto, N. Aizaki, N. Muramoto and K. Mine: *Proc. SPIE* **1466**, 538 (1991).
92. K. Watanabe, E. Yano, T. Namiki, M. Fukuda and Y. Yoneda: *J. Photopolymer Sci. Technol.* **4**, 481 (1991).
93. K. Watanabe, M. Igarashi, E. Yano, T. Namiki, K. Nozaki and Y. Kuramitsu: *J. Photopolymer Sci. Technol.* **8**, 11 (1995).
94. Y. Yamashita and M. Kajiwara: *J. Electrochem. Soc.* **137**, 3253 (1990).
95. A. Imai, H. Fukuda, T. Ueno and S. Okazaki: *Jpn. J. Appl. Phys.* **29**, 2653 (1990).
96. H. Sugiyama, T. Inoue, A. Mizushima and K. Nate: *Proc. SPIE* **920**, 268 (1988).
97. K. Nate, A. Mizushima and H. Sugiyama: *Proc. SPIE* **1466**, 206 (1991).
98. T. Noguchi, K. Nito, J. Seto, I. Hata, H. Sato and T. Tsumori: *Proc. SPIE* **920**, 168 (1988).
99. Y. Onishi, T. Ushiroguchi, R. Horoguchi and S. Hayase: *Proc. SPIE* **1086**, 162 (1989).
100. S. Sakata, T. Ito and Y. yamashita: *J. Photopolymer Sci. Technol.* **5**, 181 (1992).
101. T. Ito, M. Sakata and M. Kosuge: *IEICE Trans. Electron.* **E76-C**, 588 (1993).
102. M. Sakata, M. Kosuge, H. Jinbo and T. Ito: *Proc. SPIE* **2438**, 775 (1995).
103. M. Sakata, T. Ito, A. Endo, H. Jinbo and I. Ashida: *Digest of Technical Papres - 1983 Symposium on VLSI Technology*, pp. 147-148 (1983)
104. T. Ito, M. Sakata, A. Endo, H. Jinbo and I. Ashida: *Jpn. J. Appl. Phys.* **32**, 6052 (1993).
105. F. S. Kipping: *J. Chem. Soc.* **71**, 963 (1949).
106. R. West, L. D. David, P. I. Djurovich, K. L. Stearly, K. S. V. Srinivasen and H. Yu: *J. Amer. Chem. Soc.* **103**, 7352 (1981).
107. M. Ishikawa, T. Takaoka and M. Kumada: *J. Organometal. Chem.* **42**, 333 (1972).
108. D. C. Hofer, R. D. Miller and C. G. Willson: *Proc. SPIE* **469**, 16 (1984).
109. R. D. Miller and S. A. MacDonald: *J. Imaging Sci.* **31**, 43 (1987).
110. G. H. Taylor, M. Y. Hellman, T. M. Wolf and J. M. Zeigler: *Proc. SPIE* **920**, 274 (1988).

111. G. M. Wallraff, R. D. Miller, H. Clecak and M. Baier: *Proc. SPIE* **1466**, 211 (1991).
112. R. R. Kunz, P. A. Bianconi, M. W. Horn, R. R. Paladugu, D. C. Shaver, D. A. Smith and C. A. Freed: *Proc. SPIE* **1466**, 218 (1991).
113. M. Suzuki, K. Saigo, H. Gokan and Y. Ohnishi: *J. Electrochem. Soc.* **130**, 1962 (1983).
114. H. Gokan, Y. Ohnishi and K. Saigo: *Microelectron. Eng.* **1**, 251 (1983).
115. A. E. Novembre, E. Reichmanis and M. Davis: *Proc. SPIE* **631**, 14 (1986).
116. B. W. Smith, D. A. Mixon, A. E. Novembre and S. Butt: *Proc. SPIE* **2438**, 504 (1995).
117. D. D. Granger, L. J. Miller and M. M. Lewis: *J. Vac. Sci. Technol.* **B6**, 370 (1988).
118. M. A. Hartney, A. E. Novembre and F. S. Bates: *J. Vac. Sci. Technol.* **B3**, 1346 (1985).
119. D. A. Mixon and A. E. Novembre: *J. Vac. Sci. Technol.* **B7**, 1723 (1989).
120. H. Fujioka, H. Nakajima, S. Kishimura and H. Nagata: *Proc. SPIE* **1262**, 554 (1990).
121. S. A. MacDonald, R. D. Allen, H. J. Clecak, C. G. Willson and J. M. J. Frechet: *Proc. SPIE* **631**, 28 (1986).
122. T. Iwayanagi, T. Kohashi, S. Nonogaki, T. Matsuzawa, K. Douta and H. Hanazawa: *IEEE Trans. Electron. Devices* **ED-28**, 1306 (1981).
123. A. S. Gozdz, C. Carnazza and M. J. Bowden: *Proc. SPIE* **631**, 2 (1986).
124. L. F. Thompson and M. J. Bowden: *J. Electrochem. Soc.* **120**, 1722 (1973).
125. P. S. D. Lin and A. S. Gozdz: *J. Vac. Sci. Technol.* **B6**, 2290 (1988).
126. A. S. Gozdz and P. S. D. Lin: *Proc. SPIE* **923**, 172 (1988).
127. A. S. Gozdz, H. Ono, S. Ito, J. A. Shelburne III and M. Matsuda: *Proc. SPIE* **1466**, 200 (1991).
128. T. Higashimura, B.-Z. Tang, T. Masuda, H. Yamaoka and T. Matsuyama: *Polymer J.* **17**, 393 (1985).
129. A. S. Gozdz, G. L. Baker, C. Klausner and M. J. Bowden: *Proc. SPIE* **771**, 18 (1987).
130. K. Saigo, Y. Ohnishi, M. Suzuki and H. Gokan: *J. Vac. Sci. Technol.* **B3**, 331 (1985).
131. E. Babich, J. Paraszcczak, M. Hatzakis, S. Rishton, B. Grenon and H. Linde: *Microelectron. Eng.* **9**, 537 (1989).
132. N. Hayashi, T. Ueno, H. Shiraishi, T. Nishida, M. Toriumi and S. Nonogaki: *ACS Symp. Ser.* **346**, 212 (1987).
133. T. Ueno, H. Shiraishi, N. Hayashi, K. Tadano, E. Fukuma and T. Iwayanagi: *Proc. SPIE* **1262**, 26 (1990).

134. Y. Saotome, H. Gokan, K. Saigo, M. Suzuki and Y. Ohnishi: *J. Electrochem. Soc.* **132**, 909 (1985).
135. C. W. Wilkins, Jr., E. Reichmanis, T. M. Wolf and B. C. Smith: *J. Vac. Sci. Technol.* **B3**, 306 (1985).
136. W. C. Cunningham, Jr. and C. E. Park: *Proc. SPIE* **771**, 32 (1987).
137. T. Ueno, H. Shiraishi, T. Iwayanagi and S. Nonogaki: *J. Electrochem. Soc.* **132**, 1168 (1985).
138. H. Hiraoka and K. N. Chiong: *J. Vac. Sci. Technol.* **B5**, 386 (1987).
139. Y. Tanaka, H. Horibe, S. Kubota, H. Koezuka, H. Yoshioka, S. Aoyama, Y. Watanabe and H. Maezawa: *Jpn. J. Appl. Phys.* **29**, 2638 (1990).
140. T. Igo and Y. Toyoshima: *J. Non-Cryst. Solids* **11**, 304 (1973).
141. H. Nagai, A. Yoshikawa, Y. Toyoshima, O. Ochi and Y. Mizushima: *Appl. Phys. Lett.* **28**, 145 (1976).
142. A. S. Chen, G. Addiego, W. Leung and A. R. Neureuther: *J. Vac. Sci. Technol.* **B4**, 398 (1986).
143. A. Yoshikawa, O. Ochi, H. Nagai and Y. Mizushima: *Appl. Phys. Lett.* **29**, 677 (1976).
144. A. Yoshikawa, O. Ochi, H. Nagai and Y. Mizushima: *Appl. Phys. Lett.* **31**, 161 (1977).
145. K. J. Polasko, C. C. Tsai, M. R. Cagan and R. F. W. Pease: *J. Vac. Sci. Technol.* **B4**, 418 (1986).
146. H. Okamoto, T. Iwayanagi, K. Mochiji, H. Umezaki and T. Kudo: *Appl. Phys. Lett.* **49**, 298 (1986).
147. A. Ishikawa, H. Okamoto, K. Miyauchi and T. Kudo: *Proc. SPIE* **1086**, 180 (1989).
148. M. Green, C. J. Aidinis and O. A. Fakolujo: *J. Appl. Phys.* **57**, 631 (1985).
149. M. Green, C. J. Aidinis, F. Khaleque and Y. Feng: *Microelectron. Eng.* **11**, 539 (1990).
150. A. Scherer and H. G. Craighead: *J. Vac. Sci. Technol.* **B5**, 374 (1987).
151. A. Muray, M. Scheinfein, M. Isaacson and I. Adesida: *J. Vac. Sci. Technol.* **B3**, 367 (1985).
152. J. Fujita, H. Watanabe, Y. Ochiai, S. Manako, J. S. Tsai and S. Matsui: *Appl. Phys. Lett.* **66**, 3065 (1995).
153. G. H. Taylor, L. E. Stillwagon and T. Venkatesan: *J. Electrochem. Soc.* **131**, 1658 (1984).
154. T. M. Wolf, G. H. Taylor, T. Venkatesan and R. T. Kraetsch: *J. Electrochem. Soc.* **131**, 1664 (1984).
155. S. M. Miller, M. W. Spindler and R. L. Vale: *J. Polym. Sci.* **A1**, 2537 (1963).
156. G. H. Taylor and T. M. Wolf: *Proc. Microcircuit Eng.* **81**, 381 (1981).

157. F. Coopmans and B. Roland: *Proc. SPIE* **631**, 34 (1986).
158. F. Coopmans and B. Roland: *Solid State Technol.* June, 93 (1987).
159. K. Taira, J. Takahashi and K. Yanagihara: *Proc. SPIE* **1466**, 570 (1991).
160. K. Kato, K. Taira, T. Takahashi and K. Yanagihara: *Proc. SPIE* **1672**, 415(1992).
161. K.-H. Baik, L. Van den hove, A. M. Goethals and M. Op de Beeck: *J. Vac. Sci. Technol.* **B8**, 1481 (1990).
162. T. Alfrey, Jr., E. F. Gurnee and W. G. Lloyd: *J. Polymer Sci.* **C12**, 249 (1966).
163. H. Hiraoka, A. Patlach and C. Wade: *J. Vac. Sci. Technol.* **B7**, 1760 (1989).
164. T. T. Dao, C. A. Spence and D. W. Hess: *Proc. SPIE* **1466**, 257 (1991).
165. K.-H. Baik, L. van den hove and B. Roland: *J. Vac. Sci. Technol.* **B9**, 3399 (1991).
166. K.-H. Baik, K. Ronse, L. van den hove and B. Roland: *Proc. SPIE* **1672**, 362 (1992).
167. J. M. Shaw, M. Hatzakis, E. D. Babich, J. R. Paraszczak, D. F. Witman and K. J. Stewart: *J. Vac. Sci. Technol.* **B7**, 1709 (1989).
168. B.-J. L. Yang, J.-M. Yang and K. H. Chiong: *J. Vac, Sci. Technol.* **B7**, 1729 (1989).
169. E. Babich, J. Paraszczak, D. Witman, R. McGouey, M. Hatzakis, J. Shaw and H. Chou: *Microelectronic Eng.* **11**, 503 (1990).
170. R. A. Haring and K. J. Stewart: *J. Vac. Sci. Technol.* **B9**, 3406 (1991).
171. J. P. W. Schellekens and R.-J. Visser: *Proc. SPIE* **1086**, 220 (1989).
172. M. A. Hartney, M. Rothschild, R. R. Kunz, D. J. Ehrlich and D. C. Shaver: *J. Vac. Sci. Technol.* **B8**, 1476 (1990).
173. C. A. Spence, S. A. MacDonald and H. Schlosser: *Proc. SPIE* **1262**, 344 (1990).
174. M. A. Hartney, D. W. Johnson and A. C. Spencer: *Proc. SPIE* **1466**, 238 (1991).
175. D. R. Wheeler, S. Hutton, S. Stein, F. Baiocchi, M. Cheng and G. Taylor: *J. Vac. Sci. Technol.* **B11**, 2789 (1993).
176. T. Ohfuji and N. Aizaki: *Digest of Technical Papers - 1994 Symposium on VLSI Technology*, pp. 93-94 (1994).
177. T. G. Vachette, P. J. Paniez and M. Madore: *Microelectronic Eng.* **11**, 459 (1990)
178. M. A. Hartney, D. C. Shaver, M. I. Shepard, J. Melngailis, V. Medvedev and W. P. Robinson: *J. Vac. Sci. Technol.* **B9**, 3432 (1991).
179. R. Sezi, R. Leuschner, M. Sebald, H. Ahne, S. Birkle and H. Borndörfer: *Microelectronic Eng.* **11**, 535 (1990).

180. R. Sezi, H. Borndörfer, R. Leuschner, C. Nölscher, M. Sebald, H. Ahne and S. Birkle: *Jpn. J. Appl. Phys.* **30**, 3108 (1991).
181. M. Sebald, J. Berthold, M. Beyer, R. Leuschner, C. Nölscher, U. Scheler, R. Sezi, H. Ahne and S. Birkle:*Proc. SPIE* **1466**, 227 (1991).
182. T. Matsuo, K. Yamashita, M. Endo, M. Sasago, N. Nomura, M. Shirai and M. Tsunooka: *Digest of Technical Papers - 1994 Symposium on VLSI Technology*, pp. 91-92 (1994).
183. M. Shirai, N. Nogi, M. Tsunooka and T. Matsuo: *J. Photopolymer Sci. Technol.* **8**, 141 (1995).
184. O. Nalamasu and G. Taylor: *Proc. SPIE* **1086**, 186 (1989).
185. P. Vettiger, P. Buchmann, K. Dätwyler, G. Sasso and B. J. VanZeghbroeck: *J. Vac. Sci. Technol.* **B7**, 1756 (1989).
186. Y. Mimura: *J. Vac. Sci.Technol.* **B4**, 15 (1986).
187. K. Mitani, S. Okazaki, F. Murai and H. Shiraishi: *J. Electrochem. Soc.* **135**, 1014 (1988).
188. M. Sasago, M. Endo, Y. Hirai, K. Ogawa and T. Ishihara: *IEEE International Electron Devices Meeting (IEDM '88)*, pp.316-319 (1988).
189. T. Yoshimura, F. Murai, H. Shiraishi and S. Okazaki: *J. Vac. Sci. Technil.* **B6**, 2249 (1988).
190. Y. Okuda, T. Ohkuma, Y. Takashima, Y. Miyai and M. Inoue: *Proc. SPIE* **771**, 61 (1987).
191. H. Fukumoto, Y. Okuda, Y. Takashima, T. Okumura, S. Ueda and M. Inoue: *Proc. SPIE* **1086**, 56 (1989).
192. H. Fukumoto, Y. Okuda, Y. Takashima, T. Ohkuma, S. Ueda and M. Inoue: *J. Photopolymer Sci. Technol.* **2**, 365 (1989).
193. H. Hiraoka, W. Hinsberg, H. Clecak and A. Patlach: *J. Vac. Sci. Technol.* **B6**, 2294 (1988).
194. M. Morita, S. Imamura, T. Tamamura, O. Kogure and K. Murase: *J. Vac. Sci. Technol.* **B1**, 1171 (1983).
195. T. Matsuda, T. Ishii and K. Harada: *Appl. Phys. Lett.* **47**, 123 (1985).
196. T. Ishii, T. Matsuda and K. Hirata: *Technical Report of IEICE (The Institute of Electronics, Information and Communication Engineers)* **SSD85-106**, 23 (1985).
197. T. Ishii, S. Moriya and K. Harada: *Microelectronic Eng.* **11**, 465 (1990).
198. G. N. Taylor and T. M. Wolf: *J. Electrochem. Soc.* **127**, 2665 (1980).
199. G. N. Taylor, T. M. Wolf and J. M. Moran: *J. Vac. Sci. Technol.* **19**, 872 (1981).
200. G. N. Taylor, M. Y. Hellman, M. D. Feather and W. E. Willenbrock: *Polym. Eng. Sci.* **23**, 1029 (1983).
201. M. Tsuda, S. Oikawa, M. Yabuta and H. Nakane: *Jpn. J. Appl. Phys.* **23**, 259 (1984).

202. M. W. Geis, J. N. Randall, T. F. Deutsch, N. N. Efremow, J. P. Donnelly and J. D. Woodhouse: *J. Vac. Sci. Technol.* **B1**, 1178 (1983).
203. M. W. Geis, J. N. Randall, T. F. Deutsch, P. D. DeGraff, K. E. Krohn and L. A. Stern: *J. Vac. Sci. Technol.* **B1**, 1178 (1983).
204. J. M. Zeigler, L. A. Harrah and A. W. Johnson: *Proc. SPIE* **539**, 166 (1985).
205. A. N. Broers, W. W. Molzen, J. J. Cuomo and N. D. Wittels: *Appl. Phys. Lett.* **29**, 596 (1976).
206. T. W. Weidman and A. M. Joshi: *Appl. Phys. Lett.* **62**, 372 (1993).
207. T. W. Weidman, O. Joubert, A. M. Joshi and R. L. Kostelak: *Proc. SPIE* **2438**, 496 (1995).
208. T. W. Weidman, O. Joubert, A. M. Joshi, J. T-C Lee, D. Boulin, E. A. Chandross, R. Cirelli, F. P. Klemens, H. L. Maynard and V. M. Donnelly: *J. Photopolymer Sci. Technol.* **8**, 679 (1995).
209. C. P. Umbach, A. N. Broers, C. G. Willson, R. Koch and R. B. Laibowitz: *J. Vac. Sci. Technol.* **B6**, 319 (1988).
210. E. W. Scheckler, S. Shukuri and E. Takeda: *Jpn. J. Appl. Phys.* **32**, 327 (1993).
211. T. Yoshimura, H. Shiraishi, J. Yamamoto and S. Okazaki: *Jpn. J. Appl. Phys.* **32**, 6065 (1993).
212. M. Nagase, H. Namatsu, K. Kurihara, T. Ishiyama, Y. Takahashi, K. Murase and T. Makino: *Technical Report of IEICE (The Institute of Electronics, Information and Communication Engineers)* **SDM96-5**, 29 (1996).
213. D. R. Allee, C. P. Umbach and A. N. Broers: *J. Vac. Sci. Technol.* **B9**, 2838 (1991).
214. T. K. Whidden, J. Allgair, A. Jenkkins-Gray, M. Khoury, M. N. Kozicki and D. K. Ferry: *Jpn. J. Appl. Phys.* **34**, 4420 (1995).
215. M. Taneya, Y. Sugimoto, H. Hidaka and K. Akita: *J. Appl. Phys.* **68**, 3630 (1990).
216. H. Kawanishi, Y. Sugimoto, N. Tanaka and T. Ishikawa: *Jpn. J. Appl. Phys.* **32**, 4033 (1993).
217. Y. Ochiai, M. Baba, H. Watanabe and S. Matsui: *Jpn. J. Appl. Phys.* **30**, 3266 (1991).
218. H. Hiroshima, S. Okayama, M. Ogura, M. Komuro, H. Nakazawa, Y. Nakagawa, K. Ohi and K. Tanaka: *J. Vac. Sci. Technol.* **B13**, 2514 (1995).
219. K. Matsumoto, S. Takahashi, M. Ishii, A. Kurokawa, S. Ichimura and A. Ando: *Proceedings of the 13th Symposium on Future Electron Devices*, Tokyo, 40-43 (1994).
220. K. Matsumoto, S. Takahashi, M. Ishii, M. Hoshi, A. Kurokawa, S. Ichimura and A. Ando: *Jpn. J. Appl. Phys.* **34**, 1387 (1995).

221. K. Matsumoto: *Proceedings of the 14th Symposium on Future Electron Devices,* Tokyo, 57-60 (1995).
222. K. Matsumoto, M. Ishii, J. Shirakashi, K. Segawa, Y. Oka, B. J. Vartanian and J. S. Harris: *IEEE International Electron Devices Meeting (IEDM '95),* pp.363-366 (1995).
223. M. Ishii and K. Matsumoto: *Jpn. J. Appl. Phys.* **34**, 1329 (1995).

Appendix: Two-dimensional Diffraction

The purpose of this appendix is to provide the mathematical basis for the calculation of the intensity of diffracted light discussed in Sections 3.1 and 3.2 in Chapter 3. Here, the two-dimensional diffraction problem is rigorously solved on the basis of Huygens-Fresnel principle [1].

HUYGENS-FRESNEL PRINCIPLE

The so-called Huygens-Fresnel principle postulates that every point of a wavefront is regarded as the center of the secondary wavelet (Huygens) and the wavelets mutually interfere (Fresnel). Based upon this principle, an approximated solution of diffraction usually called "Fresnel diffraction" has been obtained. The error involved in this approximation comes from the form of the wavelet and the mathematical approximation used there. If the supposed wavelet satisfies the wave equation, and no mathematical approximation is used, the principle will lead to a correct result.

We suppose that a linearly polarized plane wave of light is propagating in free space, where we set an orthogonal coordinate system O-x,y,z with the z-axis parallel to the direction of wave propagation, and the y-axis to the electric field of the wave, as illustrated in Fig. A.1.

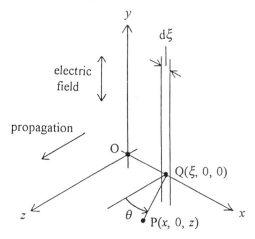

Fig. A.1. The coordinate system and definition of r and θ.

The electric field of the wave is expressed as

$$E_x = 0, \qquad (A.1)$$

$$E_y = E_0 \exp[i(kz - \omega t)], \qquad (A.2)$$

$$E_z = 0, \qquad (A.3)$$

where E_0 is the amplitude of electric oscillation, k is 2π times the wavenumber, ω the number of vibrations in 2π seconds, and t the time. We use complex representation for electric or magnetic wave as shown in eq. (A.2), where the vector component of electric or magnetic vector is regarded as the real part of an associated complex wave.

We take the x-y plane ($z = 0$) as the wavefront to which the Huygens-Fresnel principle is applied. In the present case, we exclusively deal with the two-dimensional problems which are completely independent of y. In accordance with this, we consider, as the source of secondary wavelet, an infinitely long strip with infinitesimal width $d\xi$ parallel to y-axis and passing a point $Q(x,0,0)$, as shown in Fig. A.1.

The spherical wavelets starting from all points on this strip mutually interfere to form a cylindrical wavelet. The electric field of the cylindrical wavelet is parallel to y-axis and independent of y. Therefore, we can denote this electric field by a scalar dE_y. We assume that this minute electric field at point $P(x,0,z)$ where $z \geq 0$ is given by

$$dE_y = KE_0 \frac{\cos(\theta/2)}{\sqrt{r}} \exp[i(kr - \omega t)] d\xi \qquad (A.4)$$

where K is a constant, r the distance from Q to P, and θ the angle between z-axis and the direction from Q to P taken anticlockwise (from z-axis to x-axis), as shown in Fig. A.1. This type of wave has been referred to in Born and Wolf's textbook as one of the well known solutions of two-dimensional wave equation [2]. The electric field of the wavelet expressed by eq. (A.4) is illustrated in Fig. A.2.

This wavelet is considered to be the most appropriate one in the present case by the following reasons.

First, the wavelet satisfies the two-dimensional wave equation expressed by

$$\left(\frac{\partial^2}{\partial r^2} + \frac{1}{r}\frac{\partial}{\partial r} + \frac{1}{r^2}\frac{\partial^2}{\partial \theta^2}\right)(dE_y) = -k^2(dE_y). \quad (A.5)$$

It is readily certified by differential calculation that dE_y defined by eq. (A.4) satisfies eq. (A.5).

Secondly, the wavelet satisfies the equation of continuity for the energy flow of electric wave. This is understood by noticing that the energy of oscillation, being proportional to the squared amplitude of oscillation, is proportional to $1/r$ according to eq. (A.4), while the area of normal cross-section of energy flow is proportional to r.

Finally, the wavelet is symmetrical with regard to the z-direction which is the direction of propagation of the primary wave.

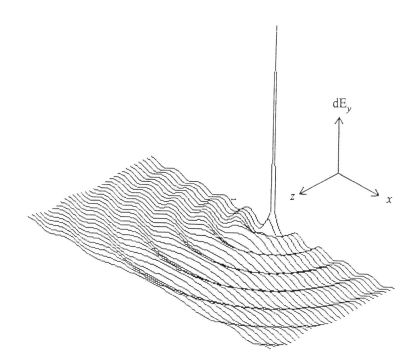

Fig. A.2. The electric oscillation of cylindrical wavelet expressed by eq. (A.4), where $t = 0$.

The Huygens-Fresnel principle applied to the present case postulates that the integration of dE_y with regard to ξ from $-\infty$ to ∞ gives the electric field of primary wave at point P, that is,

$$\int_{-\infty}^{\infty} KE_0 \frac{\cos(\theta/2)}{\sqrt{r}} \exp[i(kr-\omega t)]d\xi = E_0 \exp[i(kz-\omega t)]. \quad (A.6)$$

The integral calculation described below shows that eq. (A.6) holds if K is given by

$$K = \frac{1}{\sqrt{\lambda}} \cdot \frac{1-i}{\sqrt{2}}, \quad (A.7)$$

where λ is the wavelength of the wave.

First, we consider the indefinite integral

$$I = \int \frac{\cos(\theta/2)}{\sqrt{r}} \exp(ikr) d\xi. \quad (A.8)$$

Changing the integration variable from ξ to r, we obtain

$$I = \int \frac{\cos(\theta/2)}{\sqrt{r}} \exp(ikr) \frac{-dr}{\sin\theta}, \quad (A.9)$$

where we have used the correlation,

$$d\xi \sin\theta = -dr.$$

If $z \geq 0$, that is, $-\pi/2 \leq \theta \leq \pi/2$, then

$$\cos\frac{\theta}{2} = \frac{\sqrt{1+\cos\theta}}{\sqrt{2}}, \quad (A.10)$$

and

$$\sin\theta = \varepsilon\sqrt{1-\cos^2\theta}, \quad (A.11)$$

where ε is defined by

Appendix 307

and the sign function sgn(x) is defined as

$$\text{sgn}(x) = 1 \quad \text{for } x > 0,$$

$$\text{sgn}(0) = 0, \quad \text{and}$$

$$\text{sgn}(x) = -1 \quad \text{for } x < 0.$$

Using eqs. (A.9), (A.10) and (A.11), we obtain

$$\begin{aligned}
I &= \int \frac{\sqrt{1+\cos\theta}}{\sqrt{2}\sqrt{r}} \exp(ikr) \frac{-\varepsilon}{\sqrt{1-\cos^2\theta}} \, dr \\
&= \frac{1}{\sqrt{2}} \int \frac{-\varepsilon}{\sqrt{r(1-\cos\theta)}} \exp(ikr) \, dr \\
&= \frac{1}{\sqrt{2}} \int \frac{-\varepsilon}{\sqrt{r-z}} \exp(ikr) \, dr \\
&= \frac{\sqrt{k}}{\sqrt{2}} \int \frac{-\varepsilon \exp(ikz)\exp[ik(r-z)]}{\sqrt{k(r-z)}} \, dr.
\end{aligned} \qquad (A.13)$$

Changing the integration variable from r to φ which is defined by

$$\varphi = k(r-z), \qquad (A.14)$$

we obtain

$$I = \frac{\exp(ikz)}{\sqrt{2k}} \int \frac{-\varepsilon \exp(i\varphi)}{\sqrt{\varphi}} \, d\varphi. \qquad (A.15)$$

Secondly, we calculate the definite integral on the left-hand side of eq. (A.6). From eqs. (A.8), (A.12) and (A.15), we obtain

$$\int_{-\infty}^{\infty} \frac{\cos(\theta/2)}{\sqrt{r}} \exp(ikr) d\xi = \int_{-\infty}^{x} \frac{\cos(\theta/2)}{\sqrt{r}} \exp(ikr) d\xi$$

$$+ \int_{x}^{\infty} \frac{\cos(\theta/2)}{\sqrt{r}} \exp(ikr) d\xi = \frac{\exp(ikz)}{\sqrt{2k}} \left(\int_{\infty}^{0} \frac{-\exp(i\varphi)}{\sqrt{\varphi}} d\varphi + \int_{0}^{\infty} \frac{\exp(i\varphi)}{\sqrt{\varphi}} d\varphi \right)$$

$$= \frac{\sqrt{2}\exp(ikz)}{\sqrt{k}} \int_0^\infty \frac{\exp(i\varphi)}{\sqrt{\varphi}} d\varphi$$

$$= \frac{\sqrt{2}\exp(ikz)}{\sqrt{k}} \left(\int_0^\infty \frac{\cos\varphi}{\sqrt{\varphi}} d\varphi + i \int_0^\infty \frac{\sin\varphi}{\sqrt{\varphi}} d\varphi \right)$$

$$= \frac{\sqrt{2}\exp(ikz)}{\sqrt{k}} \left(\frac{\sqrt{\pi}}{\sqrt{2}} + i\frac{\sqrt{\pi}}{\sqrt{2}} \right)$$

$$= \sqrt{\frac{2\pi}{k}} \exp(ikz) \frac{1+i}{\sqrt{2}}$$

$$= \sqrt{\lambda} \exp(ikz) \frac{1+i}{\sqrt{2}} . \quad (A.16)$$

By using eq. (A.16), the left-hand side of eq. (A.6) is expressed as

$$\int_{-\infty}^{\infty} KE_0 \frac{\cos(\theta/2)}{\sqrt{r}} \exp[i(kr-\omega t)] d\xi = KE_0 \sqrt{\lambda} \exp[i(kz-\omega t)]\frac{1+i}{2}. \quad (A.17)$$

Therefore, if K is defined by eq. (A.7), eq. (A.6) holds. Combining eq. (A.6) with eq. (A.7), we obtain

$$\int_{-\infty}^{\infty} \frac{1}{\sqrt{\lambda}} \cdot \frac{1-i}{\sqrt{2}} \cdot E_0 \frac{\cos(\theta/2)}{\sqrt{r}} \exp[i(kr-\omega t)] d\xi = E_0 \exp[i(kz-\omega t)]. \quad (A.18)$$

Equation (A.18) is a quantitative expression of the Huygens-Fresnel principle in the present case, indicating that the primary wave represented by the right-hand side is equal to the superposition of the wavelets expressed in the left-hand side.

DIFFRACTION BY A HALF PLANE

The problem of two-dimensional diffraction by a half plane has been rigorously solved by Sommerfeld [3]. The solution has been obtained by a sophisticated method based on the knowledge of modern analysis such as the integral calculation on Riemann surface. However, the same solution is easily derived from the Huygens-Fresnel principle described above.

To deal with the diffraction by a half plane, we consider the same primary wave and the same coordinate system as described in the foregoing section. We

Appendix 309

suppose that the half part of *x-y* plane defined by $x \geq 0$ and $z = 0$ is covered with a perfectly black half-plane screen with its edge on the *y*-axis. The primary wave expressed by eqs. (A.1), (A.2) and (A.3) is diffracted by the screen. As the half part of *x-y* plane covered with the screen does not emit the wavelets supposed to be emitted in the Huygens-Fresnel principle, the electric field of diffracted wave, denoted by E_{1y}, is derived from eq. (A.18) by changing the integration range from $(-\infty, \infty)$ to $(-\infty, 0)$. The result is shown below.

$$E_{1y} = \frac{1}{\sqrt{\lambda}} \cdot \frac{1-i}{\sqrt{2}} E_0 \exp(i\omega t) \int_{-\infty}^{0} \frac{\cos(\theta/2)}{\sqrt{r}} \exp(ikr) \, d\xi. \quad (A.19)$$

The integral in the right-hand side of eq. (A.19) is transformed by using eqs. (A.15) and (A.16) as follows.

$$\int_{-\infty}^{0} \frac{\cos(\theta/2)}{\sqrt{r}} \exp(ikr) d\xi = \int_{-\infty}^{x} \frac{\cos(\theta/2)}{\sqrt{r}} \exp(ikr) d\xi + \int_{x}^{0} \frac{\cos(\theta/2)}{\sqrt{r}} \exp(ikr) d\xi$$

$$= \frac{\exp(ikz)}{\sqrt{2k}} \left(\int_{-\infty}^{0} \frac{-\exp(i\varphi)}{\sqrt{\varphi}} d\varphi + \text{sgn}(x) \int_{0}^{u} \frac{-\exp(i\varphi)}{\sqrt{\varphi}} d\varphi \right)$$

$$= \sqrt{\lambda} \exp(ikz) \frac{1+i}{\sqrt{2}} \left[\frac{1}{2} - \text{sgn}(x) \frac{1}{2\sqrt{\pi}} \cdot \frac{1-i}{\sqrt{2}} \int_{0}^{u} \frac{\exp(i\varphi)}{\sqrt{\varphi}} d\varphi \right], \quad (A.20)$$

where *u* is defined as

$$u = k(r - z) = k\left(\sqrt{x^2 + z^2} - z \right) \quad (A.21)$$

By using eq. (A.20) and defining W(*u*) as

$$W(u) = \frac{1}{2\sqrt{\pi}} \cdot \frac{1-i}{\sqrt{2}} \int_{0}^{u} \frac{\exp(i\varphi)}{\sqrt{\varphi}} \, d\varphi, \quad (A.22)$$

eq. (A.19) is transformed into

$$E_{1y} = E_0 \exp[i(kz - \omega t)] w_1, \quad (A.23)$$

where w_1 is defined by

$$w_1 = \frac{1}{2} - \text{sgn}(x) W(u). \quad (A.24)$$

The other electric vector components E_{1x} and E_{2z} are equal to zero.

The Sommerfeld's solution contains the wave reflected by the screen because the screen is supposed to be perfectly reflective. This component does not appear in the solution expressed by eq. (A.23) because the screen is supposed to be perfectly black. It is easy to derive the Sommerfeld's solution from eq. (A.23) through some steps of modification and generalization [1].

In the present case, the wave is linearly polarized, and the direction of electric field is parallel to y-axis and remains unchanged by the diffraction. This type of wave is called transverse electric wave (abbreviated to TE wave).

As the energy of oscillation is proportional to the squared amplitude of oscillation, the relative intensity of electric wave, denoted by $B_{TE}(x,z)$, at $P(x,y,z)$ normalized with that of primary wave is given by

$$B_{TE}(x,z) = |w_1|^2. \tag{A.25}$$

To deal with the diffraction of unpolarized light, we must consider another case where the electric field of the wave is perpendicular to the edge of the half plane. In this case, the associated magnetic field of the wave is parallel to the edge and expressed by

$$H_x = 0, \tag{A.26}$$

$$H_y = E_0 \exp[i(kz - \omega t)], \tag{A.27}$$

$$H_z = 0, \tag{A.28}$$

where H_x, H_y and H_z are the magnetic vector components, and the amplitude of magnetic oscillation H_0 has been replaced with that of electric oscillation because H_0 and E_0 have the same magnitude and the same dimension.

As the magnetic wave satisfies the same wave equation as that the electric wave does, the magnetic vector of the wave diffracted by the half-plane screen in the present case is readily obtained by replacing the electric vector with the magnetic vector. The result is expressed as

$$H_{1x} = 0, \tag{A.29}$$

$$H_{1y} = E_0 \exp[i(kz - \omega t)] w_1, \tag{A.30}$$

$$H_{1z} = 0, \tag{A.31}$$

Appendix

where H_{1x}, H_{1y} and H_{1z} are x-, y- and z-components of the magnetic vector of diffracted wave. As the direction of the magnetic vector remains unchanged by the diffraction, this type of wave is called transverse magnetic wave (abbreviated to TM wave).

To derive the electric vector associated with the TM wave shown above, we use one of Maxwell's equations,

$$c \text{ rot } \boldsymbol{H} = \frac{\partial \boldsymbol{E}}{\partial t}, \quad (A.32)$$

where c is the velocity of light, $\boldsymbol{H}$ and $\boldsymbol{E}$ the magnetic and electric vectors, respectively. As the time-dependent factor in $\boldsymbol{E}$ is $\exp(-i\omega t)$, the right-hand side of eq. (A.32) is equal to $-i\omega \boldsymbol{E}$. Therefore, we obtain

$$\boldsymbol{E} = \frac{c}{-i\omega} \text{rot}\boldsymbol{H} = \frac{i}{k} \text{rot}\boldsymbol{H}. \quad (A.33)$$

Using eqs. (A.29), (A.30), (A.31) and (A.33), we obtain the electric vector components of the diffracted TM wave as follows.

$$E'_{1x} = E_0 \exp[i(kz - \omega t)] w_2, \quad (A.34)$$

$$E'_{1y} = 0, \quad (A.35)$$

$$E'_{1z} = E_0 \exp[i(kz - \omega t)] w_3, \quad (A.36)$$

where w_2 and w_3 are defined as

$$w_2 = \frac{1}{2} - \text{sgn}(x) W(u) + \frac{-\text{sgn}(x)}{2\sqrt{\pi}} \cdot \frac{1+i}{\sqrt{2}} \cdot \frac{\sqrt{u}}{u+kz} \exp(iu), \quad (A.37)$$

$$w_3 = -\frac{1}{2\sqrt{\pi}} \cdot \frac{1+i}{\sqrt{2}} \cdot \frac{\sqrt{u+2kz}}{u+kz} \exp(iu). \quad (A.38)$$

The relative intensity of electric oscillation, denoted by $B_{TM}(x,z)$, normalized with that of primary wave is given by

$$B_{TM}(x,z) = |w_2|^2 + |w_3|^2. \quad (A.39)$$

In the case where the light is not polarized, the waves of light are considered to be the superposition of TE and TM waves with equal intensity. Therefore, the relative intensity of electrical oscillation, denoted by B(x,z), at P(x, y, z) normalized with that of primary waves is obtained by averaging the right-hand sides of eqs. (A.25) and (A.39). The result is shown below.

$$B(x,z) = \frac{|w_1|^2 + |w_2|^2 + |w_3|^2}{2} \quad . \tag{A.40}$$

An example of calculated result according to eq. (A.40) has been shown in Fig. 3.1 in Chapter 3.

DIFFRACTION BY A SLIT

We consider the same primary wave and the same coordinate system as described in the foregoing sections. In this section, we suppose that a perfectly black screen with a slit is placed on the x-y plane and the edges of slit are defined by $x = a$ and $x = b$ ($a < b$), as shown in Fig. A.3.

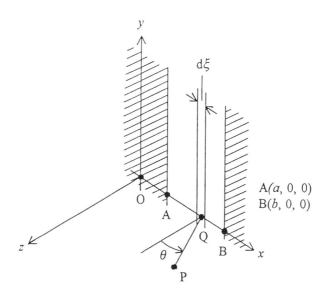

Fig. A.3. Perfectly black screen with a slit.

Appendix

The linearly polarized electric wave expressed by eqs. (A.1), (A.2) and (A.3) is diffracted by the slit. In the same way as described in the foregoing section, we can solve the problem of diffraction in the present case by changing the integration range in eq. (A.18) from $(-\infty, \infty)$ to (a, b). As a result, the electric field, denoted by E_{2y}, is expressed as

$$E_{2y} = \frac{1}{\sqrt{\lambda}} \cdot \frac{1-i}{\sqrt{2}} E_0 \exp(-i\omega t) \int_a^b \frac{\cos(\theta/2)}{\sqrt{r}} \exp(ikr) d\xi. \quad (A.41)$$

The x- and y-components of electric vector are equal to zero at any point in the space, which means that the diffracted wave is TE wave.

The right-hand side of eq. (A.41) is transformed in a similar way to that used to derive eq. (A.23) as follows.

$$\int_a^b \frac{\cos(\theta/2)}{\sqrt{r}} \exp(ikr) d\xi = \int_a^x \frac{\cos(\theta/2)}{\sqrt{r}} \exp(ikr) d\xi$$

$$+ \int_x^b \frac{\cos(\theta/2)}{\sqrt{r}} \exp(ikr) d\xi$$

$$= \frac{\text{sgn}(x-a) \exp(ikz)}{\sqrt{2k}} \int_0^{u_A} \frac{\exp(i\varphi)}{\sqrt{\varphi}} d\varphi$$

$$+ \frac{\text{sgn}(b-x) \exp(ikz)}{\sqrt{2k}} \int_0^{u_B} \frac{\exp(i\varphi)}{\sqrt{\varphi}} d\varphi, \quad (A.42)$$

where u_A and u_B are defined by

$$u_A = k\left[\sqrt{(x-a)^2 + z^2} - z\right], \quad (A.43)$$

and $$u_B = k\left[\sqrt{(x-b)^2 + z^2} - z\right]. \quad (A.44)$$

Using the function $W(u)$ defined by eq. (A.22) and defining ε_A and ε_B as

$$\varepsilon_A = \text{sgn}(x-a), \quad (A.45)$$

and $$\varepsilon_B = \text{sgn}(b-x), \quad (A.46)$$

we can transform eq. (A.42) into

$$E_{2y} = E_0 \exp[i(kz-\omega t)]w_4, \tag{A.47}$$

where w_4 is defined by

$$w_4 = \varepsilon_A W(u_A) + \varepsilon_B W(u_B). \tag{A.48}$$

It is obvious from the principle of superposition that E_{2y} given by the integration of wavelets which satisfy the wave equation (A.5) also satisfies the same equation.

Since the x- and z-components of electric vector are equal to zero, the relative intensity of electric oscillation of this TE wave, denoted by $B_{TE}(x,z)$, at $P(x,y,z)$ normalized with that of primary wave is given by

$$B_{TE}(x,z) = |w_4|^2. \tag{A.49}$$

Replacing the electric vector components with the magnetic components, we obtain an expression for the diffracted TM wave where the magnetic vector is parallel to the edges of slit. The result is expressed as

$$H_{2x} = 0, \tag{A.50}$$

$$H_{2y} = E_0 \exp[i(kz-\omega t)]w_4, \tag{A.51}$$

$$H_{2z} = 0. \tag{A.52}$$

Here, the amplitude of magnetic oscillation has been replaced with that of electric oscillation because they have the same magnitude and the same dimension. Using eqs. (A.33), (A.50), (A.51) and (A.52), we obtain the electric vector associated with this TM wave. The result is expressed as

$$E'_{2x} = E_0 \exp[i(kz-\omega t)]w_5, \tag{A.53}$$

$$E'_{2y} = 0, \tag{A.54}$$

$$E'_{2z} = E_0 \exp[i(kz-\omega t)]w_6, \tag{A.55}$$

Appendix

where w_5 and w_6 are defined as

$$w_5 = \varepsilon_A W(u_A) + \varepsilon_B W(u_B)$$
$$+ \frac{1}{2\sqrt{\pi}} \cdot \frac{1+i}{\sqrt{2}} \left[\frac{\varepsilon_A \sqrt{u_A}}{u_A + kz} \exp(iu_A) + \frac{\varepsilon_B \sqrt{u_B}}{u_B + kz} \exp(iu_B) \right] \quad \text{(A.56)}$$

and

$$w_6 = \frac{1}{2\sqrt{\pi}} \cdot \frac{1+i}{\sqrt{2}} \left[\frac{\sqrt{u_A + 2kz}}{u_A + kz} \exp(iu_A) - \frac{\sqrt{u_B + 2kz}}{u_B + kz} \exp(iu_B) \right]. \quad \text{(A.57)}$$

From eq. (A.53), (A.54) and (A.55), we obtain the relative intensity of electric oscillation, denoted by $B_{TM}(x,z)$, at $P(x,y,z)$ normalized with that of the primary wave as

$$B_{TM}(x,z) = |w_5|^2 + |w_6|^2. \quad \text{(A.58)}$$

In the case of unpolarized light, the relative intensity of electric oscillation, denoted by $B(x,z)$, is obtained by averaging $B_{TE}(x,z)$ and $B_{TM}(x,z)$. The result is expressed as

$$B(x,z) = \frac{|w_4|^2 + |w_5|^2 + |w_6|^2}{2}. \quad \text{(A.59)}$$

An example of result of calculation using eq. (A.59) has been shown in Fig. 3.2, where $B(x,z)$ is plotted against two-dimensional spatial position (x,z).

DIFFRACTION BY A BLACK STRIPE

The light diffracted by a black stripe is considered to be complementary with the light diffracted by a slit whose edges coincide with those of the stripe. Therefore, in the case where the slit shown in Fig. A.3 is replaced with a black stripe, the relative intensity of electric oscillation of unpolarized wave, denoted by $B(x,z)$, at $P(x,y,z)$ normalized with that of the primary wave is given by

$$B(x,z) = \frac{|1-w_4|^2 + |1-w_5|^2 + |w_6|^2}{2}. \tag{A.60}$$

An example of light intensity distribution calculated by using eq. (A.60) has been shown in Fig. 3.5 in Chapter 3.

In the case of TE wave, eq. (A.60) is simplified to

$$B(x,z) = |1-w_4|^2. \tag{A.61}$$

An example of light intensity distribution calculated by using eq. (A.61) has been shown in Fig. 3.16 in Chapter 3.

REFERENCES

1. S. Nonogaki, A rigorous solution of two-dimensional diffraction based on the Huygens-Fresnel principle, *Jpn. J. Appl. Phys.*, **28**, 786-790 (1989).
2. M. Born and E. Wolf, *Principles of Optics* 6th Ed. p.570, Pergamon Press (1987).
3. A. Sommerfeld, *Optics*, pp. 247-272, Academic Press (1954).

Index

Subject Index

Absorption coefficients for x-rays, 161
Acetylated poly(phenylsilsesquioxane) (APSQ), 249
Acid-catalysed reaction in chemically amplified resist systems, 101-113
    cationic polymerization, 111-112
    crosslinking and condensation, 108-112
    depolymerization, 107-108
    deprotection reaction, 101-107
    electrophilic aromatic substitution, 110
    polarity change, 112-113
    silanol condensation, 112
Acid generators in chemically amplified resist systems, 94-100
    halogen compounds, 96
    onium salts, 94-96
    sulfonic acid esters, 96-99
    sulfonyl compounds, 99-100
Acid hardening of melamine derivatives, 108-109
Adhesion enhancement of substrate, 133-136
Adhesion factor, 134-135
Aluminum fluoride, 272
APSQ, 249
ArF excimer laser, 14
ArF lithography, 24
    resists for, 115-118
Arylnitrones, 233
ASSP, 250
ASTRO, 267
Azide-cyclized rubber photoresist, 69
Azide-phenolic resin photoresists, 72-77
1-Azidopyrene, 249

Bethe energy formula, 211
Bis-acryloxybutyltetramethyldisiloxane (BSBTDS), 187
α,α'-Bisarylsulfonyldiazomethanes, 99
2,6-Bis(4'-azidobenzal)-4-methylcyclohexanone, 69
Block projection electron beam exposure system, 204-205
t-BOC group, 102-104
t-BOC-PHS, 102
BSBTDS, 187
tert-Butoxycarbonyl (t-BOC) group, 102-104

Cadmium chloride, 270
Cadmium floride, 271
Cadmium iodide, 271
Cationic polymerization, 111-112
CBR, 121-122, 224
CEL, 230-238
Cell projection electron beam exposure system, 203-204
CEM, 233-238
Chemical amplification resist systems, 92-115
       acid-catalysed reaction of, 101-113
       acid generators in, 94-100
       improvement in process stability, 114-115
Chemically amplified resists, 6, 92-115, 222-224
Chemically amplified silicone-based negative resist (CSNR), 250
Chemistry of photoresist materials, 65-123
Chlorinated poly(methylstyrene) (CPMS), 185-186
Chloromethylated polystyrene (CMS), 185-186, 218
Chrome masks, 142-143
CMS, 185-186, 218
Contact printers, 16
Contact printing, 4, 25-36, 142-143
       of black line, 29-32
       of bright line, 27-29
       of large pattern, 25-27
       of line-and-space pattern, 32-36
Contrast boosted resist (CBR), 121-122, 224
Contrast enhancement lithography (CEL), 230-238
       materials for, 233-238
       principle of, 230-232
Contrast enhancement material (CEM), 233-238

Index

Contrast of resist, 7
COP, 185, 218
Correct pattern transfer, 7-8
CPMS, 185-186
CSNR, 250
Cyclized polyisoprene, 70

DCOPA, 185-187
DCPA, 182, 185-187, 285
Deep uv hardening, 150
Deprotection reaction, 101-107
Depth of focus (DOF), 49-51, 57
DESIRE process, 275-277, 282
Development of resist, 149-150
Diaryliodonium salts, 94
5-Diazo-Meldrum's acid, 236
Diazonaphthoquinone (DNQ) compounds, 87-92
Diazonaphthoquinone-novolak posoitive photoresists, 78-92
        photochemistry of, 78-79
Diazonium salts, 233-235, 238
2,3-Dichloro-1-propyl acrylate (DCPA), 185, 285
Diffraction of light, 25-34
        by line-and-space pattern, 32-34
        by opaque half plane, 25-27
        by opaque stripe, 29-32
        by slit, 27-29
Diffraction of x-ray, 36-42
        by line-and-space pattern, 41-42
        by opaque half plane, 37-38
        by opaque stripe, 40
        by slit, 38
Dill's model for optical characteristics of positive photoresist, 89-92, 231-232
3,3'-Dimethoxy-4,4'-diazidobiphenyl, 75
Diphenyliodonium salts, 94
Diphenylsilanediol, 267
Direct fabrication of substrate, 288-290
Dissolution mechanisms of positive photoresists, 80-87
DNQ, 87-92
DOF, 49-51, 57
Dry development, 285-288

Dynamics of spin coating, 136-142

EBR-9, 216
EK-88, 185-186
Electron beam exposure systems, 200-207
Electron-beam induced polymerization, 287-288
Electron-beam induced reaction of resist materials, 213-214
Electron-beam induced vapor phase graft polymerization, 283-284
Electron beam lithography, 3, 199-225
    electron beam resists, 213-224
    exposure systems, 200-207
    resolution of, 207-209
Electron beam resists, 213-224
    negative electron beam resists, 218-224
    positive electron beam resists, 214, 222
    sensitivity of, 211-213
Electrophilic aromatic substitution, 110
EPB, 218
Epoxide compounds, 111
Epoxidized polybutadiene (EPB), 218
ESCAP, 114-115
Etching, 1-2, 5-6, 151-156
    plasma etching, 5, 152-153
    reactive ion etching (RIE), 6, 153-156
    wet etching, 1-2, 151-152
Excimer lasers, 12-15
Exposure of photoresist, 142-148
    contact printing, 142-143
    phase-shifting masks, 144-148
    projection printing, 143-144
Exposure systems for deep-uv lithography, 20-24
Exposure systems for photolithography, 11-24

FBM, 183, 216
Focus latitude enhancing optical filter, 61-63

Germanium-containing resists, 263-264
Glass pecursor resist (GPR), 258

Index

G-line, 19
GPR, 258

Halogenated polystyrene, 267
Halogen compounds as acid generators, 96
Heteropolytungstic acid, 270
Hexamethyldisilazane (HMDS), 133
History of microlithography, 3-6
HMDS, 133
HPA, 270
α-Hydroxymethylbenzoin sulfonic acid esters, 99

I-line, 19
Image-reversal resists, 99, 240, 243
Iminosulfonates, 99
3-Indenecarboxylic acid, 79
Inorganic resists, 269-272
Intensity distribution of light diffracted by:
        edge of opaque half plane, 26-27
        line-and-space pattern, 32-36
        opaque stripe, 29-32
        slit, 27-29
Intensity distribution of light in the vicinity of projected image of:
        edge of opaque half plane, 46-47
        line-and-space pattern, 52-57
        opaque stripe, 50-52
        slit, 48-50
Intensity distribution of x-ray diffracted by:
        edge of opaque half plane, 37-38
        line-and-space pattern, 41-42
        opaque stripe, 40-41
        slit, 38-39
Interactions of electrons with atoms and molecules in resist materials, 210-211
Iodinated polystyrene (IPS), 218-219, 267
IPS, 218-219, 267

Kodak Photo Resist (KPR), 3
Kodak Thin Film Resist (KTFR), 3, 69

KPR, 3
KrF excimer laser, 14
KrF excimer laser stepper, 20-23
KTFR, 3, 69

Ladder type polysiloxane, 252-255
Lambert-Beer law for light absorption, 68
Liftoff process, 241-244
Light sources used in photolithography, 11-15
Lithium aluminum fluoride, 272
Lithographic process, 1-3
LMR, 243-244

Mercury lamps, 11-13
Micralign 500, 17
Micrascan, 20-23
MRS, 72-74
MSNR, 248
Multilayer resist process, 244-268

NA, 44
Negative resists, 1
Nitrated cellulose, 286
Nitrene, 70
p-Nitrobenzyl-9,10-diethoxyanthracene-2-sulfonate, 98
2-Nitrobenzyl sulfonic acid esters, 96
Novolak, 79-87
NPR, 184, 217
Numerical aperture (NA), 44

One-to-one projection printers, 16-17
One-to-one Wyne Dyson exposure system, 19-20
Onium salts, 94
Optical pattern transfer, 25-63
    contact printing, 25-42
    projection printing, 43-46
    proximity printing, 36-42 (see also contact printing, 25-42)
    by x-ray, 36-42

Index 323

Organosilicon positive photoresist (OSPR), 257
Organotin-containing reisst (TMR), 268
OSPR, 257

PAC, 87
PAS, 254
PBS, 217
PBTMSS, 264-265
PCM, 238-241
PCMS, 185-186
P(CMS-MOTSS), 263
PEB, 148
Peroxopolytungstic acid (HPA), 270
PGMA, 218
Phase shifter grating, 34-36
Phase-shifting masks, 4-5, 34-36, 58, 60-61, 144-148
Photoablation, 286-287
Photoactive compound (PAC), 87
Photochemistry of diazonaphthoquinone (DNQ), 78-79
Photochemistry of resists, 66-69
Photolithography:
    contact printing, 25-42
    correct pattern trtansfer, 7-8
    exposure systems, 11-24
    light sources, 11-15
    photolithographic process, 1-3
    projection printing, 43-63
    resolution, 28-29, 31-36, 49-51, 55-58
Photoresist, 1, 3, 69-123
Photosensitive polysiloxane, 246-247
PHS, 72, 220-221
Plasma etching, 152-153
PLASMASK, 276
Plasma x-ray sources, 176-177
PMIPK, 286
PMMA, 182-183, 214-216
PMOTSS, 263
PMPS, 184, 217
PMSS, 253-254
Poly-N,O-acetals, 108

Poly(alkylsilane), 286-287
Poly(allylsilsesquioxane) (PAS), 254
Poly(butene-1-sulfone) (PBS), 217
Poly(*p-tert*-butoxycarbonyloxystyrene) (t-BOC-PHS), 102, 288
Poly[(*tert*-butoxycarbonyloxy)styrene-co-acetoxystyrene-co-sulfone], 121
Poly(α-chloroacrylate-co-methylstyrene) (ZEP), 216-217
Poly(4-chloromethylstyrene) (PCMS), 185
Poly(cyclohexylmethylsilane), 233
Poly(2,3-dichloro-1-propyl acrylate) (DCPA), 182, 185-187
Poly(diphenoxyphosphazene), 267-268
Poly(fluorobutyl methacrylate) (FBM), 183, 216
Poly(glycidyl methacrylate) (PGMA), 218
Poly(glycidyl methacrylate-co-ethyl acrylate) (COP), 185, 218
Poly(hexafluorobutyl methacrylate) (FBM), 183, 216
Poly(hydroxystyrene) (PHS), 72, 220-221
Poly(γ-methacryloyloxypropyltris(trimethylsiloxy)silane) (PMOTSS), 263
Poly(methylisopropenylketone) (PMIPK), 286
Poly(methyl methacrylate) (PMMA), 182-183, 214-216
Poly(2-methylpentene-1-sulfone) (PMPS), 184, 217
Poly(methylsilsesquioxane) (PMSS), 253-254
Polyphthalaldehyde (PPA), 107-108
Polysilanes, 233, 259-262
Polysiloxane, 245-258
Polysilyne, 262, 287
Poly(tetrathiafulvalene) (PSTTF), 185
Poly(trifluoroethyl-α-chloroacrylate) (EBR-9), 216
Poly(trimethylsilylpropyne), 266
Poly(*p*-vinylphenol), 72, 220-221
Poly(N-vinylpyrrolidone), 234-236
Portable comformable mask (PCM), 238-241
Positive photoresists, 1, 65, 78-92
Postbaking, 150
Post-exposure baking, 148
PPA, 107-108
Practical process in microlithography, 133-156
Projection printers, 16-24
Projection printing, 4, 43-63, 143-148
    of black line, 50-52
    of bright line, 48-50
    of large pattern, 46-47
    of line-and-space pattern, 50-59

Proximity printing, 36-42 (see also Contact printing)
P(SI-CMS), 262-263
P(SiSt-CMS), 262
PSTTF, 185-186
Pullulan, 235
PVSS, 253

Quantum yield, 68-69

Raster-scan electron beam exposure system, 201
RE-500P, 281
Reactive ion etching (RIE), 6, 153-156
Reciprocity law failure, 72
Reduction projection aligners, 17-18
REL, 282-283
Resists:
        for ArF lithography, 115-118
        definition of, 1
        electron beam resists, 213-224
        photoresists, 1, 3, 69-123
        types of, 1
        x-ray resists, 180-189
Resist coating process, 136-142
        dynamics of spin coating, 136-141
        thickness-spin speed correlation, 141-142
Resist removal, 3, 156
Resolution:
        of electron beam lithography, 207-209
        of photolithography, 28-29, 31-36, 49-52, 55-62
        of x-ray lithography, 39-42, 164-171
Resolution enhanced lithography (REL), 282-283
RIE, 6, 153-156
Rigorous solution of two-dimensional diffraction, 25-42

SCALPEL, 205-206
Scanning tunnel microscopy (STM), 289-290
SCMR, 257
Se-Ge resist, 269-270

Self-development imaging, 108, 286
Sensitivity characteristics of resist, 6-7
 threshold dose, 6
 contrast, 7
Si-CARL, 279-280
SIEL, 284
Silanol condensation, 112
Silicon-containing negative resist (SNR), 247-248
Silicone-based positive photoresist (SPP), 249
Silicon polymer containing silylether groups, 108
Si-LMR, 238
Silylated clay material resist (SCMR), 257
SNP, 249
SNR, 247-248
SOR, 174-176
Spatial coherency factor (σ), 57
Spectral irradiance of Xe-Hg lamp, 12
Spin-coatable inorganic resist, 270
Spin coating, 136-142
SPP, 249
STM, 289-290
Strontium fluoride, 271-272
Styrylpyridinium compounds, 236-237
Sulfonic acid esters, 96-99
Sulfonyl compounds, 99-100
α-Sulfonyloxyketones, 99
Superficial image emphasis lithography (SIEL), 284
Surface imaging, 273-284
Surface silylation of single layer resist, 273-280
Swelling of photoresist, 72
Synchrotron orbital radiation (SOR), 174-176

TAS, 266
Tetrahydropyranyl (THP) group, 104-105
2,3,4,4'-Tetrahydoxybenzophenone, 87
THP group, 104-105
Three dimensional silphenylene siloxane (TSPS), 256
Threshold dose, 6-7
TMR, 268
Top-CARL, 279-280

N-Tosylphthalimide, 99
Trends of lithography and resists, 66
Triarylsulfonium salts, 94
2,4,6-Tribromophenol, 223-224
Trichloromethyl-s-triazine, 96
2,4,6-Trichlorophenol, 223-224
2,4,6-Triiodophenol, 223-224
Triphenylsulfonium salts, 94
Tris(alkylsulfonyloxy)benzene, 98
1,3,5-Tris(2,3-dibromopropyl)-1,3,5-triazine-(1H,3H,5H)-trione, 96
TSPS, 256

Vapor-phase etching, 5-6
Variable shaped electron beam exposure system, 203
Vector-scan electron beam exposure system, 201

Wet etching, 151-152
Wolff rearrangement, 78, 99
WSP, 235-236

Xe-Hg lamp, 11-12
Xenon-mercury lamp, 11-12
X-ray lithography, 4, 159-193
    masks used in, 178-180
    reduction projection system, 189-192
    resist materials, 180-189
    resolution of, 39-42, 164-171
    x-ray sources, 171-177
X-ray masks, 178-180
X-ray resists, 180-189
    negative x-ray resists, 184-186
    positive x-ray resists, 182-184
X-ray sources, 171-177
    electron beam bombardment x-ray sources, 171-174
    plasma x-ray sources, 176-177
    synchrotron orbit radiation (SOR), 174-176

ZEP, 216-217

Of all the reckless, foolhardy...

Jase's condemnation of himself was total as he paced the length of the living room at the Martine villa.

From the start he had sensed that the woman wasn't being honest with him. Letting himself kiss her before he had solved the mystery was the height of stupidity. Now "The Princess and the Playboy in Secret Love Tryst" was the national breakfast-time reading.

"Secret Plot Against Playboy" was more like it, he fumed inwardly. He had fought some tangled corporate battles in his time, but they paled alongside this for deviousness. Talay's denials had sounded convincing enough, but it was too neat a scheme for him to believe she hadn't foreseen this outcome.

Valerie Parv has been a successful journalist and nonfiction writer. She began writing for Harlequin Mills & Boon in 1982. Born in Shropshire, England, she grew up in Australia and now lives with her cartoonist husband and their cat—the office manager—in Sydney, New South Wales. She is a keen futurist, a *Star Trek* enthusiast, and her interests include traveling, restoring dollhouses and entertaining friends. Writing romance novels affirms her belief in love and happy endings.

In *The Princess and the Playboy*, Valerie Parv has returned to the fictitious realm of Sapphan, which she created in one of her previous titles, *A Royal Romance*.

The Princess and the Playboy
Valerie Parv

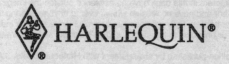

TORONTO • NEW YORK • LONDON
AMSTERDAM • PARIS • SYDNEY • HAMBURG
STOCKHOLM • ATHENS • TOKYO • MILAN • MADRID
PRAGUE • WARSAW • BUDAPEST • AUCKLAND

If you purchased this book without a cover you should be aware that this book is stolen property. It was reported as "unsold and destroyed" to the publisher, and neither the author nor the publisher has received any payment for this "stripped book."

ISBN 0-373-17416-0

THE PRINCESS AND THE PLAYBOY

First North American Publication 1999.

Copyright © 1998 by Valerie Parv.

All rights reserved. Except for use in any review, the reproduction or utilization of this work in whole or in part in any form by any electronic, mechanical or other means, now known or hereafter invented, including xerography, photocopying and recording, or in any information storage or retrieval system, is forbidden without the written permission of the publisher, Harlequin Enterprises Limited, 225 Duncan Mill Road, Don Mills, Ontario, Canada M3B 3K9.

All characters in this book have no existence outside the imagination of the author and have no relation whatsoever to anyone bearing the same name or names. They are not even distantly inspired by any individual known or unknown to the author, and all incidents are pure invention.

This edition published by arrangement with Harlequin Books S.A.

® and TM are trademarks of the publisher. Trademarks indicated with ® are registered in the United States Patent and Trademark Office, the Canadian Trade Marks Office and in other countries.

Printed in U.S.A.

CHAPTER ONE

IT WASN'T always easy being a princess, Talay Rasada thought with a sigh. There were so many rules, things you could and couldn't do, and endless protocol to be followed. 'If I were an ordinary Sapphan woman I could arrange a meeting with this Jase Clendon and tell him all the reasons his plans are totally unacceptable.'

Her friend, Allie Martine, smiled understandingly. 'But you are a princess, Talay. Your uncle is probably right. It isn't good for your public image to be seen around a man with Jase's reputation with women. What would Luc Armand think, for a start?'

Talay let her flashing eyes betray her opinion. 'Luc Armand isn't my keeper, no matter how attractive and highly suitable my uncle thinks he is.'

Allie grew serious. 'It must be rough, having so many expectations heaped on you.'

'It goes with being royal,' Talay accepted, 'although I'm so far away from the throne that I could be an old maid and nobody would notice.'

'I'd notice,' Allie said firmly. 'This isn't about being royal. It's about you and marriage, isn't it?' When Talay nodded she went on, 'When you lost your parents you lost more than anyone should have to bear, but denying yourself love for the rest of your life won't change what happened.'

'I know.'

'So why not give Luc Armand a chance, instead of exhausting yourself fighting battles you can't hope to win? From what Michael tells me, Jase Clendon does things his way so you're unlikely to sway him with emotional arguments even if you do get to meet him, which you won't.'

'You get to meet him,' Talay mocked.

Allie sighed. 'In the first place I don't belong to the royal family and, in the second, Michael has known Jase for years as they went to the same university in Australia.'

Talay was intrigued. 'But you've not actually met him yet?'

'No, and I wish I didn't have to this time. Why must he come to Sapphan now, just when Michael has arranged a second honeymoon for us?' She gave a dreamy sigh. 'Paris, imagine. I've never been there and, with Jase coming next week, it looks as if it will be a long time before I do.' She patted the soft curve of her stomach. 'Once Michael junior arrives there won't be much time for second honeymoons.'

'Paris is glorious,' Talay agreed. 'I studied jewellery design there for a year and I never tired of the museums, the galleries, the open-air cafés. We have so much French heritage in Sapphan that I felt completely at home.'

'I envy you. Instead of drinking *café au lait* at some open-air bistro next week, I'll be playing hostess here while Michael and Jase discuss business endlessly. With them both involved in the tourism industry, I probably shan't get a word in edgewise.'

'There must be a solution,' Talay said slowly, but her mind was racing a mile a minute. 'It's crazy that you're being forced to meet Jase Clendon and I can't when I'm the one who wants to. He can't possibly be allowed to open one of his three-ring-circus resorts along the Pearl

Coast. I'm all for progress, but his plan is inappropriate for one of the most beautiful and unspoiled regions in our country.'

'They aren't exactly three-ring-circuses,' Allie pointed out. 'All right, they are attractive to the mega-rich, but only because they're designed with exquisite taste. From what Michael tells me, Jase insists on the best of everything. He prefers locations like the Pearl Coast because they're remote and exotic.'

'And they appeal to the jaded tastes of people who've seen it all and done it all, like Jase Clendon himself, from what I read,' Talay stated scornfully. 'Since he announced his plans for Crystal Bay I keep seeing his picture everywhere.' Usually on the arm of some stunningly beautiful woman, she recalled. She had even read something about an ex-wife in Australia. Jase's playboy image was the main reason her ever-protective uncle, King Philippe, had thought it unwise for her to be seen in the man's company. But maybe she didn't have to be seen...

'How much do you want to go to Paris with Michael?' she asked.

Allie looked puzzled. 'I'd give anything to go, but it isn't possible.'

'Perhaps it is. You say you haven't actually met Jase Clendon face to face?'

Allie shook her head. 'He and Michael went to college together then afterwards Michael came here for a holiday, married me and never went home to Australia. Jase has been here before but, for one reason or another, our paths haven't crossed. He was supposed to attend our wedding, but became shipwrecked in mid-Pacific during an around-the-world yacht race and never made it to the ceremony. Although Michael is as disappointed as me about our second honeymoon, he likes the idea that I'll finally meet his friend.'

She wandered to a cabinet on which was displayed an assortment of framed photographs. Talay's graduation photo was among them. Allie picked up the one beside it and handed it to Talay. 'This was taken last year when Michael and Jase competed in the Sydney to Hobart yacht race.'

The photo showed the two men hauling on ropes on the deck of an ocean-going yacht and Talay felt a jolt of reaction as she looked at the man beside Michael. Allie's husband was almost six feet tall but Jase was half a head taller with a head of thick wavy hair the colour of burnt toast. Some of it fell across his forehead in a boyishly appealing look. Talay could imagine him constantly brushing it off his face but it would, no doubt, fall back again just as quickly.

Jase's hair was the only remotely boyish thing about him. In the photo he was soaked to the skin and his crew shirt was plastered over shoulders that looked as if they could carry the weight of the world without flinching. The effect was enhanced by a deeply sculpted chest and muscular arms.

He had eyes the colour of a storm-tossed sea, she also noticed. They gazed out of the photograph right into her own with a familiarity that tugged at her. Had she met him somewhere before? Or was she reacting to the sensual appeal of the man which practically leapt out of the photograph at her?

She blinked furiously to dispel the sensation. He was the enemy, the man who wanted to plunder her beloved Pearl Coast for commercial gain. How could she think of him in anything but disparaging terms? Still, it was hard to tear her eyes away from his mesmerising sea-green ones. Her throat dried as she imagined meeting him in the flesh. The thought was so overpowering that she put the photo down hastily.

'Would Michael be put out if Jase Clendon were to change his plans, arriving maybe two weeks later, so you'd have time to go to Paris and return?'

Allie's eyes narrowed. 'What are you scheming, Talay Rasada, and why do I get the feeling that what you're about to suggest is conduct unbecoming a princess?'

'Then I shan't suggest it. Let's say I have a strong feeling Jase is about to receive a message about the two-week delay.'

Allie laughed. 'I get it, you're going to pull some royal strings to delay him so he can't get here until after we've been to Paris.'

Talay hadn't thought of that but it was a good idea, and far less daring than what she actually had in mind. She smiled regally. 'What's the use of being royal if you can't occasionally use it to your advantage?' It was close enough to the truth that it didn't alert Allie's suspicions.

Her friend looked relieved. 'Sometimes it's great, having royal connections. Do you know, before I met you and we shared a room at boarding school I thought you would be stuck-up and horrible?'

'And now?'

Allie enveloped her in a hug. 'You're one of the sweetest, most caring people I know. Doesn't the king realise you're only upset about the resort plans because you care so much about this country?'

'He cares, too,' Talay said soberly, 'but he lives in the capital most of the time. And Andaman is a long way from the Pearl Coast. He's so used to going everywhere with a great entourage that he doesn't see what I see—a simple, traditional way of life which may not survive a huge influx of tourism.'

'I suppose you've pointed this out to the king?'

Talay nodded. 'Who listens to a twenty-six-year-old jewellery designer? I'm not a politician or a member of the cabinet.'

'But the king did entrust you with chairing the cultural advisory board for this province.'

Talay gave a disdainful sniff. 'A paper tiger, if you ask me. They put advisory in the title for a good reason, so we don't get to actually *do* anything but advise, and the advice isn't always listened to, as in this instance.'

'So what are you going to do?'

Talay gave a slow smile. 'You and Michael go ahead and finalise your second honeymoon plans. I'll find a way to let Mr Clendon know how I feel.' As casually as she could, she added as an apparent afterthought, 'I need somewhere quiet and private to work on some new designs. Can I use your villa while you're away?'

CHAPTER TWO

JASE CLENDON filled his lungs with the glorious, ginger-scented air that was unique to the island kingdom of Sapphan and tried to relax. It was inconvenient of Michael Martine to be called away on business at the last minute but there wasn't much either of them could do about it. The same thing had happened to Jase himself often enough.

It was strange of Michael to send a message, rather than calling direct. But it was decent of him to give Jase the run of the villa. As soon as he caught sight of the inviting pool, Jase changed into his swimming gear, intending to make the most of it. A swim was just what he needed to help him adjust to Sapphan time.

He was accustomed to luxury but this was on a scale unknown in Australia. The villa reminded Jase of a small palace, with ancient stone walls, a tropical garden studded with statuary and large, airy rooms with cool slate floors and walls panelled with aromatic eaglewood. The rattan furniture with its hand-printed silk coverings was as comfortable as it was beautiful. Michael had done well for himself, he thought, wandering around the casual living room which opened onto the pool area.

On a dresser stood a collection of family photographs, most of them meaning nothing to Jase. He considered Michael a friend but they gave each other a loose rein. Sometimes they were out of touch for a couple of years but when they got back together it was as if they'd never been apart.

His mouth twisted wryly. Male friendship was something women had trouble understanding. They wanted you *there* every minute, preferably talking or—more accurately—listening to them, or at least his former wife had. She'd never understood his need for solitude and quiet, a direct legacy of growing up in a boys' home with dozens of other children who were never quiet.

Jase shrugged off the memory and started to turn away but his attention was caught by one photo in particular. It must be Michael's wife, whom Jase had yet to meet, and it had been taken at some kind of graduation ceremony. It definitely wasn't one of Michael's photographs. For a start, unlike most of Michael's photographs, you could actually make out the subject, which meant it was Jase's first really good look at Michael's wife, Allie.

Studying her, he felt his swimming trunks growing uncomfortably tight. Not only was she gorgeous, she looked out of the picture as if she owned the world. There was something—he searched for a word—regal about her.

Her dark hair fell in a satin curtain halfway down her back. She was tall for a Sapphan woman, judging by the doorframe behind her, and she had a figure like a model, tiny of waist and full...well, full everywhere else. There was also something familiar about her that he couldn't pin down. It was probably because he'd half seen her a few times in Michael's blurred attempts at family photography.

Jase's grin was self-deprecating. Just as well she was married. Michael would laugh himself silly if he could see his friend, poring over a woman's photograph like a lovesick puppy. If he wasn't careful his reputation as a playboy would be in jeopardy and he had worked hard to create it. It served him too well to drop now.

When you were as successful and wealthy as he had made himself you were fair game for every female for miles, not to mention their fathers, mothers and ugly sisters. His one experience of marriage had convinced him he was a lone wolf, better left to hunt solo. He'd need to watch himself in Sapphan if there were many women as bewitching as Allie Martine.

If she came back early from her week-long expedition to the capital, as Michael's message had warned him she might, Jase would have to watch himself. Michael had assured him her presence wouldn't interfere with Jase's use of the villa, but it didn't solve the problem of her extraordinary effect on him.

There was another problem, too. The key Michael had sent him didn't fit the door to the guest pavilion, which he had assumed he was to use. It did fit the main house entrance so Jase had decided to move in there for the time being. If Allie came back while he was still here he would have her unlock the guest pavilion and he'd gladly move out there. Another glance at the photo in his hand warned him it might be wise to keep some distance between himself and Michael's wife.

He took another leisurely swallow of the strongly flavoured local beer Michael favoured. Jase didn't mind serving himself, but it was odd to be in such lavish surroundings without any servants. He shrugged inwardly. Maybe it was a Sapphanese custom to give the servants time off when the boss was away.

Outside his air-conditioned cocoon the air steamed. It was the end of the dry season and the humidity levels were starting to build. He finished the beer, returned the glass to the kitchen and threw open the wide doors leading to the pool and waterfall. After his reaction to Allie's photo he needed to cool off more than ever. He took a running dive into the pool.

His dive cut the water cleanly, his body knifing through the deep water like a torpedo until he surfaced on the far side of the pool, slicking his hair back and gasping for breath. This beat the heck out of cold showers.

Talay heard the sounds of someone in the pool and froze. Now the moment had arrived she was tempted to turn around and flee the house before Jase Clendon discovered her presence. He had accepted without question her message, saying that Michael would be overseas when he arrived. It wasn't exactly a lie. Allie and Michael *were* in Paris by now, enjoying their second honeymoon before their baby was due, Michael having also received a message saying Jase's arrival would be delayed for a couple of weeks.

She hadn't forged anyone's signature. She had simply 'forgotten' to append any name or signature at all. In these days of faxes and e-mail messages lots of people did. It was a sin of omission, she recognised, but she was desperate enough to try anything.

There was still time to change her mind, she assured herself as she moved cautiously towards the open French doors leading to the pool area. First she would take a look at her adversary.

He wouldn't hear her over the splashing of the waterfall, but she moved softly until she could see him without being seen. The effect was instant and electrifying. He had levered himself onto the stone rim of the pool and water streamed from muscles she had rarely seen on a male body outside the statues in her uncle's palace.

Apart from a thin band of salmon-coloured Lycra, clinging to his narrow hips, he was naked, and his Australian tan gleamed in the Sapphanese sun. Straight arms braced wide shoulders and his posture was erect, probably from his experience as a yachtsman, she guessed. His dark hair was slicked back but looked collar-length, an unusual choice for a businessman, she considered, but somehow looked right on him. Like a buccaneer from Sapphan's far past, or a modern-day pirate.

She sucked in a breath, feeling her heart race. As far as she was concerned, he was a pirate, as dangerous to her beloved coastline and its gentle people as any buccaneer in history. Still, with Jase filling her field of vision, it wasn't hard to understand how, in times past, women sometimes fell in love with pirates and ran away to sea to spend their lives with them.

Then he lifted his head and shock slammed through her so hard she had to cling to the doorframe for support. Those eyes! She had never met Jase Clendon before, yet the eyes inspecting the surface of the pool looked as familiar as her own in a mirror.

It was crazy, she told herself. Beyond the photograph Allie had shown her, she knew very little about him as a person. As far as she knew, their paths had never crossed. So why was she gripped by an unshakeable sense of familiarity, as if she had chanced across a former lover instead of a complete stranger?

She gave herself a mental shake. He was the enemy, and she had no business allowing foolish fantasies to interfere with her mission.

'It's OK, you can come and join me. I don't bite.'

Lost in a daydream of pirates and plunder, she was startled to hear his voice. It was deeply resonant with a hint of Australian accent, as tauntingly familiar as his eyes, although the source of the feeling remained equally elusive. Shock must have made her betray her presence, and panic whirled through her. She should leave now before she got herself in any deeper. She hadn't actually spoken to Jase Clendon so maybe Uncle Philippe would excuse her behaviour as female curiosity.

Of course the king hadn't specifically forbidden her to meet Jase, otherwise she would have felt duty bound to obey. He had advised against it because he considered her committed to Luc Armand. But unless she met Jase Clendon she had no hope of convincing him to change his plans. In any case, she told herself, it wasn't Her Royal Highness Princess Talay Rasada, meeting Jase Clendon, but Allie Martine, the wife of his old friend. The thought bolstered her failing courage. Gathering her flowered sarong around her, she stepped out of the shadows. 'Good afternoon. You must be Mr Clendon.'

He got to his feet and moved smoothly around the edge of the pool, coming to stand close beside her and offer his hand. 'Hello. I take it you're Michael's wife, Allie. I recognise you from your photograph.'

The touch of his fingers against her own started a chain reaction of tremors which travelled along her arm and somehow found the vein leading to her heart. Or so it felt. 'My photograph?' Even though she had put all the pictures of Allie and Michael out of sight Talay was anxious enough to try to tug her hand free, but Jase's fingers closed around hers.

He nodded. 'On the dresser inside. Some kind of graduation thing.'

To add to her pretence of being Allie, Talay had left out the picture Allie kept of her. Jase must have seen it and drawn his own interpretation. Instead of making her feel relieved, Talay was disturbed by the success of her deception. 'It was taken when I got my masters in business administration at the University of Andaman,' she said, thankful she could be honest about this at least.

'Brains as well as beauty. I'm impressed.' Very slowly he drew her hand up to his mouth, his eyes never leaving hers. When his lips brushed the backs of her fingers she felt a coil of something hot and sensual so deep inside her that it almost eluded conscious awareness. It was the most gentlemanly of greetings, perhaps even old-fashioned, but there was nothing old-fashioned about her response.

He saw the startled reaction she was unable to conceal but misinterpreted it and released her hand. 'I mustn't give you the wrong idea about me, Mrs Martine.'

'Call me Allie, please,' she invited, horrified by how shaky her voice sounded.

'And I'm Jase, Allie. No need to look so anxious. I'm sure Michael has filled you in on my...er...reputation with women, and some of it may even be deserved, but married women are strictly off limits, as Michael well knows or he wouldn't have invited me to stay here while he was away.'

'Of course.' But Jase's honeyed assurance only increased Talay's alarm. What on earth had she got herself into? She had encouraged Allie to give the servants their holidays, thinking the fewer people around who could give her away the better, but it meant she was entirely alone with Jase.

Even Sam, her devoted bodyguard, had returned home at Talay's insistence. She had told him she intended to spend the night at the Martine villa, which was true. Luckily, it hadn't occurred to Sam to check that the Martines were actually in residence. He assumed Talay was safe with their staff, as well as the villa's extensive security system, until he came to collect her the following afternoon.

She had thought that arranging the meeting would be the hardest part, but actually facing Jase himself was much more challenging than she had anticipated. Her own reaction was the problem, she acknowledged. She simply hadn't expected the magnetic power of his personality to affect her so strongly. Why hadn't anyone told her that a man could make her feel over-heated and chilled, confused and empowered, all at the same time?

'Michael's message said you were spending a few days in the capital,' Jase went on. 'You must be tired after your return journey. Why don't you join me for a swim? According to your husband, you're a real water baby who gets into the water at every opportunity.'

Allie was the true water baby. Talay also enjoyed swimming, but the thought of appearing in a swimsuit in front of Jase made her knees weaken. 'I don't think so, not today,' she dissembled.

'Then I must get dressed and join you inside. Anything else would be impolite,' he insisted.

Alarm rippled through her. With him in it, the spacious room would seem confining, the walls closer together, the ceiling lower. It was his impressive breadth and height, she accepted, as well as the sheer presence he managed to exude. It was easy to see why he was so successful in business. He radiated the same kind of easy authority as her uncle, the king.

Philippe Rasada, nicknamed the Hawk by his supporters and political adversaries alike, had the same knack of dominating a room simply by entering it. Talay forced a smile. 'In that case, I will have a swim after all,' she said around a throat gone suddenly dry. 'I don't wish to spoil your pleasure.'

His gaze lingered on her for the longest time. 'Sapphan has many pleasures. Her crystalline waters hardly compare with the attractions much closer to hand.'

He gave her no time to absorb the poetic compliment, far less frame a coherent response, before he led the way back to the pool and cut a sleek arc through the air as he dived in. She held her breath as he stayed under for a long time and only released it when he finally surfaced on the far side, treading water with powerful thrusts which he managed to make appear effortless.

Hastily she turned towards the dressing rooms, where Allie kept swimwear for her as she spent much of her free time here. She emerged, wearing a modest one-piece costume which usually felt comfortable. In indigo and white, it was a traditional Sapphan design known as 'flowing water' which showed stepped patterns representing streams, rivers and waterfalls.

With Jase's eyes on her as she walked towards the water, she was more aware of the parts the suit *didn't* cover, such as the curve of her hips, her legs—which were long for a Sapphan woman—and the way the traditional material outlined the swell of her breasts.

As a member of the royal family she should be accustomed to public scrutiny, but Jase's inspection managed to convey a far more personal interest. His appraisal was leisurely and frankly appreciative as she stepped to the water's edge. His expression seemed to say, 'If you were not a married woman...'

She dived into the water and welcomed the cool, silken feel as it closed over her. Unfortunately Jase moved while she was under water, or else she misjudged the distance, because she surfaced uncomfortably close to him. 'Michael was right—you are a real water baby,' he commented.

She smiled to hide her discomfiture. 'In Sapphan we have a natural affinity with the water. Two centuries ago many of our people earned their living as pearl divers or shell hunters.' Many were also sea-nomads and pirates but she didn't point this out. 'During the early eighteen hundreds many pearl divers from Sapphan worked along the north-west coast of Australia.'

'With the pearling luggers, based in Broome,' he confirmed. 'At first the divers were aboriginal, then they came from Sapphan and later the Japanese took over.'

'You know your history, Jase.'

He smiled wryly. 'I should. I was born in Broome. I built my first resort there.'

It was the opening she'd hoped for but she hesitated, before taking advantage of it. Something about Jase Clendon warned her he would make a formidable enemy. He would also make a formidable friend, she suspected, which was probably why Michael Martine was so loyal to him.

Everything about Jase suggested he would also make a formidable lover, but Talay pushed the thought away. She wasn't likely to find out. Nor did she want to, she added hastily to herself. They had other business and delaying it would only make it more difficult. As it was, she had only these two days in which to try to change his plans.

She side-stroked to the edge of the pool and clung to it, her feet just touching the bottom. 'How many resorts do you own?'

'Crystal Bay will be the fifth.'

'Provided something—or someone—doesn't change your mind about going ahead with it,' she said, unable to stop her tone from sharpening.

He levelled a long look at her until she wondered if

he sensed her disapproval of his plans. Before she could answer he shook his head, shedding water like a tiger having drunk at a watering hole. 'Why would they want to try, Allie? My resort is needed to give the Pearl Coast an injection of new commercial life. The place is in danger of stagnating, otherwise.'

Despite the coolness of the water, her blood felt heated. How dared he call her beloved Pearl Coast stagnant? 'Surely there's a difference between tradition and stagnation?' she demanded.

He looked startled by her vehemence. 'You sound as if the area is important to you, Allie.'

'It is. My mother was born there,' she snapped.

She realised her mistake as soon as the words escaped her mouth. He frowned. 'Michael told me your people come from the Jarim islands in the Andaman Sea.'

'*Oh, what a tangled web,*' she thought furiously. Her mother had come from the Pearl Coast. According to Sapphan law, royalty could not marry another member of the royal family so her father, the king's brother, had taken as his bride a woman from a pearl-farming community. A blue-blooded woman, true enough, with vast land holdings and pearl farming interests of her own, but still a commoner under the law.

Bitterness rose in Talay as she thought of her parents' lives cut cruelly short by a terrorist bomb attack ten years ago as they had boarded a plane for a visit to a neighbouring island. Talay, sixteen at the time, had been about to board the plane and had survived with horrific scarring to her face. Only the devotion of her grandfather, Leon, and the skills of Australian cosmetic surgeons had repaired the damage. But, however deep her gratitude towards his people, she wasn't about to let this arrogant Australian dismiss her mother's way of life as stagnant.

'My family is scattered,' she supplied diffidently. 'Many of them come from the Pearl Coast. They're a hard-working, fiercely proud people with strong ties to the province. The historical name for Crystal Bay even translates as "mother place". It is said to whisper to anyone who leaves it, the voices only ceasing when they return to stay.'

The pool was barely large enough to contain her growing anger. He didn't understand anything. Tremors shook her as she levered herself onto the stone coping and stood up. She had hoped they could discuss rationally the unspoilt beauty of Crystal Province, its historic and cultural uniqueness. Instead, she had allowed emotion to get in the way. She was as annoyed with herself as with him for letting him provoke her.

She was unaware of footsteps on the stone behind her until he took her arm and spun her around. The contact triggered a maelstrom of sensations inside her. She tried to tell herself it was because, as a member of the royal house, she was seldom touched other than by her maid and closest friends. It couldn't have anything to do with finding Jase a hair's breadth away, his arm extended towards her so every detail of his long-fingered hand burned itself into her awareness.

He had followed her out of the pool in such a hurry that water streamed from him, steaming gently in the hot air to create a misty halo around his body.

Her attention was captured by the contrasting tenderness in his gaze, and a totally unexpected warmth surged through her. Physically, he had no equal in her experience, but she sensed something more, a connecting of souls she hadn't anticipated and couldn't possibly allow with this man. Her every instinct warned her against such foolish indulgence.

The heart-stopping moment ended when he said, 'I apologise for whatever I said to offend you.'

She shook her head. 'You don't understand why I'm angry, do you?'

His mouth twisted wryly. 'No doubt you're going to tell me.'

'Pearl Coast Province is the last remnant of a way of life which has existed unchanged for thousands of years. The people are pearl farmers, shell hunters and sea-nomads, not innkeepers.'

He folded his arms across his broad chest. 'What was the population of the province ten years ago, Allie?'

'About five thousand. Why?'

'And two years ago?'

She had to think. 'Maybe three thousand.'

'And today?'

She saw what he was getting at. 'All right, I'm well aware the people are growing older and the younger ones are moving away to the cities to work.' The whispering voices of the mother place couldn't alter the fact that there was little work for young people in the province.

He nodded. 'If they had a future at home they might not be forced to leave. A Clendon Resort is not only a playground for the rich. It's also a training ground for the young, a nursery for endangered plants and animals and a monument to the past as well as the future. I'm proud of the concept, which is rare in the tourism business.'

It was hard to think rationally around the thunder of her own heartbeat. She wished they could have had this discussion in the air-conditioned living room, preferably fully dressed. While he talked her swimsuit had dried, and she was disturbingly conscious of the way it was moulded to her figure. She took refuge in annoyance. 'I hardly think token eco-tourism can compensate for what will be lost.'

Fire snapped in his eyes. 'You obviously know little about how I do things. Why don't I take you with me to the site tomorrow and show you why you're wrong?'

Given the way he made her feel, going anywhere with him was reckless. It was also impossible, without giving away her true identity. 'I can't.'

'Afraid, Allie?'

His lowered tone stole over her like a caress. Musical voices were a characteristic of her people but his defied such a mundane description. It was as deep and rich as volcanic soil. The sound vibrated through her. She *was* afraid, but not in the way he apparently thought. Visiting Crystal Bay with him could only strengthen her conviction so it must be her reaction to his company she feared.

The surreal nature of today's experience crashed over her. Today she wasn't Princess Talay Rasada, she was Allie Martine, commoner and married woman. It was alarming how readily *her* Allie entertained fantasies which were forbidden to a princess or even to a married woman. It would have to stop. 'I have other plans tomorrow—sorry.'

'A lover, perhaps?'

She stared at him in shocked surprise. 'What an extraordinary thing to say.'

He met her look levelly. 'You're an extraordinary woman, not at all the way Michael describes you. This thing between us, for instance...'

Tension gripped her. 'There's nothing between us.'

'Oh, yes, there is. We both felt it from the moment we set eyes on each other. It suggests to me that you're not as faithful to Michael as he thinks you are. Which is why I asked if you're seeing someone else.'

She drew herself up, regal hauteur in every line of her bearing. At some level she was intrigued by his willingness to confront her on his friend's account. It suggested a capacity for loyalty on an enviable scale—provided he considered you his friend. For her own sake she was furious at being so unfairly suspected. 'I can only assume you speak from your own experience. It's said we suspect others of our own misdeeds.'

'Quite possibly.' His tone was mild but his eyes burned into her. 'I don't deny my marriage was a spectacular failure, as Michael would have told you. Nor do I deny having seduced many women but they were all willing, not to mention enthusiastic, at least at the time. And they were all available.'

She recalled his vow that he considered married women off limits, and felt the merest flaring of regret. She resisted it but couldn't stop herself from asking, 'Why did your marriage fail, Jase?'

'The question should be: why did we get together in the first place? The answer is that she got pregnant—on purpose.' His expression hardened. 'Don't look so shocked. I'm sure women in Sapphan do it to snare men, too. She talked me out of using precautions, swearing she was protected, then used her pregnancy to put a noose around my neck.'

None of Allie's talk about Jase had mentioned the existence of a child, and something clenched inside Talay as she pictured him with a tiny baby cradled against the hard wall of his chest. 'Where is your child now—with the mother?'

'There's no child any more,' he said in a voice laced

with bitterness. 'The pregnancy didn't last beyond the fifth month. By then we were stuck with each other.'

It was as cynical an opinion of marriage as she had ever heard. 'With such a sad experience of marriage, no wonder you're quick to jump to conclusions about me,' she said. 'I don't know what you think you sense between us, Mr Clendon, but you're wrong. I would never cheat on the man I love.'

'Then there's no reason why you can't come with me to Crystal Bay tomorrow.'

Hooked as neatly as a fish on a line, she thought furiously. She would have to watch herself around him if she was to have any hope of winning her battle. That it might be lost already, she couldn't afford to consider. 'Very well, I'll go,' she conceded. Keeping up the fiction of being Allie Martine wouldn't be easy, but she would find a way.

Keeping up the fiction that Jase had no effect on her—now *there* was the real challenge.

CHAPTER THREE

It was easier than she had anticipated for the simple reason that Jase declined her offer to prepare a meal for them and went out for the evening. Royal or not Talay was perfectly able to cook, having been taught at boarding school. The king himself was an enthusiastic cook and had taught Talay some of his favourite recipes.

So she felt more than a little piqued when Jase announced he was attending a business dinner that evening. He seemed almost eager to escape the villa, and she couldn't shake off the feeling that she was part of the reason.

Had he somehow guessed her identity? She didn't think so and she had looked forward to the evening to provide her best chance to impress upon him the uniqueness of Crystal Bay. Now she would have to wait until he took her to visit the site to spend more time with him.

Frustration gnawed at her. As a princess, she could have requested his company at dinner and he would have felt bound by protocol to accept, no matter what other engagements pressed him. However, as Allie Martine she had no such influence.

'I'd invite you to join me but it's mainly business,' he explained.

She pretended indifference. 'Please don't concern yourself. I'm looking forward to a restful evening at home.'

'After your long journey,' he said.

Her blank look almost betrayed her until she remembered the trip she was supposed to have taken. 'It's a three-hour drive from the capital. No wonder I'm all in.'

His eyes narrowed speculatively. 'Which reminds me. I'm surprised Michael let you drive back alone. Didn't he insist you have a driver?' He looked around as if seeking evidence of one. But the longest 'journey' Talay had taken was to the Martine villa from her residence, a mere twenty minutes' drive away, where her bodyguard had returned with the car after dropping her here.

'Michael is a husband, not a keeper,' she said tartly, aware that her ill humour had a lot to do with Jase deserting her for the evening. She pushed the feeling away. 'He doesn't *let* his wife do anything. She makes up her own mind. I'm well able to drive myself wherever I wish to go.' She winced inwardly as a betrayingly regal note crept into her tone.

He didn't appear to notice because his attention was fixed on something else. 'Is it a peculiarly Sapphan custom to talk about yourself in the third person?'

'Sometimes,' she said warily. It wasn't, but it enabled her to stick to the truth as much as possible.

'I see.' He straightened his tie and the simple act drew her gaze upwards, back to the hawk-like planes of his face. Stripped off to swim, he had looked awesome. It was hard to believe he could look even more prepossessing in a maroon tuxedo with a blindingly white dress shirt which showed off his Australian tan to perfection. 'I'd better get going. It's a shame the guest villa isn't available. I don't want to wake you if I return late.'

The message was clear—*don't wait up*. She felt a quick flaring of anger but controlled it. What he did was no concern of hers, except as it affected her beloved province. 'I'm not your keeper, either, Jase. Return as late as you wish. My bedroom suite is sufficiently far from the front door that you're unlikely to wake me.' Unlikely because he had disturbed her so much she was sure she would have trouble sleeping at all tonight.

'Then I'll wish you a good evening. If Michael rings give him my thanks and best regards.'

She inclined her head. 'Of course.'

Then he was gone and the villa echoed with emptiness. Having spent many nights at her uncle's vast Pearl Palace at Andaman, she wasn't troubled by the emptiness. But she had never been so conscious of it before, as if some vital force had been drained from the rooms.

She started to pace then checked the action. She wasn't bothered by Jase's unexpected departure, only that it had robbed her of the chance to discuss his plans with him, she told herself. Nothing else explained the sensation that she would explode if she didn't move.

The feeling almost drove her back to the swimming pool, but Jase had stamped his presence on it too indelibly. It wouldn't help to be reminded of what a narrow band of Lycra could do for the male physique, in his case at least.

She resisted the vision, knowing the link between them was more than physical. Some of her more spiritual friends would say they had known each other in a previous life. She had certainly known him *somewhere* but more probably in this life. But where and when? Men like Jase Clendon were not easily forgotten. It would come to her in time.

In the meantime, she had told Allie that she wanted to use the villa to work on some new jewellery designs for her collection so that was precisely what she would do. Beyond his involvement with Crystal Bay, Jase meant nothing to her. She wouldn't even miss him this evening.

As she rounded up her drawing materials she wondered why she found herself remembering Allie's favourite English phrase about pigs flying.

Jase's fingers drummed impatiently on the armrest as his driver negotiated the busy streets of Alohan, capital of Pearl Province. Traffic here was nowhere near as bad as in Andaman but it was bad enough.

He wished fervently that he had elected to drive himself, instead of letting his associates send a limousine for him. The traffic would have served as a distraction from thoughts he had no business thinking, such as how exotically beautiful Allie Martine was. No wonder Michael had fallen headlong for her, giving up his Australian citizenship to live permanently in Sapphan. For a woman like Allie, it wouldn't be a sacrifice, Jase thought.

His stomach muscles tightened as he remembered how she had looked in a swimsuit. It was modest enough, covering far more of her body than the garments Australian women wore back home on Bondi Beach. But, in Allie's case, the sensuous fabric hinted at secrets which practically invited exploration.

Lord, it was hot in here, he thought, reaching to turn up the air conditioning in the passenger compartment. The collar of his dress shirt felt tight suddenly and he hooked a finger into it, knowing the collar had fitted perfectly well when he had left the villa.

It came to him that Allie hadn't been pleased to hear he was going out for the evening. The thought of spending the evening alone with her in the villa as the sun set and darkness gathered around them had him tugging at the collar once more. He hadn't actually planned this business dinner until he had met her but it was the only sensible option. If he had stayed with her tonight...

He slammed one fist into the other palm so hard that pain vibrated all the way to his shoulders, shattering the image before it could take form. Allie was married, for goodness' sake. She knew what could happen when a man and a woman struck sparks off each other the way they did. She should be grateful he had taken the initiative and removed himself from temptation.

Another thought occurred to him. She was married, but she wasn't wearing a ring. Odd. He tried to remember if couples exchanged rings in Sapphan. They had some unusual customs, such as declaring two people legally married as soon as they formally agreed to the union. There was no concept of an engagement, simply, 'Do you? I do.' Any ceremony came later but it was purely a formality. The marriage existed from the time they agreed to be married. So rings were probably optional. All the same, Michael was Australian-born. Surely he would have wanted to give Allie a wedding ring, even if local custom didn't demand it?

Jase frowned at his own thoughts. What business was it of his whether the Martines had exchanged rings or not? Ring or no ring, he was well aware of her status and it screamed 'hands off' at him. No trappings were needed, only a good deal of self-restraint, enough to leave him feeling shaken.

'Did you sleep well last night, Allie?' Jase enquired politely when he joined her for breakfast next morning. She had set out a traditional local repast of fresh papaya, pineapple and mango slices, croissants and an assortment of sliced cold meats. He was glad to see there was coffee. Tea was more common in Sapphan but it wouldn't help his head this morning.

She smiled but he saw a hint of censure in her eyes. 'Better than you, from the look of you.'

He massaged his forehead. 'It was a heavy night. Lots of business to discuss.'

'Naturally.'

He didn't add that his business could have been concluded at the restaurant. He had had no need to continue to a nightclub where the music had pounded at him and the drinks had been at stellar prices. He didn't normally drink to excess but last night he had needed the distraction for some reason. Unfortunately he was paying for it now.

The drink she offered him was a vile orange colour, and she persisted even after he shook his head, a shudder taking him. 'It's a local remedy for late nights and heavy business discussions,' she said, with the merest trace of sarcasm.

He took a cautious swallow then another. After the first bitter taste it was curiously refreshing. 'What is this stuff?'

'Mostly tropical juices with herbs and a dash of pepper,' she explained. 'What you would call "hair of the pup".'

'Dog,' he corrected. At her puzzled look, he added, 'It's called "hair of the dog" but this doesn't qualify. The complete phrase is "hair of the dog that bit you" so, strictly speaking, it should be alcoholic.'

She started to rise but he stayed her with a hand on her arm. 'This is fine, thanks.'

The effect of the contact was instant and electrifying. He felt it all the way to the soles of his shoes. She felt it too, from the way her pupils enlarged and she trembled ever so slightly under his hand. He hastily withdrew it and finished the juice.

'Are you still coming with me to Crystal Bay?' Even as he said it he knew he should have withdrawn the invitation, giving some excuse to go alone. Instead, he held his breath as he waited for her answer.

'I wouldn't miss it for the world,' she assured him. 'After you've shown me the site of your resort, I want to show you a Crystal Bay which outsiders seldom see.'

He felt a frown etch his brow. 'Carting me around some picturesque village won't make me change my plans, if that's what you're hoping.' It didn't take a genius to work out that she didn't favour the resort, which was strange, given her husband's involvement with tourism.

Her look was mild but her hands wove together in her lap, he noted. 'Somehow I doubt if anyone *makes* you do anything, Jase.'

She didn't exactly say, 'So who am I to try?' And she didn't bat her eyelids. But both were implied. He got a sense of performance in her behaviour today, as if she were acting a part. Probably the submissive Sapphan woman, he decided. Feminism wasn't exactly rampant here but neither were the women especially submissive. They owned property, ran businesses, held government office, exactly as they did in his own country. Maybe she had some notion of using feminine wiles to influence his plans. 'Why didn't you go with Michael to Europe?' he asked, suddenly suspicious.

She shrugged. 'He didn't ask me to.' It was the absolute truth.

'You didn't remain behind precisely so we could have the discussion we're having now?'

She drew herself up. Regal was the only word which fitted her bearing, as if she wasn't accustomed to having her word doubted. 'What are you implying, Jase?'

He folded his arms across his chest. 'You don't like the idea of a resort at Crystal Bay.' It wasn't a question. Her behaviour had already given him the answer.

'I make no secret of it,' she confirmed. 'Today I mean to show you my reasons.'

Honesty at last. He nodded slowly. 'This should prove fascinating.'

The road to the resort site at Crystal Bay was a winding dirt track, littered with fist-sized stones. Jase kept the car windows wound up against the gritty dust blowing against the glass. 'This road is the first thing I plan to upgrade,' he said through clenched teeth.

The daunting road also deterred outsiders from intruding on the villagers' way of life, but Talay kept the thought to herself, reluctant to invite another lecture about the dangers of stagnation.

Another jolt threw her sideways against Jase and she was forced to cling to him until she could lever herself upright. About the only benefit she could see in a smooth road was to save her the indignity of constantly being thrown into contact with him, she thought, feeling her face flame. In the driving mirror she glimpsed amusement dancing in his eyes. The wretched man was enjoying this.

Fortunately, he blamed the jolting ride rather than the intimacy of the contact for her discomfiture. She would die before admitting that every move, every touch between them, sent her senses haywire. She had never experienced anything remotely like his effect on her, and it took her breath away. Keeping her mind on her mission was becoming more and more of a challenge.

'I gather you're a jewellery designer,' he surprised her by saying.

Her startled look flitted to him. 'How did you know?'

'This morning I saw some sketches you left lying on the coffee-table. From the look of them, you have a lot of talent. I'm surprised Michael never mentioned it.'

Her thoughts raced. 'It's something I studied as a single woman.'

Jase nodded, his lean hands flexed around the steering-wheel as he controlled the powerful car over the tortuous road. 'And now you've decided to go back to it.' He shot her a sidelong look. 'Are you and Michael having some trouble?'

Her eyebrows lifted involuntarily. 'Why do you ask?'

'I can add up. He didn't ask you to go with him to Europe. You're reviving an interest in a former career. And you obviously don't see eye to eye on the resort plans.' Without warning he stopped the car and turned to her. 'What are you playing at?'

Confusion ripped through her. His instincts had warned him she wasn't being honest with him but he had reached a totally unexpected—and wrong—conclusion. 'What do you mean?' she hedged.

'Michael didn't sign the message inviting me to use the villa in his absence. Did you send it, Allie?'

'What makes you think—?'

He seized her wrist and turned her to face him. 'First things first. Did you send the message?'

White-hot anger seared her veins. 'You forget with whom you're dealing.' In dismay she realised her imperious tone belonged more to Princess Talay than Allie Martine.

His hold didn't slacken but thankfully he misunderstood the source of her rage. 'It's a bit late to remind me of your married status now, isn't it? You should have remembered it before arranging for us to be alone.'

'I have no idea what you mean.' In truth, she didn't.

His hard gaze bored into her. 'Don't you? You may think you know what kind of man I am—but think again. In spite of my reputation, I have no interest in providing a fling for a woman whose marriage has gone stale.'

Horror gripped Talay. She had never dreamed he would misinterpret her actions so completely. She couldn't let him believe Allie would do such a thing. 'You're wrong,' she stated emphatically. 'There's nothing amiss with the marriage.'

His eyebrows climbed. '*The* marriage, Allie? It's a peculiar way to describe a love match, surely? What about, "I still love Michael"? Say it and I'll admit I'm out of line.'

The silence in the car became deafening. Lies argued against Talay's nature. Already she wished with all her heart that she had never pretended to be Allie. She couldn't bring herself to compound her crimes by telling Jase she loved Michael.

An impatient breath whistled past Jase's lips. 'I rest my case. So there's only one thing left to prove.'

Something in his tone set her senses on full alert. 'What do you—?'

Before she could finish the question he slid an arm around her neck and pulled her towards him, the suddenness of the movement driving the air out of her lungs. His mouth crashed down on hers and she was enveloped in a sensation like drowning.

It was drowning of the most sensuous sort and the protest she tried to make forced her lips apart, exactly the way he wanted them she found out when his tongue sought hers in a sinuous dance. She had been kissed before, but never so compellingly that she could hardly think.

Then, somewhere between her attempted cry of protest and his invasion of her mouth, something changed. All the pent-up emotion of the last few hours forced its way into her response until she found herself returning his kiss with all the passion in her soul.

Her arms came up and wound around his neck, her fingers threading through his wonderful long hair which felt like silk. The firmness of his scalp was another source of sensory wonder and she explored it with fingertips as sensitised as a surgeon's. She felt hungry for something beyond food, thirsty for something beyond water. Blackness fringed the edges of her vision and she wondered fleetingly if you could pass out from an overload of sensual pleasure.

She never found out. As abruptly as he had begun the kiss, he ended it by drawing away from her, imposing a yawning chasm of space between them as he folded his arms and stared grimly out of the window.

As the sensual heat subsided Talay's muscles ached, as if from a mile-long run. 'What is it?' she asked. Why had he stopped before they had barely begun?

He heard the question, without her having to say it aloud. 'There's no need to go on. I've proved my point. If you want more you'll have to find some other man to provide it. Michael is my friend.'

Fighting a crushing sense of disappointment, she decided to tell him the truth. She couldn't let him go on thinking Allie would ever seek an affair outside her marriage. Not only were she and Michael blissfully happy, they were planning for the arrival of their first child at this very moment.

'This isn't what you think,' she began. 'I'm not—'

'Save it, I'm not interested in a litany of Michael's shortcomings as a husband,' he cut across her savagely.

In spite of her turmoil, Talay was mesmerised by the way his breathing came and went, came and went, as if he, too, had been affected by the kiss. But his voice was steady as he said, 'If I had a choice I'd turn this car around and take you home right now, but my foreman expects me at Crystal Bay in a few minutes so I have to show up. Once we get there I'll have someone else drive you back. Until then I don't want to hear another word out of you.'

As a princess she should have found the injunction shocking. No one spoke to her in such a demeaning way. But in her present guise she understood and even admired his loyalty to his friend. But he had to let her tell him the truth. 'Please let me—'

With the swiftness of a king cobra, he moved to clamp a hand over her mouth, silencing her. 'Not a word, understand?'

Over the warm pressure of his fingers she saw the determined glint firing his eyes, and she had no choice but to nod. His palm tasted salty against her lips which felt swollen from his kiss. It came to her that he would be stunned when he found out how he'd treated a member of the royal family. It was a pity Sapphan no longer imposed the death penalty for lese-majesty, she raged inwardly. She would take great pleasure in making Jase pay for his callous treatment. At her nod he slowly removed his hand.

Trying to explain would only invite more punishment so she sat in mutinous silence while he restarted the car and drove the remaining distance to Crystal Bay. When they arrived she was too stunned by the scene which met her eyes to say anything.

He noticed her stillness. 'It looks worse than it is,' he assured her calmly. 'We have a complete reafforestation plan in place to ensure that every tree removed and more are put back before we're done. I have thousands of baobabs, coconut palms, tamarinds, frangipani and flame trees on standby for this area alone.'

'Big of you,' she muttered. She was still smarting from the physical way he had ensured her silence.

He gave her a searching look. 'Sulking, Allie? Or simply annoyed because I wasn't taken in by your devious scheme?'

She forced herself to meet his eyes unflinchingly, wishing fervently that looks could kill. 'It must be a terrible burden, always having to be right.'

'Then you admit I am right about you?'

'I admit nothing of the sort.' During her enforced silence she had decided he would find out soon enough whom he had mistreated. The longer it took the sweeter would be her revenge when it came. In the meantime, since he wasn't prepared to listen to her explanation, she would maintain a dignified aloofness.

'I'll say this for you, you don't lack courage.' His tone was grudgingly admiring. 'And you're sufficiently beautiful and sexy to make me wish you weren't married. But you are, and to a man I like and respect too much to indulge you. I'll arrange for someone to drive you home before I deal with my business here.'

'Afraid, Jase?' She deliberately used his own words against him.

'Afraid of what? You? I thought I demonstrated my resistance to your charms rather effectively on the way here.'

At the reminder of his forceful kiss her insides clenched but she managed to remain outwardly calm, blessing years of royal training which enabled her to disguise her inner turmoil. 'I was referring to your obvious fear of learning anything that might not fit your preconceptions about this place,' she said, pleased that her voice hardly shook at all.

He gave a grunt of annoyance. 'You're still determined to show me the error of my ways.'

'You did a good job of pointing out mine. I only wish to return the favour.'

His eyes glinted ferally. '*Touché*. Very well, I'll deal with my business here then you can show me around. But if your behaviour is less than exemplary I'll return you home so quickly your head will spin.'

It was already spinning but she bit her tongue. She was determined not to give him the satisfaction of revealing how strongly he affected her. Even if he knew he would probably assume it was a new ploy on Allie's part to seduce her husband's friend.

Silently Talay asked for Allie's forgiveness. Now she had come this far she had to continue playing the part long enough to convince Jase to change his plans. Afterwards she would take great pleasure in setting the record straight for Allie and herself while she watched Jase choke on his ugly suspicions.

CHAPTER FOUR

JASE had chosen his location well, Talay admitted reluctantly to herself. The resort rested on nearly eighteen hectares of swaying palms and tropical foliage against a backdrop of lush green rainforest.

A white sandy beach stretched around two sides of the area and curved into deep, secluded bays. A small coral reef at one end of the crescent would provide snorkelling in Sapphan's crystal-clear waters, Jase explained. From the shore it was possible to watch shoals of brilliantly coloured fish, playing among the coral gardens almost at her feet.

The buildings would blend with the lush jungle, waterfalls and mist-covered mountain peaks, Jase assured her. He borrowed plans from the foreman overseeing the site preparation to show Talay sketches of how the finished resort would look. Built from traditional materials, using timeless Sapphan carpentry techniques, it could easily be mistaken for the abode of an island chieftain from her country's history, she concluded, impressed almost against her will.

The resort would comprise several low-lying main buildings and a dozen thatched cottages, known as bures, facing the tropical sunsets. She noticed Jase took special care to describe the honeymoon bure to her. 'It will have a hand-carved four-poster bed curtained with mosquito netting, an outdoor shower for two,' he emphasised, 'and a private spa and sundeck where the newlyweds can entertain themselves, without having to set foot in the main complex. Tropical fruit and champagne will be provided and even a yacht to take them to the resort's private island, complete with gourmet provisions, if they wish to spend a night entirely alone.'

A wave of sensual heat swept through her as his words painted a vivid picture in her mind. No need for clothing on an uninhabited island. No need for anything except the company of the man you loved and endless hours to enjoy each other, free of all commitment and restriction. Her throat felt dry but she refused to swallow.

'I see the notion appeals to you,' Jase drawled, shattering the fantasy. 'Maybe you can convince Michael to reserve it for the two of you so you can direct those passions of yours where they belong.'

'If you've finished lecturing me it's my turn to show you around,' she snapped, uncomfortably aware of how easily she could imagine sharing the private island paradise with Jase himself. Having tasted the heady pleasure of his kiss, she trembled at the thought of what other sensual treats might await his partner in such a place. It would, indeed, be paradise on earth.

She froze in horror at her own thoughts. He was the last man with whom she should dream of spending nights in paradise, or anywhere else for that matter. He had as good as admitted that his playboy reputation was deserved. His refusal to seduce his friend's wife, as he believed she wanted, was honourable but hardly redeeming, given the number of other fish in the sea. And he was still the enemy. Nothing he had shown her today had altered her opinion of him as an interloper here.

She was so lost in thought that she didn't notice the foreman approaching them to consult the plans spread out on the car bonnet in front of Jase. The man froze as he caught sight of Talay, and immediately brought his palms together at chest height in a gesture of respect. 'Your Royal Highness, forgive the intrusion. I didn't recognise you from a distance.'

Fortunately he addressed her in Sapphanese and she answered in kind, assuring him she was not seeking special treatment but was here as Jase's guest to inspect the site. 'Go about your work and take no notice of me,' she urged with a smile. She had deliberately dressed in a western-style white shirt and slim-fitting cream linen trousers to avoid being recognised too easily. So far it had worked, but the foreman was spreading the word, she saw, as the distant workers began to look their way.

A deep V of interest furrowed Jase's brow. 'Your presence seems to be causing a stir.'

'I'm patron of some local charities so I'm well known around here,' she answered truthfully.

She wasn't sure if he accepted the explanation or not but he said no more about it as he concluded his business at the site. Unfortunately, the problem was even worse at their next port of call mere minutes away, the village she wanted very much to show him. There was nothing she could do about it except be glad few of the villagers spoke English and when they used her title it was in Sapphanese.

The village comprised a collection of bamboo and thatch buildings clustered around an arc of white sand where turtles came to lay their eggs between November and February. Behind the village was a forest of casuarina trees. Overhead the palm fronds waved and the rest of the world could have been on the moon.

Sea-nomads, shell hunters and pearl divers had lived here for centuries. At night they strapped battery-operated lamps to their foreheads and walked in the tidal shallows to where jutting rocks hissed and popped as they dried. Wielding hooks of bent iron, they pried up the rocks and tipped them over to expose slimy shells which, when cleaned, were breathtakingly beautiful.

A thatch-roofed cottage served as a trading post for the shells. In their raw state they wore thick rubbery coats which washed away to reveal key scallops, nautilus shells, cowries, olives, cones and the delicate, spine-tipped Venus's combs.

There were pearls, too, not the perfect farmed variety but the bizarre baroque shapes created by wild oysters in the open sea. Jase picked up a specimen which was amazingly heart-shaped. Its rainbow colours glistened in the sunlight spilling through the cottage door. 'Ask the trader how much she wants for this pearl,' he told Talay.

The woman, having recognised Talay, wanted to press the gem on her as a gift, and it took a lot of gentle persuasion to convince her to name a price, which was still ridiculously low. Jase paid for the pearl with a large note and walked away before he could be given change. Too late Talay remembered that Jase came from Broome and was bound to know the pearl was a bargain.

'Either these people are dangerously naïve or unusually generous,' he commented outside the trader's hut.

'Is generosity a crime in Australia?' She evaded the issue.

He ignored it and his searching gaze swept the area. 'No building rises higher than the palm trees.'

She welcomed the change of subject. 'It's their idea of a planning code.'

He glanced at the palm-leaf-wrapped package in his hand. 'How do they live? Educate their children?'

'There's a snake farm nearby where they milk poisonous snakes of their venom to make snake-bite serum,' she explained. 'The women also make silk on traditional hand looms. In spite of the rustic appearance, this village is prosperous and its members happy and healthy in their isolation.'

His eyes narrowed. 'In other words, hands off. Point taken. Where can we get lunch around here?'

Her tension escalated rapidly. There were several thatch-roofed cafés where they could eat and her presence would be considered an honour. Which was the problem. She had managed to explain away the reaction to her so far, but over a meal it would become obvious that she was more than a respected charity-worker.

'I thought you couldn't wait to dump me back at home,' she reminded him sharply, not liking to bring up his earlier suspicions of 'Allie', but seeing no other option.

'Since I made my position clear you've been on your best behaviour. There's no need to let you starve.' His searching appraisal took in her slender waist and hips, their narrowness emphasised by the cut of the pants. 'There's nothing of you as it is. I'd hate to have you fainting from hunger on me.'

'Sapphan women are naturally slender. We eat like horses,' she snapped back, as his disturbingly slow appraisal sent waves of warmth flooding through her, try as she might to prevent it.

'All the more reason to take you to lunch now. This one looks good.' He indicated an unpretentious little restaurant on a bluff with a superb view of Crystal Bay.

Talay's heart sank. She had hoped he would ask for her recommendation. She would never have chosen this café, where she was well known. The restaurant was famous for its local lobsters, marinated duckling and seafood steamed in a crab shell, and was also a favourite of her uncle, the king, when he visited the province on his way to the royal retreat at Chalong.

Jase saw her hesitate. 'Isn't the food any good here?'

It wasn't fair to the owner to deny it. 'It's outstanding,' she admitted, with a heavy sigh.

Inside things were as difficult as she had feared. They were shown to the best table with a spectacular view of the ocean, the owner fussing over her like a prodigal daughter. Luckily most of the fussing was in Sapphanese, but she was glad when they were left alone to study the enormous menus laboriously hand-lettered in Sapphanese script.

Jase's mouth curved wryly when he saw the menu. 'Translations, please.'

She was achingly aware of his eyes on her as he leaned forward, waiting for her to interpret the menu. Their eyes met over the top of the unwieldy parchment. 'Most of the dishes are seafood,' she began, but stopped, the intensity of his gaze burning into her. 'What's the matter?'

'I think you know—*Your Royal Highness*.'

Panic flared through her and she half rose from the table, but his fingers curled around her wrist, easing her down again gently but irresistibly. She glared at his hand but he took his time about removing it. She rubbed her wrist ostentatiously, although the pressure had been calculated not to do damage. 'How did you find out? From the owner's behaviour?'

'From yours. To paraphrase an old Australian idiom, you can take the princess out of the palace but you can't take the palace out of the princess.'

'How long have you known?'

'Since we arrived at the resort site and the foreman fell all over himself to please you. You acted as if it was perfectly normal, which it probably is for you. It happened again in this village.' He glanced around at the café owner and his wife, watching them from the entrance, their expressions anxious. 'Look at those two. Every time you crook a finger they rush to do your royal bidding.'

She felt a flush rush up her neck and face. 'It's more than you do.'

Sensation speared through her as his index finger traced a lazy line around the inside of her wrist. It took an effort not to pull her hand away, as if burned. 'Ah, but my companion is Allie Martine, not Princess... Talay, isn't it?'

'You even know my name.' Her voice sounded infuriatingly hoarse. Now the charade was over why didn't he bundle her off home and be done with it?

'It isn't surprising since we've met before, a long time ago.'

Shock jolted through her. Maybe now she would find out why his eyes haunted her so. His face wasn't familiar but she recognised his voice and the searing look in his sea-green gaze, although she still didn't know how or why.

'We met when you were in hospital in Sydney, Australia, ten years ago,' he went on. 'I came on a goodwill visit as part of the Australian Olympic yachting crew.' He picked up the menu and held it across the lower half of her face so that only her eyes glared at him over it. 'This is what reminded me of Her Royal Highness, Princess Talay Rasada. The rest of your appearance may have changed but I'll never forget those eyes, glaring at me as if I were the lowest form of animal life.'

Memory clicked into place and with it an anger

stronger than anything she'd felt for a long, long time. '*You're* Jason Carter?'

He nodded. 'It's officially Clendon now. I competed in the Olympics under my foster family's name, but even then I was in the process of tracing my birth parents. When I did, I reverted to my real name of Jason Clendon, Jase to my friends.'

'You wore a surgical mask when you came to the hospital.' It explained why his eyes were the only part of him she recognised, and only then because his visit was branded so vividly and humiliatingly into her memory. *That* was why she'd glared at him—not out of arrogance, as he evidently believed.

'We all had to wear them, even the press photographer, to avoid passing on germs to people like yourself who were recuperating from serious burns and vulnerable to infection.'

Memories flooded through her, as distressing as they were vivid. Ten years or a hundred would never dim them. She had been sixteen at the time, fluent in English, fortunately, but still almost alone in a foreign country and struggling to come to terms with her parents' death and her own horrific injuries.

The Australian doctors had been kindness itself, performing wonders to restore her damaged features through skin grafts and other miracles of medicine. But when Jase had seen her her face had been a mass of scars and her hair a wispy mess. No wonder he hadn't been able to recognise her as she was now. Jase couldn't know it, but he had provided a turning point in her recovery.

She would never forget his visit to the ward. As royalty, she had been allocated a private suite of rooms but her doctors, concerned that the isolation was impeding her progress, ordered that she spend some time each day among the other teenaged patients. She hadn't wanted to mingle, not out of disdain, as many of them had thought, but because she was ashamed of her damaged appearance.

She could hardly bear to look at her reflection in a mirror, seeing the surgical scars, the discoloured tissue and her once-beautiful hair growing back so slowly it made her want to weep. How could anyone not look at her with horror and revulsion?

When Jase Clendon paid his fateful fund-raising visit to the ward with the other gold medallists, fresh from their triumphs in the Olympic Games, the other patients were ecstatic over meeting their heroes, but Talay, the foreigner, felt shy among the boisterous strangers. They were all so big and muscular, with that peculiarly Australian ruggedness that made them look more handsome than movie stars.

She would never forget her first sight of Jase Clendon. He was one of the youngest medal winners and moved with easy confidence and charm among the patients, posing to be photographed by the trailing press entourage, the surgical mask lending him an almost operatic air of glamour.

The photographer preferred people with obvious injuries, she recalled. He passed up one young woman with a seriously burned back because her scars weren't on public show in favour of another girl in a wheelchair. It was cynical but a fact of public life, Talay noted. It would have been no different in Sapphan.

Talay had taken no interest in the visitors, and had continued reading to a tiny child who had pulled a saucepan of boiling water over herself. The little girl belonged in the children's ward but had attached herself to Talay and spent long hours in her ward instead. The nurses encouraged it because they said it helped both of them to heal. In turn, Talay felt most alive when she was helping the little one's recovery.

Because of the visitors, the nurses decided to take Talay's tiny friend back to her own ward so Talay was alone when Jase reached her chair near the window. The photographer was still talking to the girl in the wheelchair. Jase looked down at Talay from his impressive height and smiled but the strain in it wasn't lost on her. 'Hello. Who are you?'

'Talay Rasada,' she answered simply. He obviously had no idea of her true status and she wasn't sure it would have impressed him. He looked distressed at the sight of her, and pain shafted through her. Why wouldn't he be uncomfortable with someone as damaged as she was?

Something twisted inside her. She could only see his eyes and the shock of hair above the surgical mask, but she found herself longing for him to stay and talk to her. If he held her hand, hers would be lost in his powerful grip, she thought, as a thousand nameless longings washed through her. Around him she felt less child than woman and wanted desperately for him to treat her as one.

He seemed to be searching for something to say. 'You're not Australian?'

Her injuries impeded her shake of the head. 'I come from Sapphan.'

To her surprise, he nodded. 'In the Andaman Sea close to the Thai peninsula.'

'You've been there?'

'Once or twice, racing yachts.' He moved from one

foot to another, as if searching for something more to say.

Involuntarily she reached out and their fingers brushed. The electric sensation shook her, the new and wondrous feeling making her ache for closer contact. He jerked his hand away as if stung. 'I'd better get back to the others.'

But the photographer had caught up and urged Jase to pose with the princess for the camera. Jase shook his head and she heard his aside about, 'not a good idea', although it was plainly not meant for her ears. 'I hope you get well soon,' he said to her as he steered the photographer away.

Next day the nurses brought around extra copies of the newspapers, carrying the story of the hospital visit. Jase was pictured on the front page with his head close to a blonde teenager from another ward, her only disfigurement an arm and shoulder in plaster. The caption identified her as a Miss Australia entrant who had been injured in a riding accident.

Talay understood what had happened. Jase was barely able to look at her, far less pose with her for a photograph. He wasn't publicity-shy, as she'd tried to tell herself, at least not around the vivacious beauty queen. Talay's scars had repelled him too much for him to want to pose with her. The realisation had hurt almost as much as the pain of her injuries.

She had Jase to thank for one thing. His reaction had provided her with a powerful incentive to get well, both physically and emotionally.

Soon after Jase's visit, Talay was introduced to a woman who saved her sanity and was to become queen of her country as well. At the memory, a tiny smile hovered around Talay's mouth. Little had she known that dear Norah Kelsey, as she was then, would not only use her skill as a beauty therapist to restore Talay's shattered self-esteem but would accompany Talay back to Sapphan and fall in love with Uncle Philippe.

Norah's romance was the greatest good to come out of her stay in Australia. Dear Norah assured her that all the teenagers in the unit understood her suffering. In fact, she said, most teenagers experienced a crisis of self-confidence at her age, whether injured or not.

As soon as Talay was sufficiently healed Norah spent hours teaching her how to use special make-up to conceal blemishes, making the most of her undamaged features. Slowly, slowly, Talay emerged from her shell to smile with the other teenagers, and convince herself that her world hadn't ended with the terrorist bomb that killed her parents.

More affected by Jase's reaction to her than she shared with anyone, Talay threw herself into her therapy and Norah's beauty lessons with a ferocity that astonished her doctors. As a result, she had returned to Sapphan much sooner than anyone expected. No one but Talay herself suspected that some of the healing was no more than skin deep.

It was all in the past now, she decided, dragging her mind back to the present. She should thank Jase for doing her a favour, however inadvertently. He had shocked her into getting well, and she had set herself to become a confident, assured woman who now had hardly a trace of scarring, thanks to the surgeons and Norah, the woman who was now her beloved aunt and queen.

Yet she didn't feel thankful as she sat opposite him in the rustic café. She felt furiously angry, as much with herself for still feeling the tug of attraction between them as at him for his treatment of her today and ten years ago.

'I suppose this is some kind of game with you, letting me think Allie was on the make,' he fumed, sounding as angry as she felt. 'I gather there's nothing wrong between her and Michael?'

She shook her head, willing herself to calmness. 'They're enjoying a second honeymoon in Paris before their first child is born. Posing as Allie was the only way I could arrange to meet you.'

His mouth tightened into a mocking slash. 'You have the rank to command my attendance at a meeting. Why didn't you use it?'

She looked down at the table. 'My rank, as you call it, is precisely why it wasn't considered appropriate for us to meet.'

His cynical chuckle of understanding brought her head up. 'Well, well. My reputation precedes me even here. What were your advisers afraid of? That I'd ravish their precious princess?'

Recalling how close he'd come to doing exactly that this morning, she felt her colour heighten but she willed it to subside. 'My uncle, King Philippe, has chosen a suitable husband for me, Luc Armand. He doesn't want me to be...distracted.'

Jase's eyes sparked fire at her. 'And were you distracted?'

More than she was willing to admit. She took refuge in annoyance. 'It would take more than a few kisses, especially from a man I dislike as much as you, Mr Clendon. Socialising with you is purely in the line of duty.'

'It was Jase before,' he corrected quietly. 'And, forgive me, Your Highness, but I don't believe you.'

He managed to make her title sound almost insulting. 'You don't believe I would do my duty?'

'I don't believe kissing me *was* a duty. You may have hoped it would manipulate me into seeing things your way, but your response held more passion than politics. Would you like me to kiss you again to demonstrate?'

The heat pouring through her was like a physical warning of danger. To her chagrin, she wasn't so much shocked by his offer as aroused by it when it was the last thing she should be. His last kiss had been bestowed on the woman he thought was Allie Martine, to show her the folly of seeking the affair he thought she wanted, but it was Talay Rasada who had responded. It was Talay whose pulses raced and whose heart hammered against her ribs now.

It was an effort to speak normally. She picked up her handbag. 'There's no need. With the experience credited to you with women, you could probably create any effect you chose. But it would be nothing more than a physical response which wouldn't touch my emotions in the slightest.'

His lazy look rolled over her and panic set in. Had she gone too far? She had practically dared him to try to wring from her the emotional response she denied was possible. Then he stood up and moved her chair back for her, his expression shuttered. 'Permit me to drive you home, Your Highness.'

Disappointment coiled through her. She didn't want him to meet her challenge, yet she hadn't expected him to concede so easily, either. What *did* she want from him? She realised it hardly bore thinking about. Another thought occurred to her. They hadn't discussed his plans for the resort or his feelings about the village. She hated to think this whole charade had been in vain.

It seemed likely because he said nothing more to her as he paid their bill, over the protestations of the owners, and escorted her from the restaurant. On the wooden steps she almost slipped and Jase's arm came around her, supporting her. Before he could remove it they were blinded by a flash, which she took to be lightning until she realised it was a photographer's flashbulb.

She recognised the photographer as one of the virulent paparazzi who loved to stalk celebrities in hopes of obtaining sensational pictures. 'Oh, no,' she gasped, as the man gave her a cheeky thumbs-up sign, and disappeared around a nearby building. Moments later she heard the sound of a car receding into the distance.

Jase looked puzzled. 'What was that all about?'

She explained who the man was. 'He must have gone looking for you at the resort site and been told where you were.'

'And who I was with,' Jase added grimly. 'It seems your secret assignation won't be a secret for much longer.'

'It's worse than that,' she said miserably, picturing the scene as the photographer had captured it. 'He'll make it look as if we were arm-in-arm like lovers and it will be all over the newspapers tomorrow.'

'Being photographed with you is not a problem for me,' Jase stated. 'However, I assume it will cause trouble for you with the palace.'

'But it is a problem for you,' she pointed out. 'Since the king has nominated Luc Armand as my future husband, under our laws, you are the one who will have to answer to Luc.'

CHAPTER FIVE

So MANY people in the village wanted to greet Talay that several minutes passed before Jase could retrieve the car and give chase. By then the paparazzo was nowhere in sight.

Talay expected to be black and blue from jolting along the rough road at high speed while Jase tried to catch the photographer, but the man had too great a head start. 'So the picture of you and me is likely to be front-page news tomorrow. We simply ride it out,' Jase said, slowing to a more suitable speed for the terrain and pointing the car towards Alohan.

She shook her head. 'I wish it were that simple.'

His swiftly assessing gaze raked her. 'What did you mean—that I would have to answer to Luc Armand? Who is he?'

'The man my uncle, King Philippe, has chosen to be my future husband,' she explained.

'Do you love this Luc Armand?'

'We've known each other since childhood.'

'But do you love him?'

She held herself stiffly against the washing-machine motion of the car. 'The question really doesn't arise.'

She had been read like a book, she saw from his quick nod. 'You don't love him but you'll marry him because the king decrees it.'

'Not decrees, exactly. I could refuse but it would be foolish, given my uncle's greater wisdom in these matters. His own bride was chosen for him by my grandfather, as is customary in Sapphan.'

In the driving mirror she saw his eyebrows cocked ironically. 'Wisdom greater than your own heart?'

'The heart isn't always the most reliable judge,' she said unhappily, aware that in her reaction to Jase it was certainly true. He had only to take her in his arms to sweep away all common sense. It would be all too easy to get in over her head and set herself up for inevitable heartbreak. She had ample forewarning that Jase didn't love women, he made love to them—quite a different thing.

Jase's breath whistled between his teeth. 'You may have a point. Your divorce rate is only a fraction of most country's, including mine, so the system must work.'

She heard the censure in his voice. 'You don't approve?'

'It seems wrong. What if you fall in love with someone else?'

Normally she wouldn't have had the opportunity, spending most of her time with Luc until they had known each other well enough formally to agree to marry. Part of the reason the king hadn't wanted her to spend time with Jase Clendon was to shield her from possible temptation. She didn't like to admit how wise her uncle had been. 'It could be...awkward,' she admitted.

Jase glanced assessingly at her. 'How did Allie—the real Allie—manage to marry an outsider? Wasn't a husband chosen for her by a family elder?'

'There are exceptions. I was in Paris when Michael actually proposed, but I imagine he paid off Allie's intended groom—or fought him.'

'Fought him?' Jase's tone was frankly incredulous. 'As in a duel?'

'Not as you know it,' she denied, thinking of the western films she'd seen where competing suitors confronted each other with pistols or swords. 'Usually it takes the form of a contest of strength or skill. The tradition arose centuries ago when two men wanted to prove to a woman that each would make the better husband for her. A test of strength was arranged and she married the victor.'

'The men do get a choice in this, don't they?' Jase growled.

She nodded, seeing him arrive at the conclusion she had been avoiding sharing with him. 'The one refusing the challenge can leave in disgrace.'

He brought the car to a shuddering halt but kept his stony gaze on the road ahead. 'Such an outcome never crossed your mind, I suppose?'

She gave him a startled glance. 'What are you implying?'

This time he did look at her and the contempt in his eyes turned her blood to ice. 'I'm not implying anything. I'm stating facts. You knew you were promised to this Luc Armand when you arranged to spend time with me. You also knew if we were discovered I'd have to face your intended on some archaic field of honour.'

Her heart hammered. 'You're wrong. I didn't—'

'I haven't finished.' His cold voice sliced through her protestations. 'It was a fair assumption that I wouldn't fight him for your hand since, presumably, that would mean I get stuck with you—an outcome I was unlikely to welcome, right?'

His choice of words was galling but she was forced to nod. 'It would be logical *if* I had any such plan.'

He ignored this. 'So there was only one possible outcome. I would have to leave Sapphan and you would be the victor. You'd marry this Luc with no hard feelings. In fact, he'd be a hero in everyone's eyes for facing me down. And I'd be out of the country and out of your hair.'

As if the car could no longer contain his anger, he slammed out and stalked to where the road dropped away towards the sea. On one side of the sheltered, curving bay sheer limestone cliffs shaggy with jungle growth rose hundreds of feet out of the translucent waters, while on the other side a series of white-sand beaches, studded with rock formations, stretched towards the horizon.

Jase seemed to be intensely interested in the breathtaking view but Talay felt sure he wasn't seeing it at all. His back was rigid and his broad shoulders were set in an angry line. From her perspective the shadows gave the hard planes and angles of his face a thunderous look which made her quail.

She had to admit that the plan he believed she had hatched was ingenious, and she wished she had been devious enough to think of it. But she hadn't, and for some reason his suspicions wounded her more deeply than they had any business doing.

Unhappily she got out of the car and approached him on the cliff top. Her footsteps crunched on the gravel road shoulder, but no change in his demeanour showed he had heard her approach. 'I only meant to convince you with words,' she offered quietly. 'It never occurred to me that this would happen.'

He swung around his eyes spitting fire. 'Under the circumstances, I can hardly call you a liar, can I, *Your Highness*?' His tone dripped sarcasm.

Her anger rose. 'You didn't feel the need to defer to

my status when you thought I was Allie. You may feel free to speak your mind now.'

'It's hardly the same when it's by royal decree.'

She drew herself up. 'Then pretend I'm not royal. Treat me as you did before.'

The moment the words were out of her mouth she wished she could snatch them back. How he had treated her before didn't bear thinking about. At the same time she felt a rising excitement. In spite of all the reasons against it, she had enjoyed the way he had treated her before. By issuing such a rash invitation, was she actually seeking more of the same?

Jase's scorching look told her she was about to find out. 'Now that,' he drawled, 'will be a pleasure.'

The moment his arms closed around her she felt as if she would melt. Desire licked through her tentatively at first like the flames of a brush fire, building and building with each tender caress until it was like wildfire tearing through her veins.

His mouth hovered inches from hers, vexingly close yet nerve-searingly distant, his breath tantalising her heated skin until she ached for him to bridge the small gap and kiss her as she dreamed of being kissed. Instinctively she raised herself on tiptoe, her slight body taut as a bowstring, as everything in her reached towards him, like a sunflower responding to the sustaining radiance of the sun.

Then he did kiss her and a moan of surrender surged in her throat, to be instantly swamped by the fusion of his lips with hers. By turns, she felt starved of breath and giddy as if from too much oxygen. Tremors of pure sensual response ripped through her, blazing a trail of desire all the way to her core.

In vain she reminded herself of his callous behaviour

towards her in the hospital in Australia ten years ago. She had every reason for disliking him and none for allowing this sweetly torturous invasion of her mind and spirit.

She should also remember his playboy reputation. He was an expert at drawing from a woman whatever response he desired. No woman in her right mind could resist the provocative way his hands slid over her hair and skin, as if he wanted to know every inch of her by feel alone, or the sensuous exploration of her face and throat with lips that burned where they touched. There was no shame in feeling so pliant, so wanton. It was exactly what his caresses were intended to make her feel.

Which was why it felt so wrong.

All the same, she was tempted to close her eyes and abandon herself utterly to the intoxicating feelings invoked by the pressure of his mouth against hers. She shuddered and mustered the last ounce of resistance she possessed to push her hands up between them, flattening her palms against his chest. 'No, please.'

Instantly he stilled and his hands dropped to his sides. A breath of cool sea breeze whispered between them and she shivered. But it was more with the sense of being abandoned than actual coolness. Perversely she wanted the warmth of his arms around her, instead of the chill of his furious look raking her from a heartbeat away. 'Do you have any other royal decrees, *Your Highness*?'

He hadn't kissed her because she decreed it, she could swear. This close she could see the rapid rise and fall of his chest and feel the measured beat of his heart under her palms where they rested against him. A practised lover he might be, but he was not immune to her either, no matter how much he evidently wanted her to think so. It was small consolation for the animosity she could feel simmering between them.

Not even sure why she did it, but acting on pure instinct, she let her hands drift down with deliberate slowness and explored the muscular contours of his chest through the gauzy fabric of his shirt. At his indrawn breath she slid her fingertips lower, over the ridge of leather belt at his waist, then lower still over the hard angles of his hips.

His breathing quickened and she had the satisfaction of feeling his stomach muscles clench beneath her questing fingers. No, not immune at all. Then she let her hands drop away, appalled at herself. He had an uncanny ability to make her forget entirely who and what she was. 'Take me home, please,' she whispered, not remotely in the tone of a royal decree. More like a plea, actually.

Jase dragged in a lungful of the ginger-scented air. 'As you wish, Your Highness.' For the first time he didn't make her title sound insulting. The air was charged between them but he kept a careful distance, she noticed. When he helped her into the front seat and fastened her seat belt, there was nothing but solicitousness in his touch. 'Where is home? Not the Martine villa,' he said when they were back on the road again.

She shook her head. 'Allie is my closest friend so I spend a lot of time at her home. It's one place where I can be myself.'

'Which explains your familiarity with the place,' he agreed. 'I suppose you live in that grand edifice known as the Garden Palace in the centre of Alohan? I drove past it the other night.'

She nodded, wishing she didn't feel as if she should apologise for where she lived. 'The Garden Palace is traditionally the home of the king's brother. Long ago, it was considered prudent to have the main rival for the throne live some distance from the capital. After my parents died in the blast which resulted in my hospitalisation in Australia the palace passed to me.'

'Rough deal,' he said gruffly, his nod acknowledging the pain in her voice. Then he surprised her by asking, 'Do you like living in such splendid isolation?'

She bridled at the implied criticism. 'I only occupy an apartment at the northern end. The rest is open to the public as a living museum of our culture and history.'

She hated sounding so defensive but something in Jase's attitude provoked her. 'I suppose you live in a shack on the beach at Broome?'

His eyebrow tilted ironically. 'I'm sure it won't surprise you to hear I have substantial homes in several places. In any case, I didn't mean the actual building so much as the ivory-tower nature of your existence.'

What was he getting at? 'I'm hardly in an ivory tower now.' In many ways it would be safer than having to cope with her feelings around him.

He nodded as if she had proved a point. 'Precisely. It's probably the first time in—how old are you, anyway?'

'Twenty-six,' she supplied tensely.

'In twenty-six years,' he continued without missing a beat. 'Except for a hospital stay in Australia, you've hardly ever ventured outside your royal enclave without a retinue to experience life like the ordinary people in your realm.'

The last two days had been a round of firsts, she thought furiously. Among them the first time she had been kissed so passionately it still stirred her blood to think of it. Also possibly the first time anyone had called her a liar, manhandled her as intimately as if she were a commoner, and deserted her for an evening when she would much have preferred his company.

She tossed her head, taking refuge in denial. 'If this is life as an ordinary person, it's hardly to be recommended. Besides, I've seen and experienced nothing I couldn't have seen with my bodyguard in tow.'

Jase's scorching look all but called her a liar again. 'Nothing, Princess?'

Some things she would prefer not to have Sam around to witness. She ran her tongue across tender lips. 'All right, almost nothing. But you're wrong about me. I don't live in an ivory tower, as you call it. I work for a living. Those jewellery designs you saw at Allie's villa are mine. They're for my next collection, which will feature Sapphan pearls.'

His fingers drummed an impatient tattoo on the steering-wheel. 'Perhaps you work, but hardly for a living. I'm sure it doesn't depend on whether or not you can sell your collection to put food on the table?'

'Perhaps not, but...'

'Well, it does for those people in that village we just left. That's what I mean by coming down from your ivory tower.'

Her temper flared and she fired her most regal glare of displeasure at him. He caught it in a sidelong glance but refused to quail beneath it. 'Save the royal contempt for someone who hasn't kissed you,' he snapped.

She drew herself up, the seat belt straining against her breasts which felt almost unbearably sensitised. 'How dare you?'

He grinned without humour. 'I dare, all right. You told me to treat you the way I did when I thought you were

Allie Martine. It's precisely what I am doing. Unless you'd prefer me to fawn and grovel?'

The very idea melted some of her resistance. 'I can't imagine you fawning or grovelling to anyone.'

'Not if I can help it. Australians are rebels by nature. We don't readily go down on bended knee to anyone, except the winner of the Melbourne Cup.'

Mystified, she shook her head. 'This Melbourne Cup is a contest of the higher athletic arts, perhaps?'

'You could say so. It's the most famous horse race in Australia. On the first Tuesday in November each year the whole country comes to a standstill to watch it. People who normally don't bet on anything usually have a flutter on the Cup.'

Fluttering was what her heart was doing now, she realised in alarm. It wasn't at his description of a horse race, but at the way his voice dropped to a warmer register when he spoke of his own country. Jase sounded so much more attractive when he wasn't angry that a lump the size of a lychee stone rose in her throat.

He dispelled it by going on. 'So, you see, you need to live in the real world to understand the importance of having a job and a secure future for your children.'

Unwelcome tears stung the backs of her eyes and she blinked to dispel them. They were back on the original footing, it seemed. 'Your resort will provide them, naturally.

'If you stop fighting me and let me do my job they will.'

She clenched her hands tightly together. However attractive Jase might be, he was still out to change her beloved Pearl Coast. 'Does it have to be located at a beautiful, unspoiled bay?'

His impatient breath hissed between them. 'Haven't

you heard a word I've said? Or does royalty only deal in preconceptions around here? I plan to replace every tree and blade of grass removed during the building, and the resort itself will blend in with the rainforest so thoroughly you'll be hard-pressed to know it's there until you fall over it. Why do you think King Philippe allowed me to go ahead?'

Jase evidently underestimated his own powers of persuasion. Talay had the greatest respect for her uncle's wisdom but he was not infallible. He would find the argument about jobs and futures a compelling one. And if the king believed in Jase's project, could there be merit in it after all?

She stole a look at Jase. Was she being swayed by his logic or the power of his kisses? She made an effort to override the rush of sensations that threatened to swamp her at the very thought. 'Why do you choose such remote locations for your resorts?' she asked, remembering Allie's observation that he preferred exotic sites.

Was it her imagination or did his fingers tighten around the steering-wheel? 'It's none of your business.'

'Is it because their people are more innocent and easily bent to your will?' she persisted. His expression warned her she was being reckless but she couldn't stop herself. 'You say that Australians don't readily accept authority. Do you come here because you can't get your own way at home?'

'That's enough, Talay,' he thundered. It was the first time he'd used her first name and the sound of it sent quivers along her spine. 'You have no idea what you're talking about.'

Her chin jutted out. 'Because I'm right?'

'Because you're so wrong it's ridiculous. You remind

me of the sign which says, "Rule one: the boss is always right. Rule two: if the boss is wrong, refer to rule one." You don't know the first thing about me or my motives.'

'Then tell me,' she said, on a note of desperation. To her dismay, she found she wanted to know, not only for her people but also for herself. Jase Clendon was the most enigmatic man she had ever met, and the urge to dig beneath the surface until she knew everything about him was overwhelming.

There was a long moment during which she imagined him wrestling with some inner demon. She had almost decided he was going to tell her to mind her own business when he stated, 'Peace.' She was baffled until he went on, 'Places like Crystal Bay represent peace and tranquillity, something I never knew when I was a boy.'

'But you come from Broome,' she observed. From what she knew of Broome, it was a peaceful enough place on the far north coast of Western Australia. The main industries were pearl farming and tourism. It was also the gateway to the Kimberley region, rich in aboriginal heritage and, more recently, fabulous wealth in diamonds. But it was still one of his country's last frontiers.

'I was born there but I didn't grow up in Broome,' he corrected, his voice vibrant with leashed emotion. She wondered if she was rash to delve into what was clearly sensitive territory, but it was too late now. 'I never knew my father and my mother died having me. Until I was five I lived with a man I called uncle, but he was a foster-carer, no blood relation at all. His wife didn't even like kids. As I grew older she claimed I was unmanageable and I was sent to a series of children's homes, the last being in Perth where I lived with twenty-three other boys between the ages of seven and fifteen.'

She tried to imagine it. 'Twenty-three boys all living together?'

His look impaled her. 'The house we shared would have fitted into the gatehouse of your palace. We shared rooms, meals, entertainment, even clothes. There wasn't a single place where you could be alone. If you stopped long enough to think, somebody found something for you to do in case the devil made work for idle hands.'

'It sounds like a terrible existence,' she murmured.

In the rear-view mirror his eyes brightened, dismissing the moment of revelation. 'Save your sympathy for those who need it. My upbringing gave me the drive to better myself. As soon as I was old enough I went to work aboard a trawler, saved every cent I made and bought my own fishing boat. One day I hauled in some artefacts from a seventeenth-century shipwreck. The reward paid for my first land purchase. From then on, I was answerable to no one.'

More easily than she wanted to, she could picture Jase at the helm of a deep-sea vessel. It explained his far-sighted air, as if he was accustomed to scanning distant horizons. Then she recalled their first meeting and her heart lurched at the memory. It made her voice cold as she pointed out, 'You became an Olympic yachtsman. It's hardly a solo sport.'

He nodded tautly. 'At first I sailed solo but to race bigger vessels I had to become a team player. It wasn't difficult once I figured out I could spend my off-hours alone if I chose.'

'Didn't the crowds bother you when you competed in the Olympics?' she asked.

He shook his head. 'People aren't a problem unless they encroach on my private space. As long as I have somewhere to call my own, the world can camp on my doorstep.'

'Now you build resorts in out-of-the-way places to give other people somewhere to be alone as well,' she concluded. It made perfect sense but it also painted a picture of a solitary man. No wonder his marriage had failed. Sharing his private life with anyone argued against everything that he was.

'So you see why I'd be the last person to want to spoil the very attractions which brought me here,' he went on, driving his point home.

It was a perspective she hadn't considered and didn't want to now because it would mean admitting that he was right and she was wrong. Her reluctance bothered her. Did she really adhere to the rule one and rule two concept, expecting to be right simply because of her royal status?

'You aren't going to say it, are you?' he provoked.

She pretended ignorance. 'Say what?'

'That you could be wrong about me, Princess.'

His mocking tone made her temper flare. She hated being read so transparently. 'This isn't about you and me,' she flung at him.

Either by accident or design—quite possibly the latter, she suspected later when she had time to think—the car struck a rock in the road and she was thrown against him hard enough to drive the breath from her body. Her slight frame impacted with his thickly padded one until she was caught and held by one strong arm. He kept control of the car with the other as his fingers dropped over her shoulder and met the swell of her breast.

The touch could have easily been explained away as helpful, steadying her against the jolting movement. But she couldn't explain away her own reactions so easily—as if he had kissed her again instead of merely touching her. Her stomach muscles knotted but she tried to keep her features carved in stone and inject a tone of royal command into her voice. 'I'm all right, you may release me now.'

'Of course, Your Highness.'

His ready acquiescence was the first thing she needed and the last thing she wanted. As his arm slid away she fought to keep her inner turmoil from showing on her face. But he sensed it anyway, with the almost telepathic link she had sensed between them from the first.

He laughed.

She felt the colour stain her cheeks. 'What's so amusing?'

'You, Talay.' No title now. 'Acting so regal, when all you want is for me to stop this car and take you in my arms properly.'

'You couldn't be more wrong.' Her voice came out as an unconvincing whisper.

'No?' He stopped the car in the middle of nowhere and turned to her, reaching to brush the back of his hand down the side of her face. A brushfire ignited inside her. 'You don't want me to do this?'

'No.' Even less conviction.

'Or this?' His eyes searched her face as he caressed her nape under her long hair, his touch feather-light but so delicious that it was an effort not to arch her back under his hand.

Abruptly he took his hand away. 'Rule one or rule two?'

'Neither,' she snapped, aware that her anger was as much at herself for her weakness as at him. 'I can't afford the luxury.'

'Then you admit it is a luxury?'

How could she deny what her trembling body made all too obviously to him? 'Yes.'

He jolted upright. 'For me, too. Unfortunately, I can't afford it either.' Because of the risk to his business, she assumed, fighting disappointment. She had hoped—

'We'll have enough complications when that newspaper photo appears tomorrow,' he cut across her unspoken thoughts. He was all business now, restarting the car and aiming it towards Alohan.

Horrified, she realised he had made her forget the damning photograph. 'What are we going to do?'

His mouth tightened into line of fixed purpose. 'I have a plan.'

CHAPTER SIX

HIS plan was simplicity itself, Talay found out after Jase returned her to the Garden Palace. Sam raised an eyebrow when he saw who her escort was, but the bodyguard was too well trained to comment. To ward off any gossip, she introduced Jase in his official capacity as the owner of Crystal Bay, trying not to wince as she said it.

'I've persuaded Her Royal Highness to become my official adviser on the project,' Jase added, presumably for the benefit of her staff.

Jase kept a professional distance as she led him through the palace to her office, but his comments about her ivory-tower lifestyle echoed in her mind as she showed him along a hall, hand-painted in Pompeian red and crowded with Italian baroque giltwood furniture from the late seventeenth and early eighteenth centuries. Was she isolated from the very people she had been born to serve?

'Most of the furniture is museum-quality with a history behind it,' she said, hating the defensive note which crept into her voice. She had always accepted the palace as home and had never before felt the need to justify her existence.

He smiled dryly. 'So I notice.'

Wishing she had somewhere simpler to entertain him, she had no choice but to take him to the suite which served as her office. By palace standards it was minimally furnished and the room was flooded with sunshine, but the desk was still Chippendale, the carpet an eighteenth-century Savonnerie and the picture on the wall a 'lost' Degas rediscovered during the 1970s.

He offered no reaction but perched on a spindly Chippendale chair, crossing an ankle over one knee as if he were completely at home. She had a feeling Jase could make himself at home anywhere.

He refused her offer to ring for tea. Welcoming the barrier between them, she sat behind the desk and locked her fingers together, long practice enabling her to appear the picture of composure despite her agitation. 'Tell me about your plan.'

'You've just heard it,' he stated. 'I'm officially inviting you on board as my adviser on environmental matters at Crystal Bay Resort.'

It was as cynical an invitation as she'd ever heard and she said so.

He shook his head. 'Not at all. Your royal position, combined with your love of the area and understanding of the local culture, means you would be a great help on the project.'

But not to him, she noted with an unwelcome surge of disappointment. Her tone sounded flat as she said, 'It would serve as a cover story for why we were together today.'

'It's also true,' he said shortly. 'I care about the places where I build my resorts and I want the best for them, including Crystal Bay. Will you accept?'

'What would I have to do?' she asked warily.

His smile mocked her. 'More of what you did today. Tag along and point out the error of my ways.'

She bristled. 'This isn't a real appointment. You're laughing at me.'

He sobered immediately. 'Never, Your Highness. It's probably high treason in Sapphan.'

There was no longer any such crime on the statute books but she felt safer not pointing it out to him. 'A hanging offence,' she agreed seriously.

His eyes lingered on her mouth. 'So what is the penalty for kissing a princess?'

Her throat dried but she refused to swallow. 'Quite possibly, you turn into a prince.'

He did laugh then, and the sound vibrated through her until she gripped the edge of the desk for support. Jase was definitely no frog and, with Luc Armand in the wings, there was no chance of the big Australian turning into a prince. But for a moment she was preoccupied with a vision of she and Jase ruling Pearl Province side by side, sharing the king's delegated authority. Sharing everything.

She shook the vision out of her mind. Luc Armand was the one she should be imagining at her side. Yet Jase's image refused to give way. She made herself focus on the problem at hand. 'Do you think making me your adviser will be sufficient to explain away our presence at Crystal Bay?'

He was all business now. 'I'll make sure of it. If you lend me some staff I'll issue a press release, announcing your appointment and describing how you decided to make a surprise inspection of the site today.'

'It might serve.' Another thought fuelled her excitement. It also gave her a reason to spend more time in Jase's company. However unwise it might be, she couldn't make herself dislike the idea. She vacated the desk. 'My office is at your disposal.'

'Good.' As he traded places with her their bodies brushed in the space behind the desk, and she had to bite back a gasp of amazement. The slightest contact with him was enough to set her senses on fire. Was it sensible to agree to work more closely with him, given his effect on her?

With the prospect of the photograph which would appear in tomorrow's newspapers she probably had no choice. She also suspected she had no will to turn him down.

Disaster struck early next morning in the form of a telephone call from the capital. 'Uncle Philippe,' she said sleepily, immediately recognising the king's vibrant voice. 'Is something the matter?'

'You may well ask,' the king thundered down the telephone. 'You obviously haven't seen this morning's papers.'

The photograph! She sat up quickly. 'Uncle, I can explain. Jase Clendon has appointed me to advise him on his resort. We were—'

'Kissing in public,' Philippe finished for her. 'Is this how he usually treats his advisers?'

She must be still asleep and dreaming. 'Kissing? I don't understand.'

'Then I suggest you look at the newspaper and reacquaint yourself with the meaning of the term,' the king instructed. 'Luc Armand is already demanding an explanation and I'm hardly the one to provide it for him.'

In confusion she groped for the neatly folded newspaper, lying on a tray beside her bed. One of her few indulgences was glancing through the pages in bed before she launched herself into the day. Another ivorytower weakness, came the unbidden thought, but it vanished as soon as she saw the photograph taking up most of the front page.

It wasn't the expected shot of Jase helping her down the steps of the village café, but a completely different one, taken with a long lens from some distance away. Many times enlarged, the grainy picture showed her unmistakably in Jase's embrace, her arms linked around his neck and one foot drawn up behind her in a coquettish pose. Jase's head was bent and his mouth covered hers.

Her stomach muscles clenched at the graphic reminder of her abandonment to the pleasure of his kiss. It had never occurred to her, or apparently to Jase, that anyone could be spying on them. The photographer must have watched her arrive at the Martine villa, camped outside overnight, then trailed them to the resort site and the village.

'Oh, no,' she gasped aloud. It was bad enough to see it in Sapphanese, but this paper was published in English as well so it would be all over the country by now.

The king heard her. 'I see you now understand the meaning of the word "kissing".'

She always had done, but had never expected to see herself demonstrating it so publicly. 'I don't know what to say, Your Majesty.'

'It's still Uncle Philippe,' he said more kindly. 'I'm not angry about this as your King, but as the only father you have, Talay. I thought I'd taught you the meaning of discretion.'

'You did,' she said miserably. 'It was a lonely road and no one else was around, or so we thought.'

'To a princess, no road is *that* lonely,' he observed sarcastically. 'I assume Jase Clendon wasn't forcing his attentions on you?'

She couldn't let herself off the hook by blaming Jase. 'He would never do such a thing.'

The King drew an audible breath. 'I'm gratified that you're willing to take responsibility for your actions. At least I've taught you something. Unfortunately, it is Jase himself who must bear the greater responsibility under our laws.'

'Luc intends to challenge him?' Even saying it out loud didn't make the prospect any less alarming.

'I tried to talk him out of it on the grounds that Jase is a foreigner who doesn't know our ways.'

'Was he persuaded?' She was afraid she already knew the answer. Luc Armand was a hot-headed young aristocrat who loved adventure. He thought nothing of abseiling down treacherous cliffs or hiking through primitive jungles. Not so long ago, Talay had thought his exploits romantic and daring. Now he seemed foolish and reckless. She also knew the change had begun to come about from the moment she had met a real man in Jase Clendon.

'What do you think? He wants the challenge to take place as soon as it can be arranged.'

She was afraid to ask, but it couldn't be avoided. 'What form does he want the challenge to take?'

It was the challenger's right to choose. 'A dive off the Malakai Cliffs.'

Horror crept over her. 'He can't be serious!' Malakai was one of the most dangerous places in Sapphan. Located a few miles north of Alohan, along one of the wildest parts of the Pearl Coast, it was nicknamed Lovers' Leap for good reason.

Her uncle paused. 'I'm afraid he is. I hope Jase Clendon has the sense to leave before he has to see this through.'

She was afraid she already knew how Jase would react. Running away wasn't in his nature even if, on this occasion, it would be the better part of valour. 'Can't you do something? Forbid Luc to go ahead? Issue a decree, outlawing the challenge? Jase could be killed.'

At the desperation in her voice the king's tone became kind but implacable. 'If this challenge takes place they could *both* be killed.' Belatedly she heard her uncle's recognition that her first concern was for Jase. The King went on, 'I cannot alter the law to suit myself or my family. You should have considered this outcome before you met Jase Clendon against my advice.'

Uncle Philippe was being remarkably forbearing on this point, she noted belatedly. He probably thought she would be punished enough by having to live with the outcome. And he was right. She only hoped Jase would also be able to live with it—and through it. The alternative was unthinkable.

Of all the reckless, foolhardy... Jase's condemnation of himself was total as he paced the length of the living room at the Martine villa. The room was the size of a tennis court, but it was barely enough to contain his explosive movements as his long-legged strides ate up the priceless carpet. He let the newspaper drop from his hands and it landed with the photograph of him and the princess uppermost.

He slammed one fist against the palm of the other. From the start he had sensed that 'Allie' wasn't being honest with him. Letting himself kiss her before he had solved the mystery had been the height of stupidity. Now 'The Princess and the Playboy in Secret Love Tryst' was the national breakfast-time reading in Sapphan in two languages.

'Secret Plot Against Playboy' was more like it, he fumed inwardly. He had fought some tangled corporate battles in his time, but they paled alongside this for deviousness. Talay's denials had sounded convincing enough, but it was too neat a scheme for him to believe she hadn't foreseen this outcome, particularly as Jase had sensed from the first that she didn't favour the resort.

If her regally sanctioned suitor challenged him—and Jase knew that if his woman had been pictured in such a compromising pose, he wouldn't hesitate—Jase would either have to accept or leave the country for good. Tough choice.

Impossible choice.

He was still in reasonable trim from his days as an Olympic yachtsman, and he sailed regularly, which required peak physical fitness, but until he knew what he was up against he shouldn't consider accepting a challenge. Talay Rasada was twenty-six so her suitor was probably around the same age. Jase himself was thirty-two, hardly a vast difference in age but light years of difference in physical performance. The yacht crew joked about it all the time.

Jase stopped pacing abruptly. He *was* considering accepting. It was only a resort site. The world had many more. Hell, he owned half a dozen potential sites in other countries already. There was no reason to risk getting injured or killed because of this country's archaic notion of honour.

He looked down at the paper on the floor. The reason was right there in his arms. If he won the challenge Talay Rasada wouldn't have to marry a man whom she clearly didn't love.

Jase winced. Whatever she felt for him wasn't love either, but there was *something* powerful between them. She was easily the most beautiful woman he'd met in a long time, possibly a lifetime, managing to remain in his thoughts long after he should have forgotten her existence. He'd actually been pleased to discover she wasn't married to Michael Martine. Or anybody else.

No matter what it cost him, he couldn't let her be forced into a marriage she didn't want if there was a way to prevent it.

He only hoped he could live with the decision. Literally.

'He wants me to do *what*?' he demanded when he arrived at the palace after breakfast. Talay had called and said they needed to talk urgently. She had made it clear it wasn't a royal summons, in case he refused on principle. Having seen the front-page story, he wouldn't have done, but she was learning.

She looked away, misery in every line of her slight body. 'Luc told my uncle he's challenging you to dive off the Malakai Cliffs.'

'I surveyed that area before I settled on Crystal Bay,' he said evenly. 'It's the most rugged, dangerous part of the northern coastline.'

'I know. Historically it was a rite of passage for teenage boys to dive off it to prove their manhood. Some still die trying it even today.' Her tone carried a weight of concern.

'My manhood doesn't need any proof,' Jase bristled. Her colour rose, showing she was well aware of the fact, which gave him a grim kind of satisfaction. It didn't last long. 'Was Luc Armand one of the boys who tried the dive?'

She shook her head. 'Not that I know about, but he is very athletic. Diving is one of his skills. He has represented Sapphan internationally in diving contests.'

'No wonder he chose his best sport.' Jase knew he

would have done no less. 'Then I'll have to brush up on my diving, won't I?'

Her look of horror was too transparent to be faked. 'You can't mean to accept his challenge, Jase? You could be killed.'

'I could also win,' he said evenly. 'I realise it isn't what you want.'

'How can you think such a thing?' Her eyes brimmed, although she held the tears back with a visible effort. Something clenched inside Jase in response, catching him by surprise. 'I don't want you to get hurt.'

'Maybe not, but you do want rid of me and my resort. What better way than if I refuse Luc's challenge and have to leave with my tail between my legs?'

'At least you'd be alive.'

He was almost disappointed when she didn't argue the point more strongly. He hated the thought of being manipulated, even by someone as lovely as the princess. 'I intend to live, anyway. You'd better show me where this dive is to take place. I'll need to assess it carefully.'

'I can't...'

For one blinding moment she wasn't the princess. She was 'Allie' and he was entitled to touch her as he badly wanted to. His hand latched around her wrist and he hauled her closer. Enveloped in a cloud of her seductive jasmine perfume, his mind reeled. Her delectable mouth was so inviting, inches from his own. It took an almost superhuman effort to restrain himself and say, 'If I can see this through, you can. You're coming with me.'

At his touch, her breathing quickened and he registered the fast beat of her heart against his chest. 'We would have to take Sam, my bodyguard,' she said in a hoarse whisper, which told him she was fighting a similar battle with herself.

He released her before he followed through on the need coursing through his veins and kissed her senseless. 'It's probably just as well,' he agreed on a heavy outrush of breath.

'Yes,' she said, sounding numb.

What had she done? she asked herself over and over during the drive north. If only she had listened to her uncle, Jase wouldn't be in this terrible fix now. 'Why are you doing this?' she asked, finding she was holding her breath as she waited for his response.

There was a long pause. 'If I refuse the challenge I have to leave the country in disgrace—correct?'

There was no avoiding it. 'Yes.'

'What other reason do I need?'

She had been afraid business would be his main motive and it weighed heavily on her, although it shouldn't. Despite the awareness that crackled between them like summer lightning, she knew he wasn't the type to care about her as a woman, even if it had been appropriate. This morning's newspaper article had taken malicious delight in cataloguing his long string of liaisons, including his failed marriage. It wasn't news to her and she hadn't enjoyed reading about them, finding she had to resist the urge to tear the pages to shreds. But perhaps she needed the reminder.

'Turn right here.' Her gesture indicated the way to the cliffs. They fringed the northern boundary of the Leon Rasada National Park. The area was breathtakingly beautiful, popular with hikers, cliff climbers and deep-sea divers.

Millennia ago the area had been dry land, studded with lofty limestone mountains, until the glaciers melted and the sea invaded the resulting bay, leaving only the tops of the mountains rising from the water like shaggy, jungle-coated monsters.

The area abounded in natural wealth from the many caves, yielding birds' nests for gourmet cooking, to the seas, which teemed with dolphins and the elusive dugongs, and the shallows, which were home to reef egrets, herons, and hornbills.

A network of river estuaries webbed the shore, with dense forests of mangrove and palms growing down into the dark waters. Some of the mangroves were raised above the water on aerial roots like stilts. The local villagers harvested the heavy wood for building and to make charcoal, as they had done for centuries.

Normally one of Talay's favourite places, Malakai looked dark and forbidding today. She could hardly take her eyes off the headland cliff which was the height of a four-storeyed building. It was hollow, the cave said to be the home of a mythical sea princess who ensured a plentiful supply of fish. A trail led up the side of the cliff, but her eye was drawn to the rope guide anchored to the rock. The rope ended after a dozen metres, leaving climbers exposed to the lashing wind and sea spray on the arduous climb to the top.

She was aware of Sam's car, pulling up at a discreet distance behind them. He had followed them all the way from the palace and this time there was no question of evading him, even if it had been prudent. He had used his considerable training to help them elude the media, clustering outside the palace seeking her reaction to the morning's story.

Now she was almost grateful for the bodyguard's inhibiting presence to remind her of who and what she was, otherwise the temptation to taste more of Jase's kisses might have been too strong to resist. Already the need to feel his arms around her clawed at her like a hunger. She told herself it was because of the danger facing him.

'You can't really mean to go through with this?' she said, around a throat which felt painfully parched.

He looked up at the forbidding cliffs, shading his eyes with one hand. 'Does Luc Armand mean to go through with it?'

'I'm afraid so.'

Only a tiny muscle working in Jase's neck contradicted his certainty. 'Then it's settled.' He reached for her hand. 'Come on, I want to get a close look at what I'm up against. One of the first rules of business is "know your enemy".'

Did he mean the cliffs or her? she wondered fleetingly. He was approaching this as dispassionately as he might have approached a hostile takeover of one of his companies, she noted as he quartered the base of the cliff, his eyes missing nothing. She wished there was some way to convince him that she had never intended this to happen. She could hardly believe how much it hurt to have him think it was a devilish plot on her part to have him exiled from Sapphan.

For the first time it occurred to her that if he left, some vital part of her would leave with him. From the first, his impact on her had taken her breath away. Never before had she encountered such strength of character and purpose. Unlike Luc, Jase needed no macho posturing to project his certainty that the world would yield because he required it.

Would *she* yield if he required it? She was suddenly thankful that, under the present circumstances, the question wouldn't arise because the answer hardly bore thinking about.

CHAPTER SEVEN

THE going was easy enough at first, although Talay was glad she had worn sensible, rubber-soled shoes. As the cliff became steeper Jase had to take her hand to assist her over the jagged surface. She could hear Sam panting along in their wake and nearly ordered him back to the car, then thought better of it. Even if he would agree to go, she wasn't sure it was wise to be alone with Jase more than she had to.

Her flesh crawled as she imagined photographers with long lenses trained on them at this very moment. Would she trust herself to be unobserved ever again?

Jase felt the tremor sweep through her. 'Cold?' he asked.

The concern in his voice warmed her. 'No, I was wondering if we're being watched right now.'

Jase glanced wryly at Sam then back to her. 'It's a reasonable assumption.

She shook her head. 'I meant photographers, paparazzi.'

The fingers in hers tightened while his other hand splayed across a boulder which looked as if it might tumble into the sea at any moment, but was as firmly anchored as the cliff itself. 'Is this the first time this has happened to you?'

She jerked her hand free. 'I don't make a habit of kissing strangers, if that's what you're implying.'

He reached for her and drew her back to his side. 'I wasn't implying anything so you can come down off that high horse, Your Highness. I meant is this the first time you've been subjected to malicious headlines?'

Striving to ignore the fast beating of her heart, which insisted on accompanying his touch, she nodded. 'Naturally, the media are interested in everything we do, but usually they are respectful in what they write.'

He frowned. 'I noted the byline on this morning's story. The name didn't sound Sapphanese. In fact I could swear...'

She regarded him curiously as his voice trailed off. 'What?'

'I'm sure I've seen this guy's work in the Australian newspapers, the ones specialising in scandal and gossip.'

His logic became clear. 'Could he have followed you from Australia?'

'He might, or someone could have tipped him off about my movements. I certainly intend to find out which it was.'

He would—she didn't doubt it for a minute. Once Jase made up his mind there would be no stopping him. It hurt to have him think she could do such a thing as bring the Australian photographer to Sapphan to take the compromising picture. The more thorough Jase's investigation was, the better. He couldn't uncover a connection which didn't exist so eventually she would be vindicated.

Why it mattered what he thought of her, she wasn't sure. She only knew that her spirits lifted for the first time since she had opened the newspaper this morning.

'Where does this lead?' Jase asked, gesturing towards a cliff path.

'I don't know. I've never come this way before.'

It was true. She had picnicked in the park many times and swum in the sea off the isolated beaches, but she had never climbed the treacherous cliffs because one slip could mean injury or death. If her uncle had impressed nothing else upon her, it was that she had a duty to take care of herself and cause her subjects as little worry on her account as possible. No wonder the king was so disappointed with her for flouting his advice. She was doing it now, she realised unhappily. What was it about Jase that made her forget her responsibilities so readily?

She dismissed the thought. What was done was done. But she was perversely glad to have the distraction of the uncertain footing to drive her own problems from her mind. Now she actually needed Jase's helping hand to negotiate the rugged terrain.

She told herself the feel of his strong fingers curling around her own had no more effect beyond reassurance. The times when his arm came around her waist so he could help her over an outcropping of rock didn't really make her heart beat in double time. It was the exertion, nothing more. At the same time she knew she had never felt more vibrantly alive and aware of herself as a woman in her life.

'Please stop, I need to catch my breath,' she begged, after they had climbed steadily for a time.

'In a minute. We're almost there,' he promised her.

The summit was still a short distance away. 'Almost where?'

'Here.' He tugged her into what appeared to be a cleft in the rock face but turned out to be the concealed entrance to a cave.

As the sheer beauty of the place thrust itself into her awareness, she caught her breath in astonishment. 'This must be the princess cave.'

'Where else would you bring a princess?' he asked, laughter in his tone belying his serious expression.

In fact, the cave turned out to be two interlinked caverns, each illuminated through a rock chimney which directed shafts of golden sunlight into the gloom. The cave contained some of the most beautiful limestone formations Talay had ever seen. She could hardly believe it had taken a foreigner to bring her to this place. 'How did you know it was here?' she asked in wonder.

'I researched the area extensively before I chose Crystal Bay,' he explained. 'Even so, I couldn't recall the cave's exact location until we were almost on top of it. This cave is said to be the grand palace of the sea princess, while the lower one that everybody knows about is supposedly her summer palace.'

The reminder of how he had come by his knowledge rankled and she turned away to hide her disappointment. He would have none of it. His hand cupped her chin and he turned her gently but irresistibly back to meet his searching gaze. 'Still determined to see me as the enemy, Talay?'

'No, I...'

'It's written all over your face,' he insisted, something very like disappointment colouring his own tone. 'Well, it isn't true. I was offered the next bay as the site for my resort but I turned it down because of the natural and cultural significance of this area. Not that I expect you to believe me.'

Oddly enough, she wanted to but his plans for Crystal Bay argued against it. When she stayed silent he let his hand drop. 'No, of course you wouldn't. But it's easy enough to check if you care to.'

The derision in his voice tore at her and she looked away. 'I don't need to check. I believe you.'

He swore softly under his breath. 'Apparently, there is a first time for everything.'

All at once she became disturbingly conscious of their situation. She could hear her bodyguard moving around outside and she pictured him casting around for the entrance to the cave. He would find it soon, but until then she and Jase were entirely alone in this hidden place. There was no possibility of photographers having long lenses trained on them here.

The awareness and Jase's closeness made the cavern seem claustrophobic, and she felt an urge to flee outside. She resisted the urge, forcing herself to show an interest in their surroundings.

'I can't believe I've never been here before,' she murmured, thankful for the dappled sunlight which camouflaged her heightened colour from his searching inspection.

'Maybe it's just as well,' he commented wryly.

She felt her eyebrows arch in surprise. 'Why?'

'Come here. I'll show you.'

He took her hand again and tugged her into the second of the caverns. Its main feature was a sparkling quartz formation which looked for all the world like a crystal waterfall. Sunlight from another chimney in the rock turned it to gold. She drew a sharp breath. 'It's beautiful.'

Drawing her with him, Jase moved closer and let his fingers trace an inscription in the rock face above the waterfall. She recognised it as an ancient Sapphanese script. 'Can you translate this?'

Protected from the effects of wind and rain, the inscription was as fresh as if it had been carved yesterday, although it was probably centuries old. She had no trouble reading it, but every difficulty sharing it with Jase. 'I'm not certain.'

He traced the markings again. 'This is the sign for

two and this is the sign for one. The hieroglyphs clearly represent a man and a woman. It isn't difficult to decipher,' he insisted, sounding amused at her hesitation.

Her anger rose. 'If you already know then why ask me?'

'I was guessing until your reaction told me I was on the right track,' he observed. 'This is some kind of honeymoon place, isn't it?'

'Yes,' she agreed, wishing she didn't sound so much like a child having her lessons dragged out of her. 'If you must know, it's where two lovers became one in ancient times. There are several such caverns in Pearl Province but these days they are of no more than historical interest.'

He would have none of it. 'For a place that hasn't been used in centuries, it's remarkably tidy. The cavern floor looks as if it's been swept recently.'

'It probably has,' she snapped, wishing she could think of a graceful way to change the subject. 'Even today some couples like to spend their first night together in such a place. It's not so much belief any more as purely for fun.'

She wished she had expressed it differently as soon as she caught the gleam in his eye, reflected off the sparkling golden quartz. 'You have some intriguing customs in Sapphan. As my adviser, you must tell me more about them.'

The churning inside her warned her how dangerous this would be. 'Considering you asked me to advise you on environmental matters, our wedding customs are hardly likely to be of interest to you.'

He shook his head. 'You're wrong. They are of great interest to me. I might decide to create a lovers' cave on the resort's private island. Can you imagine the appeal of such a place to a couple seeking a truly romantic honeymoon?'

In the cavern of the sea princess, surrounded by reminders of the lovers who had come here to set the seal on their relationships, she found it all too easy to imagine and her pulses went haywire at the thought. It was only a cynical marketing ploy on Jase's part, but she couldn't shake off the wish that he might support the romantic custom for its own sake.

'I didn't know you were such a romantic,' she said, annoyance at her own reaction sharpening her tone.

'There's a lot you don't know about me,' he asserted, 'otherwise you would have thought twice about involving me in your little scheme.'

They were back to that again. 'You're wrong, I did no such thing.'

He caught her wrists and hauled her to him until her breasts ground against the hard wall of his chest. 'You still say being photographed with you was pure coincidence and not a set-up engineered by you to drive me out of Sapphan before I could build my resort?'

'I do say it because it's the truth, as you'll find out when you investigate,' she insisted with difficulty, as a surge of purely physical response tore through her.

Ordering her thoughts was almost impossible with the fast beat of his heart thundering in time with her own. His fingers were like steel bands around her wrists, the pressure lifting her onto her toes and her body into unwilling alignment with his.

Struggling would only increase the contact so she schooled herself to remain still, although it took an almost superhuman effort. It was bad enough to be aware of her own uncontrollable reactions. She would not give him the satisfaction of seeing how readily he could arouse her with a touch. 'You forget who I am. Release me,' she said in her most imperious tone.

She used a voice of royal command, which should have led to instant obedience, but Jase simply laughed. 'I can hardly forget who you are, with your bodyguard ferreting around outside.'

'He'll find the entrance any minute,' she warned, much more tremulously this time. Keeping her royal distance was almost impossible while his touch forced her body to telegraph messages of pure delight all the way to her brain. No one had ever held her against her will as Jase was doing.

But was it really against her will? Then why were the messages so delightful? She should be screaming the place down by now, bringing Sam crashing into the cave to her rescue. If only she could shake the suspicion that she didn't want to be rescued.

Jase's wide, inviting mouth hovered inches above her own. They were parted in a mocking smile and she squirmed at the thought of kissing that smile into submission. Except she was afraid who would end up submitting.

As she moved against him his smile gentled and a gleam lit his intense gaze. 'From the look of you, I'd say you want your bodyguard to find you—but not just yet.'

'No, I can't...'

'Can't what? Think, feel, react like a normal human being? You *are* a normal human being, Talay. You may be a princess of the royal blood, but you are also a very beautiful woman with all the normal needs and desires. If you need proof...'

The realisation that he found her beautiful threw her off balance long enough to silence her protestations as his mouth descended on hers. Although he still held her wrists, there was nothing of force in the kiss and everything of physical exploration on a level she had never experienced before.

For one shining moment she wasn't a princess and he wasn't a playboy, as she gave herself up to the power and passion of his embrace. Dizzy with the sensual delight of his touch, she could hardly bear the tangle of sensations coiling through her as his fingers left her wrists and slid across the sensitised mounds of her breasts.

She gripped his shoulders so tightly she could feel bone through the muscle. Some distant corner of her mind remarked on the strength in him even as her mouth absorbed the gentleness of a kiss that threatened to ravage her soul.

Slowly, slowly, he cupped her hips against him, making her tormentingly aware that the cave was in danger of fulfilling its ancient purpose. The awareness shredded the curtain of sensual pleasure veiling her inhibitions, and she forced her eyes open. 'Jase, we mustn't.'

He nodded but his lips left hers with a strange reluctance, as if he had done more than prove his point. Somehow he had become lost in the foolish demonstration and he looked as shaken as she felt. 'You're right. There'll be time enough after the challenge.'

The last of her languor drained away and she stiffened her spine. 'What do you mean?'

'Exactly what I said. Last night I spent some time going through Michael Martine's library. He has some interesting material on Sapphanese customs, including the marriage challenge.'

Her fingers leapt to her mouth. 'Oh, no.'

He seemed to take great satisfaction in anticipating

her thought. 'I discovered that the woman at the centre of a challenge is deemed to belong to the winner.' His eyes glinted ferally. 'Fascinating thought, isn't it?'

Almost unconsciously she wound her arms around her slight body as if to protect herself. She had hoped there would be no need for him to know this part. 'It's ancient history, no longer practised,' she denied, managing to keep the quiver out of her voice with an effort.

'Just as the challenge itself is ancient history, "no longer practised"?' he echoed in a mocking tone. 'Then precisely what are we doing here?'

The answer which sprang to mind hardly bore thinking about so she answered the literal question instead. 'The challenge is rarely involved. Luc Armand is a hot-headed fool,' she snapped. 'And you're an even bigger fool for accepting. Why don't you—?'

'Leave Sapphan and take my resort plans with me?' he finished for her. ''No chance, Princess, so you may as well get used to the idea.'

She had meant to say, 'leave before you are hurt or killed.' Either prospect filled her with raw panic. But she let him believe what he chose. At this moment, with the memory of his touch enveloping her in primal need, it seemed safer than telling him the truth—that she wanted him to live. Even the thought of him bringing tourism to her beloved Pearl Coast was easier to contemplate than the island kingdom without his dominating presence. Even so, she would sooner bear his departure than see him killed or injured.

Admitting it would be far too revealing so she drew herself up. 'Believe what you like. But the King may have something to say about handing over his favourite niece to a foreigner.'

He grinned. 'From your behaviour up to now, I'd say

he'd be glad to have you safely off his hands. And what better solution than to hand you into the care of a husband who is willing to fight for your honour?'

Distantly she registered his awareness that defending her honour was the whole purpose behind this mess. Knowing he hadn't forgotten surprised her, even as it warmed her slightly. But the other word he'd used threatened to close her throat. 'Husband? You can't mean...'

'Why not? Think about it, Princess. It could be the perfect solution for both of us. You wouldn't be committed to marrying Luc Armand. You get a husband you'd barely see because my work takes me all over the world so you could continue living here as a prim and proper wife, with the protection of my name and fortune but none of the burdens of marriage. And I get all those socialites' daughters off my back. As a married man, I'm no longer fair game and can get on with my life in peace.'

It was a solution she had never considered. 'You'd marry me, even though you don't care for me?' she whispered.

He shook his head. 'Caring for you is the easy part. As I said, you're beautiful and intelligent—a touch too headstrong, but taming you will be a challenge and I'm not one to shy away from a challenge.'

The thought disturbed her beyond all common sense until she reined in emotions which threatened to run away with her at any moment. Coming so soon after the experience of being in his arms, the prospect of marrying him strung her nerves wire-taut. Her anger was rapidly dissipating and she found herself clutching at it as if to an anchor. 'I may not prove as easily tamed as you think.'

'Then you agree, a marriage between us is worth considering?'

She *was* considering it, she realised with a frisson of panic. She pushed it away. 'First you have to win the challenge.'

His level gaze settled on her, allowing no possibility of failure. 'And when that's out of the way?'

'As you point out, I will no longer have the luxury of a choice. I will belong to you whether I will or not.'

'Do you will it, Talay?'

This conversation was rapidly getting out of hand. How had she progressed from hating him for treating her so cruelly when she was a scarred teenager in Australia to actually contemplating marriage to him? However logical his reasoning, the step yawned like a chasm in front of her. She looked away. 'How can I answer such a question? I barely know you.'

'You know more than you're admitting. You react like a startled fawn when I touch you, yet I can feel your heartbeat racing and taste the hunger for what I offer on your lips. That kind of knowing doesn't depend on background details. It's instant, like chemistry.'

'Maybe my so-called reactions are nothing more than chemistry,' she said, hating to admit how accurately he read her response to him.

'Don't underestimate chemistry,' he cautioned dryly. 'Empires have been built on it.'

They may well be built on this one, she thought with a shudder. The king's two children made her a distant heir to the throne, but an heir nonetheless. And she was under no illusions that a man of Jase's reputation would tolerate a marriage of convenience. The very thought unleashed a ravening torrent of emotion which required all her self-control to tune down to a thread.

The thread was enough to deny her the power of speech for a heartbeat. She almost collapsed with relief when Sam finally barged into the second cavern. After the brightness of the cliff face he stood blinking in the gloaming. 'Your Highness, is everything all right?'

She inclined her head regally, aware of Jase watching her, his expression amused. *Talk your way out of this*, his dancing gaze seemed to invite. Very well, she would oblige him. 'I'm perfectly all right, thank you, Sam. Mr Clendon found these caverns during a coastal survey and was anxious to show me his discovery. I hope you haven't been inconvenienced.'

'Not at all, Your Highness,' the bodyguard hastened to assure her. 'I would have appreciated some warning before you disappeared.'

As rebukes went it was mild, but Sam was entitled to be upset. 'You're right,' she acknowledged contritely, taking some satisfaction from Jase's lifted eyebrows. 'The entrance came as a surprise to me, too. I'm sure Mr Clendon will promise not to spirit me away again without warning you.'

'Certainly not while she's in your charge,' Jase contributed. The meaning wasn't lost on her. If they married he would have much more to say about it. She was thankful she had managed not to give him an answer, although it could become academic if he won the challenge. If he didn't... She refused to consider the option. Even if it meant she was bound to him, she would rather that than face having him injured or killed.

As she followed the two men out into the sunlight it occurred to her to wonder when she had begun to think of her freedom as a fair exchange for his life.

When they reached the cliff face Jase paused, shading his eyes with a hand as he scanned the terrain above them. 'You'd better wait here. It gets dangerously steep from here on.'

The rock from which the young men dived loomed above them. She lifted her head. 'If you can make it, I can.'

'More climbing?' Sam kept the groan out of his voice but she was sure it took an effort. He was commando-trained and superbly fit, but looked as if he was finding the cliff climb taxing. Yet Jase wasn't even breathing hard.

Jase looked as if he would object to her continued presence on the climb, then cursed under his breath and held out a hand. 'A challenge, indeed.'

Sam had no idea what he meant but she knew. Jase was accepting that taming a princess might be tougher than he had thought. She ignored his hand. 'I can manage.'

With a 'suit yourself' shrug, he applied himself to the jagged path, although 'path' was hardly the word. It was more of a knife-edge, slippery-sheer on the cliff side, dropping away to surf-ravaged rocks on the other. She was soon sorry she had rejected his help but she was not going to swallow her pride and say so. She ploughed doggedly on, knowing she would be a mass of aches and pains tomorrow. And this was only the climb.

Jase was the one who had to dive from this eyrie, she thought, horror creeping over her as the reality of the undertaking came home to her. They had finally reached the outcrop of rock traditionally used as a diving platform, and up here the full extent of the danger Jase courted was shockingly apparent.

'You can't go through with this,' she told him, not caring that he heard the quiver in her voice.

'If I don't, Luc wins by default,' he pointed out. 'You know what that means.'

They both knew it meant she would have no escape from her commitment to Luc and her eyes brimmed at the prospect, but she couldn't put her own feelings before Jase's life. 'It means you don't have to break your neck on my account. Surely the resort isn't worth your life?'

The sea breeze tugged at them so he hunkered down on the ledge, pulling her down and sheltering her with his body. 'What are you really afraid of, Talay? That I'll come to grief, or that I'll win and you'll belong to me?'

Both prospects sent tremors through her for vastly different reasons. She couldn't answer, without revealing how important he had become to her in the space of a few days. Even honouring her uncle's wish that she marry Luc Armand seemed preferable to what could happen to Jase if he insisted on going through with this.

'I don't want you to be hurt,' she insisted quietly. 'I'd sooner agree to marry Luc.'

Something unfathomable churned in Jase's expression. 'I believe you mean it.'

She lifted her head, steeling herself to remain outwardly calm although every fibre of her being revolted against what she was offering. It was her life or Jase's, but at least her choice didn't involve hurling herself off a cliff into treacherous water fringed by jagged rocks. 'I do. Will you agree to withdraw from the challenge?'

He shook his head. 'Sorry, Princess. You've just given me the best reason in the world to see it through.'

CHAPTER EIGHT

THE day of the challenge dawned almost obscenely bright and clear. The sun shone out of a cerulean sky and a breeze played with the palm trees around the Garden Palace, as it would play with the mangroves fringing the white-sand beaches below the Malakai Cliffs.

As she got ready to go to Malakai Talay prayed that the temperate weather would make the dive less treacherous than usual. She would give much not to have to watch Luc and Jase dive from the cliff, but honour demanded her presence. The two men were competing because of her, after all.

She was interrupted by her maid's discreet knock. 'You have a visitor, Your Highness.'

Talay tensed. Could Uncle Philippe have come personally to witness the challenge? She could think of no other likely caller at her inner sanctum today of all days. 'Who is it?' she asked testily. She was so on edge she didn't feel like being civil to anyone.

'It's me.'

At the sound of Jase's voice Talay spun around. He must have dismissed her maid and taken the woman's place at the door. Shock and surprise sharpened her voice. 'You shouldn't be here, Jase. How did you gain access to my private quarters?'

His taut grin threatened to turn her insides to jelly. 'Your bodyguard, Sam, is a decent bloke once you get to know him. He proved unexpectedly sympathetic to our cause. He arranged for me to drive in through a back entrance where I wouldn't be seen by the press, then personally escorted me here.'

She felt her eyes widen. During the endless night, when sleep had been an impossible dream, she had wished over and over that she had never pretended to be Allie and gone to meet Jase. If she had listened to her uncle the challenge wouldn't be happening. She would never have known the sweetness of Jase's kisses or the transcendent joy of his embrace, but that could have been for the best, too. What you had never had you didn't miss.

She *did* miss him, she admitted to herself, although they did almost nothing but disagree when they were together. In spite of all her reasons for disliking him, starting with his treatment of her as a teenager, when he wasn't there she missed the fires he ignited inside her. He made her feel more alive than ever before. Seeing him in her private apartments—knowing it might be for the last time—was almost more than she could handle. She masked her distress with anger. 'How dare Sam let you in, without checking with me first?'

Jase was at her side in two strides, clamping her arms against her body before she could reach for the bell to summon her staff. 'Sam isn't to blame—I am,' he insisted. 'Because of the challenge, he thinks I'm madly in love with you and I didn't disabuse him.'

Jase had interrupted her in the act of winding her traditional sarong around her slender body. Now the silken material bunched in her clenched fingers. Could part of Jase's attraction be because he was the first man to treat her as a normal human being, with no regard for her royal status? She hated to think it was true but was afraid it was so. She had always taken for granted the deference shown to her by others until she had met a man with no such inclinations, arguing with her more often than not and touching her how and when he felt like it, as he was doing at this moment.

She tried to stay angry at his cavalier behaviour, but knowing what lay ahead of him replaced her anger with a heightened awareness of him. His grip was so masculine and assured, and the feel of his body so provocative against hers, that her arousal was as instinctive as it was complete, despite her total lack of experience in such matters.

Jase was so worldly-wise he would probably laugh if he knew she was still a virgin. One of the drawbacks of being royal was having her life open to greater scrutiny than most people's. They might not judge another woman badly for sleeping with a man, but for a princess any indiscretion could create a scandal, as the photograph of her and Jase had proved.

In spite of the uncertainty her experience made her feel around Jase, Talay did not regret her virginal state if it meant waiting for a man she loved above all others. Such a man would understand and treasure her gift to him. It would not be a curiosity, as she feared it would be to Jase. His playboy reputation suggested he would have little patience with the fumbling of inexperience.

She took a deep breath and mustered her voice. 'Very well, I'll allow you to explain your intrusion before I call my staff.'

He took his time about releasing her, allowing his hands to slide down to her hips in a caress meant to send her internal temperature soaring. He had the satisfaction of seeing her breathing quicken as she looked away, but not before he glimpsed the intensity of her reaction. In a voice much less steady than it should have been, she said, 'I'm waiting.'

Desire for her tore through him so powerfully that he clenched his fists to keep them away from her silk-cocooned body. If today's dive went wrong this could be their last private moment—his last of any kind, he thought grimly, almost embracing the thought.

It gave him the strength to resist, despite every masculine instinct urging him to make such sensational love to her that she would never forget him. But, in case he failed today, letting her forget him was the one kindness he could do for her. He made himself laugh, recognising the hollowness of the sound. 'Still up on that high horse, Princess? When will you learn that it's wasted on me?'

'It seems most codes of civilised behaviour are wasted on you.'

He spread his hands, palm upwards, in a gesture of innocence. 'I'm being perfectly civilised. Heroic, almost, given what my instincts urge me to do whenever we're together.'

She was intrigued in spite of herself. 'Explain.'

His glance travelled from her to the sumptuous bedroom visible through a pair of half-open eaglewood doors. 'We're both adult enough to work it out.'

'Oh.' Warmth surged up her neck and into her face as she realised what he meant. The image which sprang to her mind was so sensual and vivid that she went hot and cold by turns, then swayed as the room tilted around her.

He caught her before she fell, and eased her gently onto a silk-covered chaise longue. She waved away his concern. 'I'm all right, really.'

He frowned. 'So it appears. Have you had breakfast yet?'

After her disturbed night she hadn't felt like eating anything. She shook her head, earning a frown of disapproval. 'I thought not. You'd better ring for your maid again and let her fetch you something to eat. It promises to be a long day.'

Longer for him than for her, she thought with a sharp stab of anguish. She looked up at him in genuine contrition. 'Jase, I'm really sorry I got you into this.'

His expression gentled. 'You didn't get me into anything. I could have refused the challenge.'

'And given up your dream,' she acknowledged softly.

He lifted his head. 'I've given up more important dreams in my time.'

The intensity of his gaze on her as he said this made her wonder if the dreams had somehow involved her. The thought brought a lump to her throat, but now wasn't the right time to ask. Jase needed to focus on the task ahead. 'Why did you come here?' she asked, as an afterthought.

'I know this day will be hard on you. I wanted to be sure you're all right.'

She almost choked. He was facing mortal danger and he wanted to make sure *she* was all right. He was definitely a man unlike any other. 'I don't know what to say,' she murmured.

'You could say thank you—and wish me luck.'

She grasped his hand, feeling some of his strength flow into her. 'Thank you, Jase. I...I'm glad you came.'

He brought her hand to his lips, kissed her fingers with heart-rending gentleness, then released her with what she could swear was reluctance. 'I'd better go. I'll see you at Malakai. I gather the villagers have created quite a carnival atmosphere down there.'

Her staff had reported the same thing. 'A marriage

challenge is a rare event these days. Just as well Malakai is off the tourist track or you'd probably have busloads of tourists there as well.'

He grimaced. 'I'm thankful for small mercies.' He gave her shoulder a reassuring squeeze. 'Don't look so worried. I don't intend to lose today.'

Surprisingly, she didn't want him to, no matter what the consequences for her own future. There would be time enough to worry about herself later. For now all her thoughts were for Jase. Her heart felt tight in her chest as she watched him walk towards the door. Somehow she found her voice. 'Good luck.'

He turned and sketched a salute of thanks. 'Have something to eat, and I'll see you out at the cliffs.'

Then he was gone and the room seemed suddenly desolate. The breeze from the closing door made the folds of the sarong billow around her like a silken cloud until she secured the garment with a gold clasp, bearing the Rasada crest. With a curious sense of detachment she noticed that her fingers were shaking. Jase was right, she should eat something before she left. She only hoped the churning inside her wouldn't make it impossible.

Her inner turmoil wasn't exclusively due to lack of nourishment, she admitted to herself. It had a lot to do with Jase's unexpected visit, which had left her exhilarated and furious by turns. He was impossible to fathom—one minute manhandling her and ordering her around as if she was a servant instead of a princess and the next setting aside his own concerns to ensure she was all right.

His admission that he was tempted to make love to her heated her blood to an alarming degree, although imagining him as a lover was not only futile but reckless, she told herself. Nor was she fooled by his suggestion that they should marry. If he won the challenge he wouldn't want to burden himself with her when there were many more fish in the sea. Monogamy was hardly Jase's style. So he would win—he *had* to win, she thought, choking off any other possibility—and he would set her free of obligation to him, returning to the life her foolishness had interrupted.

He would have his victory and his resort because she could hardly continue to oppose a man who had risked his life for her. And she would have nothing but memories. She would no longer have to marry Luc Armand, but it was small consolation since she wouldn't have Jase either.

What on earth had possessed him to call at the Garden Palace? Jase asked himself as he steered the rental car towards Malakai. Until then he had felt surprisingly composed, considering what he had to do today. He wasn't nervous, or at least no more so than when he had raced before the eyes of the world in the Olympics. Of course, he had been quite a few years younger then, he thought wryly.

Obviously those years hadn't taught him much sense or he wouldn't be doing this at all. He wasn't kidding himself it was going to be easy. Visiting the cliffs with Talay had forewarned him that the dive would be a true test of skill and courage. His opponent, Luc Armand, possessed both, with the added advantage of local experience.

On the other hand, Luc had never faced Olympic competition and won. Nor did he have Jase's notorious bulldog determination. Pitting yourself against odds you knew you could beat was hardly a challenge. Using Michael Martine's computer, Jase had researched his competitor's background and discovered that Luc preferred to enter competitions where he knew he stood a fair chance of winning.

It was the one edge Jase felt he had today. He knew Luc Armand, but he would bet his last dollar that Luc was so sure of himself that he hadn't troubled to check out Jase Clendon. If he had, he would know that Jase was also a competent diver. He'd had to develop the skill to crew aboard the big racing yachts. More than once he'd had to dive into a heaving ocean to aid a fellow crew member. The experience would stand him in good stead today.

He took a quick inventory of himself. His hands on the steering-wheel were rock steady, his nerves on edge but under control. He had refused to allow the possibility of defeat to enter his consciousness in the slightest degree.

Until he had seen Talay Rasada again for what they both knew could be their last time alone.

The sight of her, wound in that silky sarong-thing, reminded him of a butterfly about to emerge from a cocoon. And what a butterfly she would make. He grinned at himself in the driving mirror. There was something delightfully pure about her which defied examination, as if she had barely tasted the delights of the flesh. Every time he kissed her—and he was surprised how many times he'd managed to find an excuse—he got the feeling it was a novelty for her.

'Crazy man,' he muttered to himself. Such innocence had to be skin-deep. No woman reached twenty-six without *some* experience these days, even a princess. Maybe he was a first for her, making her forget all previous encounters. It was a satisfying thought but also terrifying. Her attractiveness raised today's stakes almost beyond tolerating. She thought he had accepted Luc's challenge out of pride, but she was the real reason. No matter what the outcome for themselves, Jase couldn't let her marry a man she didn't love so he had to win today.

'It's time to leave, Your Highness.'

Sam's voice on the intercom told her there could be no more delays. She had managed to eat some toast and felt better as a result, but was still far from ready to face the day. 'I'm on my way. I'll meet you at the east gate,' she told her bodyguard.

It was the entry Jase had used this morning. She didn't feel up to running the gauntlet of the press contingent Sam had warned her were staking out the main palace gates. Thank goodness her uncle had forbidden live television coverage of the event, she thought. He hadn't wanted to encourage copycat challenges. The King had also mentioned he intended to introduce a law banning future challenges. It wouldn't help Jase but at least no one else would suffer the hell of such a day.

Jase was right. The normally deserted beach at Malakai was crowded with local villagers, treating the occasion as a holiday. Food stalls had been set up near the line of mangroves and children played in a bowl-shaped lagoon formed by a cleft between two cliffs. She was led to a thatch-roofed shelter where a carved wooden chair awaited her. Jase and Luc sat under the shelter, barely acknowledging each other's presence.

Seeing them together came as a shock. She had always thought Luc handsome, with his rippling muscles and skin the colour of mahogany. His hair and eyes were a devilish black and, at twenty-eight, he was in his prime.

Alongside Jase, however, Luc looked what he was—

a hot-headed young aristocrat for whom life's way had always been smoothed by money and connections. Her heart swelled as her eyes met Jase's. Physically larger than Luc, he looked tougher, more seasoned by his hard upbringing, the picture of a self-made man.

'Hello, Luc, Jase,' she managed around a dry throat.

Both men got to their feet as she approached and Luc reached for her hand, pressing the back of it to his lips. 'Today is for you, my princess.'

She resisted the urge to say, 'grow up', but the phrase screamed through her mind. 'Do you still insist on going through with this?' she asked. Luc alone had the power to withdraw his challenge.

Luc glanced at Jase and nodded his head. 'This foreigner has sullied your good name. My good name, since they are linked. The challenge is the only answer,' he said in Sapphanese.

Jase probably understood the intent, if not Luc's actual words. Since she had been pictured kissing Jase with as much abandon as he had kissed her, she could hardly be called a victim, so that left Luc's good name to be avenged. 'Jase, are you prepared?' she asked in English.

He made no extravagant gestures or speeches but his eyes lingered on her, resting for the longest time on her mouth which felt as if his kisses were imprinted there. She resisted the urge to lick her lips. He merely set his shoulders in a 'get-on-with-it' gesture.

'Then let the challenge proceed.'

A sigh rippled around the small crowd as if they had feared she would call off the event and deprive them of their entertainment. She only wished there was some way she could manage it. But the law was the law.

The two men drew lots. Jase won and elected to dive first. It was so typical of what she knew of him that she almost smiled. Any such inclination vanished as he started to climb. She pressed her hands together to stop them from shaking. Without the diversion to the Princess Caverns the climb was swift, but, to Talay, it felt like an eternity.

By the time he reached the rocky outcrop he was a distant figure against the sky. He paused for a moment, assessing wind and sea, she assumed, then with no further hesitation launched himself into the air.

Like a great bird plummeting into the ocean to feed, Jase arrowed through the air in a sleek line. She held her breath. She had heard that high divers hit the water at over thirty kilometres an hour. The waves would feel like concrete when Jase impacted with them at such a speed.

She could hardly bring herself to look but forced her eyes to stay open. Down, down, down, Jase flashed until he cut the waves in a surge of spray. A collective gasp went up from the crowd.

When he surfaced and waved, the cheers were deafening. At her side, Luc was sullenly silent. Jase had survived. If Luc did the same a committee of the elders of the province would have to decide which contestant's diving technique was the more skilful. Jase's had been faultless. She could almost hear Luc wondering if he could do as well.

Luc had already begun his climb when Jase padded back to the shelter, water streaming from his magnificent body in ribbons. His chest heaved with exertion but he managed a smile as he accepted a towel and slung it around his broad shoulders.

She should do Luc the courtesy of watching him, but it was hard to tear her gaze away from Jase. His movements were assured as he towelled himself off but his eyes were guarded, as if the dive was only the first challenge awaiting him. It came to her that he was more vulnerable than he allowed the world to see, and her heart tightened in her chest in involuntary response.

She managed to look away in time to see Luc step to the edge of the outcrop. Like Jase, he paused to assess the conditions then lifted his arms and knifed through the air. His training showed in the arrow-straight line of his body as he flew towards the ocean. Even to Talay's unpractised eye, Luc's dive was the more skilful.

He entered the water as cleanly as Jase had done and she held her breath. Unlike Jase, Luc didn't surface immediately. 'What's taking him so long?' Jase muttered, his eyes fixed on Luc's point of entry.

The crowd was silent, scanning the waves for signs of the athlete. The seconds ticked by until Jase stood up. 'Something's wrong. I'm going after him.'

'Jase, no.' He had survived one encounter with the savage sea. It was tempting fate to try again. One of the pearl divers could go to Luc's aid. Before she could order it Jase sprinted to the rocks and took a running dive into the heaving ocean.

Oh, God, she couldn't bear to lose him now, she thought as he disappeared beneath the waves. Watching him dive off the sheer cliff had been torture enough but this was unendurable. She was barely aware of following him to the water's edge until Sam's arm pulled her back from the brink.

Suddenly a dark head broke the surface. Her knees weakened. It was Jase, and he had Luc in a lifesaver's hold, swimming sidestroke with the other man braced in one arm. Luc looked as if he was barely conscious.

Willing hands helped Jase out of the water and took his burden from him. A doctor was already on hand and went to work on Luc as soon as he was safely ashore. Jase returned to Talay's side, streaming water, his chest heaving. He looked exhausted. 'He must have hit his head when he went into the sea,' he said, sounding as if speaking pained him.

Hungry for assurance that he was all right, she let her eyes devour him. 'Are you all right?'

He shook his head dismissively. 'Never mind me. What about Luc?'

She relayed the question to the doctor, who nodded. 'He probably has concussion. I'll admit him to hospital for observation as a precaution.'

When she approached Luc to thank him he closed his eyes, still groggy or pretending in order to salvage his pride, she couldn't tell. She rested a hand briefly on his shoulder. 'Thank you for defending me, Luc.' He had been defending his own honour, but it cost nothing to be gracious. As he was carried off the beach she resisted a guilty feeling of relief because he had no further claim on her.

A robe was fetched for Jase, and the rest of the morning passed in a blur. Jase was the hero of the hour, honoured and fêted with song, dance and traditional foods cooked in beach ovens. He ate almost nothing of the delicacies put before him, she noticed. Was it her imagination or did he look grey under his Australian tan? It hurt to think he might be regretting today because it meant that she belonged to him under the law. He must know she wouldn't hold him to such an archaic principle.

He held himself stiffly to accept the accolades but shook his head when she asked if he wanted to swim. 'I've done enough of that for one day.'

She was sensitive enough to him to register his fleeting grimace of pain and an echo of it knifed through her. 'Jase, what's wrong?'

He hesitated, as if debating whether to share it with her. 'As I fought to get Luc to the surface the current slammed us both against the rocks. I'm afraid it has dulled my enthusiasm for the festivities.'

When he sagged in his chair alarm bells rang in her head. 'You're hurt. I'll have someone fetch the doctor.'

She could see the effort it cost him to force himself upright. 'No, that would only complicate everything. Is there any way we can get out of here gracefully before I ruin everything by collapsing?'

CHAPTER NINE

'YOU'RE coming back to the palace with me.'

Jase looked as if he wanted to argue but lacked the energy. Swiftly Talay gave orders to Sam to fetch her car and arrange for someone to return Jase's rental car to Alohan. Then she helped Jase to his feet and straightened her slight frame under his right arm, making it look as if he was supporting her rather than the other way around.

The unusual arrangement attracted curious glances as they moved painfully back to the car. Let people think what they liked, she told herself. Jase had fought for her and won so he was entitled to his prize, if that was how it appeared.

She didn't relax until they were safely back in her car, heading back towards the city, with Sam at the wheel. She would have preferred to get there with all speed but had asked Sam to drive carefully, jolting Jase's injuries as little as possible. On touching him, she was shocked at the clamminess of his skin, despite the heat of the day. 'You're burning with fever,' she said. 'I wish you'd agree to see the doctor.'

He settled his head against the leather headrest but kept his eyes open. 'And undo all the good I did today? Not a chance. Drop me at the Martine villa. All I need is somewhere quiet and private to rest and I'll be fine.'

'I doubt it. You could have punctured a lung when you hit that rock.'

He shook his head. 'As a yachtsman, I've cracked

enough ribs to know what it feels like. I can breathe well enough.' He demonstrated but went white and pressed an arm against his side, breathing shallowly until the pain passed. 'Maybe not well, but I'll live, sorry to say.'

Her eyebrows arched. 'Why should I be sorry?'

His head lifted and his eyes taunted her. 'Aren't you? I did win the challenge.'

He didn't need to say it. They both knew she was bound to him by custom until he released her, but in her wildest dreams she had never expected him to exercise the right. 'You can't mean...' She couldn't bring herself to say it.

He regarded her with infuriating certainty. 'To possess you?' he said. 'It's quite a bargain. One dive off a cliff and a few bruises to claim a princess.'

She held herself away from him, the audacity of his statement leaving her breathless. 'It may be the letter of the law, meant to deter frivolous challenges, but it was never intended to deny the woman freedom of choice.'

'You'll have plenty of freedom...during the daytime,' he promised lazily, sounding as if he was enjoying himself in spite of the pain.

A wave of primal heat surged through her. She couldn't meet his eyes. 'You can't be serious. The King will never permit me to become your consort under these conditions. You don't even intend to remain in Sapphan.'

Jase's gaze travelled the length of her body, as if he was committing every detail of her to memory and relishing the exercise. 'Staying here is becoming more attractive by the minute. And I know King Philippe. He considers himself bound by your laws like everyone else. He won't fight me on this.'

'But I will,' she said through clenched teeth. 'I will not become your possession in any sense of the word.'

He stirred and caught his breath as the movement pained him. His suffering stabbed at her, too, but she fought the urge to minister to him. She wasn't sure yet what he proposed but it seemed he wasn't willing to release her under any but his own terms. His next words confirmed it. 'Are you sure it isn't already too late?'

Her head snapped up. 'It's far from being too late. I foolishly allowed you to kiss me and I can hardly deny that I enjoyed it. But I've been kissed before and enjoyed that, too. It doesn't mean you have any more claim over me than any other man.'

'The law disagrees with you,' he insisted, the steely undercurrent in his voice belying his quiet tone. 'Are you saying you're above the law, Princess?'

'Of course not. And stop calling me Princess. You make it sound insulting.'

'Talay, then.' The intimate way her name rolled off his tongue, the syllables broadened by his Australian accent, sent a tremor through her. She should have stuck to Princess. 'It makes sense if we're going to be husband and wife.'

Weariness hit her like a wave. 'Why are you doing this, Jase? To prove that you can? All right, I concede you have the right. But you don't want to marry me. I'd only cramp your style with other women.'

It was surprising how much the admission hurt but she refused to explore the reasons. He hadn't denied his playboy reputation so why on earth would he want to saddle himself with a woman who didn't even l-like him, she finished the thought, aware that she had almost used another word altogether, one which had no place in the strange relationship between them.

'I've already told you why a wife would be... convenient. I can be faithful to you within your borders, without having to become a monk elsewhere in the world.'

Her derisive look was meant to scorch him although she doubted its effectiveness. 'You think I'd agree to such an arrangement?'

'It should suit you, too. You wouldn't have the King matchmaking for you any more, and when I'm away you could live your life as you chose.'

'Except for being married to you.' The admission brought a lump to her throat at all the term implied. The thought of marrying Jase repulsed her, she told herself, at the same time afraid that what she felt was the very opposite of repulsion. 'What would happen if...if we have children?'

His expression hardened. 'I'll make sure we won't. No child of mine will be brought up in the kind of loveless environment I endured.'

'Just because a marriage is made for reasons other than love, it doesn't have to be loveless,' she pointed out, everything in her resisting his cynicism. 'There are other kinds of love such as caring, trust, a common history...'

'None of which applies to us,' he cut in, pain making his voice deeper than usual.

She was about to protest until she remembered that his first wife had trapped him into marriage by becoming pregnant. 'You're right, a business arrangement hardly creates an ideal climate in which to bring up children.'

His eyebrows arched. 'You're actually agreeing with me?'

'No, but you don't seem to require it as a condition of marrying me.'

His sharp laugh turned into a gasp of pain as he braced his damaged ribs, but he waved away her concern. 'I'm OK. You're the first woman I've known who gives me as good as she gets. This promises to be interesting at least.'

And he was the first man who took no account of her royal status, she thought. He treated her as he would any other woman. If only he didn't need the other women... She reined in her thoughts. Where had that idea come from? It was just as well they were turning into the palace gates because this was dangerous territory indeed.

When he saw where they were Jase looked annoyed. 'Why are we here? I told you to drop me at the Martine villa.'

She pressed a finger to his lips, silencing him. 'It may not be for long, but for the moment I rule. You're seeing my personal physician and spending the night in the guest wing where you can be looked after.'

'Shrew,' he murmured around her finger, then closed his eyes against what she guessed was a wave of pain. She gasped as his lips tightened around her finger and delicious sensations tore through her. As she tried to pull free his tongue teased her fingertip. Her breathing gathered speed.

Then he released her and grinned. 'What was that about ruling me?'

She swallowed hard but made her tone light. 'It's time somebody did.'

'Somebody meaning you?'

How did they invariably end up discussing the personal when all she wanted to do was keep things businesslike between them? 'Perhaps I will,' she said on a regally dismissive note.

It was a measure of his suffering that he didn't argue

the point, although his expression said 'later for this discussion'. Sam drove in through the secluded east gate and straight up to the portico leading to her private apartments, where any number of willing hands waited to help Jase into the palace.

Word had spread about his victory in the cliff dive and no one found his presence surprising. Everyone but her took it for granted that she now belonged to Jase, she noticed irritably. Even the King, when he telephoned a short time later.

'I understand Jase acquitted himself spectacularly well. I almost regret not allowing television coverage. I would have liked to see his dive.'

Talay shuddered. She never wanted to see it again as long as she lived. 'You were right not to allow filming, Uncle Philippe. Luc was almost killed. If Jase hadn't dived in a second time to rescue him...' Her voice tailed away.

'But he did, and I understand from my call to the hospital that Luc is recovering.'

She felt a twinge of guilt that Uncle Philippe had already checked with the hospital when the same thought hadn't yet occurred to her. 'The doctor who assisted Luc after the dive said he would be held for observation and should be released tomorrow,' she said in her own defence.

The King paused. 'No regrets over the outcome, Talay?'

There were too many to count but she said, 'No, Uncle. I couldn't marry Luc. I don't love him.'

'Do you love Jase Clendon?' The King's voice rang with surprise.

'Of course not.' She couldn't possibly love him. 'He has some ridiculous idea of us staying together,' she added uncertainly, waiting for the King to laugh and agree it was impossible.

'He won the challenge. He has the right under the law,' the King astonished her by saying.

She went cold. 'You mean you'd agree to him marrying me? Not long ago you didn't want me meeting a man with his reputation.'

The King's outrush of breath whistled down the phone. 'The matter was decided when he bested Luc this morning. What happens between you is entirely up to Jase now.'

'I thought you'd be on my side, Uncle Philippe.'

He laughed. 'I am, my darling child. I want only the best for you.'

Which hardly included marriage to Jase, she thought furiously. She had counted on the King's support to convince Jase that his proposal was inappropriate. She had never expected her uncle to side with Jase against her. But he refused to discuss it further, wishing her well as he ended the call.

She was finally forced to contemplate a future as Jase's wife. It repelled her, she told herself, wondering why a contrary thrill ran through her at the same time. His kisses promised so much. They made her ache to know more, to know everything there was to know between man and wife.

She only hoped he would have patience with her inexperience. The thought that it might drive him back to the other women in his life was a sobering one, but she couldn't deny the possibility. It was just as well she didn't love him. If she did, such an arrangement would be beyond bearing.

The doctor had been and gone by the time she changed and joined Jase in the guest wing. She could already see an improvement in him. His colour was better and, although he held himself stiffly, every movement no longer appeared to cause him pain.

In the living room of the suite, he rested against a bank of tapestry pillows on a chaise upholstered in Fortuny fabric from Venice, managing to look completely at home in the ornate surroundings. Most first-time guests at the palace tiptoed around the valuable furnishings as if they could bite, but Jase's attitude seemed to say, 'furniture is furniture, to be used.'

His towelling robe was open to the waist after the doctor's examination. Her lungs felt as if steel bands were clamped around them as she took in the wide expanse of tanned chest with the faintest scattering of dark hair, arrowing down below the loosely tied belt. She had expected to find him bandaged, telling herself it would be easier to deal with him as a patient than a whole man. It didn't help that the thin material of the robe emphasised just how much of a man he was.

'What did the doctor say?' she asked, trying not to sound as breathless as she felt.

He gave her an I-told-you-so look. 'Just as I expected, I cracked a couple of ribs which were best left to heal by themselves. I can be on my way tomorrow.'

Her eyebrows rose suspiciously. 'Tomorrow?'

He had the look of being caught out. 'All right, within the next few days, as soon as I can manage without painkillers.' He passed a hand across his eyes, hinting at a weariness she knew he would never admit to. 'It's good of you to have me here. I'll be out of your hair as soon as I'm fit enough.'

If he made good his promise he would never be out of her hair, she mused, savouring the expression.

'There's no hurry. You can remain here as long as you wish.'

'My, aren't we formal all of a sudden? Is it the palace atmosphere making you go all royal on me, Princess?'

His comment stung. It wasn't the palace so much as his wish to marry her, which caused tension to stretch like steel filaments between them. 'I am royal,' she snapped. 'You may as well get used to the idea if we're to remain together.'

'And you may as well get used to the idea that you don't rule me.'

She held herself stiffly. 'You've made the fact perfectly clear. For what it's worth, King Philippe agrees with you.'

Jase caressed her with his eyes, provoking a shiver of response. 'Well, well. So we have the King's blessing?'

'He approves of us marrying, but only because winning the challenge gives you the right.'

'And you—will you marry me because it's the custom or because you want to?'

She wrapped her arms around herself. 'Must we discuss this now?' If he forced the issue she was terrified of what answer she would give.

He nodded. 'You're right. I'm too tired and doped-up with the doctor's potions to think rationally. Maybe in a couple of days I'll see the sense in letting you go and getting the hell out of Sapphan. Right now all I can think of is how much I want to pull you down on this couch and make love to you until you're dizzy.'

It was probably the painkillers talking but she found the idea more exhilarating than it had any right to be. He had the excuse of his injury and medication. What was her excuse? 'I'd better let you rest,' she murmured.

She was almost at the door of the suite before his voice stayed her. 'Talay?'

At the low tone her heart picked up speed automatically. 'Yes?'

'However things work out between us, I'm glad I won today.'

Because it meant he could go ahead and build his resort unimpeded, she assumed. Her vision blurred. How would she feel if his reasons had been...more personal? They weren't but she couldn't be less than honest. 'I am, too.'

Watching her go, Jase wondered at her last remark. She seemed genuinely glad that he had won today. Speaking for himself, he was glad as hell, but his reasons gave him plenty of pause. He had told himself he wasn't going to be driven out of any country, not by Luc Armand or anyone else.

Showing that blue-blooded young fool what a real man was made of was undeniably satisfying. What man didn't relish winning a test of courage and skill? But there was a lot more going on here. Jase had been glad to save Talay from being forced to marry a man she didn't love. He had never meant to hold her to this ridiculous notion that she now belonged to him.

So why didn't he tell her so, and let them both off the hook?

Through a haze of painkillers it was an effort to think straight, but at some level he knew he didn't *want* her off the hook. He wanted her in his arms and, if he was honest, in his bed.

But did he have to marry her? Princess or not, she responded to him more passionately than any woman he'd ever known. If he went about it the right way he could probably take her without the formalities. Love her and leave her, as he'd done to so many women before, including his ex-wife.

Damn it all, he knew it wasn't what he wanted with Talay. Not because she was a princess. Princesses were humans with feelings and passions just like everybody else. Mainly because Talay Rasada was special. At some level, he'd known it ever since his first sight of her in the hospital.

He stretched, grimacing as a stab of pain broke through the medication. He was hardly in any condition to argue this through now. In the morning he would probably laugh at his own craziness. Marry a princess, indeed. He might as well put a leash around his own neck and be done with it.

That it might be a velvet leash and extremely appealing was a possibility he didn't care to examine.

When she looked in on Jase next morning, Talay found him sprawled on a sun lounge on the terrace adjoining his suite, tapping notes into a laptop computer. He had exchanged yesterday's beach wear for a silk dressing-gown, loosely belted over matching pyjama trousers. 'What are you doing?' she demanded, concern lending an edge to her voice.

He looked up and smiled, creating the absurd impression that the sun had come out. 'I had one of your staff fetch my things from the Martine villa. I may be outvoted by the combined weight of the medical profession and the Royal Family, but I refuse to spend the next few days in swimming trunks.'

'You're supposed to spend them resting,' she said, glaring at the computer.

'My mind wasn't injured, only my ribs.'

Contrition gripped her. 'How do you feel this morning?'

He shrugged but his indrawn breath wasn't lost on her. 'I'll live, but I won't be diving off any cliffs for a few days.'

'You shouldn't have had to dive off one.'

He tapped a couple of keys and closed the computer, setting it on a table, then looked at her seriously. 'I won. Honour is satisfied. There's no point blaming yourself for something that's over and done with.'

Was it? She had spent most of last night awake and wondering what she wanted to have happen next. She had reached only one conclusion: she didn't want him to leave. But if he stayed it meant pursuing an involvement she wasn't nearly ready for. So what *did* she want?

Jase homed in on her uncertainty like a guided missile. 'My offer's still open, you know.'

'Offer?'

He straightened confidently but she noticed he braced his damaged ribs with his left arm. Despite his assurance that she wasn't to blame, a guilty feeling overtook her at the sight. 'Marry me,' he urged. 'It's the ideal solution for us both.'

She gave a regal toss of her head. 'For you, perhaps. You gain everything, without losing your freedom to enjoy a life of your choosing outside Sapphan's borders.'

'You're forgetting the most important part,' he said quietly. 'I gain you.'

If he only cared it would make a difference, but she knew he didn't. 'A trophy wife?' she asked, hurt turning to scorn in her tone.

His expression darkened. 'Perhaps, but one I'll treat with the greatest care and respect of which I'm capable. I'm hardly penniless, you know. I may not be royal, but

I can match your fortune and raise you a million or two. But this isn't about status, is it? It's about fear.'

She drew herself up. 'I'm not afraid of you, Jase.'

He shook his head. 'I didn't say you were. But you are afraid of love.'

When had he come to understand her so completely? She was fully dressed in a watered-silk house sarong but she felt naked in front of him, as if he could see all the way to her soul. 'You're wrong,' she denied in a tortured whisper.

'I researched what happened to your family,' he went on relentlessly. 'You've lost everyone you've ever loved. Friends and relatives in the major earthquake which put your Uncle Philippe on the throne, then your immediate family to a terrorist bomb which almost killed you, too, and finally your grandfather, Leon Rasada, to a series of strokes. He took care of you after your parents died, didn't he?'

She nodded, her vision blurring. Leon had taken her to Australia for cosmetic surgery and introduced her to the woman who not only restored her appearance but also became her queen, thanks to Leon's machinations. 'Why bring the past up now?' she asked in a strangled voice.

'For you it isn't past. It's why you live here alone, afraid to trust to love in case something terrible happens.'

White-hot anger flared through her. How dared he analyse her motives and reach conclusions which were none of his business? 'What if I do?' she demanded haughtily.

'Then you allow part of yourself to die with the others.'

She felt the colour drain from her face and whirled to

the edge of the terrace, grasping the parapet so hard that her knuckles whitened. Without looking at him, she said, 'By custom, winning the challenge gave you some rights over me, but they don't include the right to pry into my feelings. You can have your resort and all the hero-status you want in my people's eyes. But you can't have me.'

'Are you sure?'

He had moved so swiftly and quietly she had no idea he was behind her until he pulled her against his chest, his arms tightening around her. If she struggled she risked aggravating his injured ribs so she held still, the pounding of her heart the only sign of the havoc wrought by his touch the instant she felt his hard body against hers. 'Let me go, please,' she said as levelly as she could.

His lips moved against her hair and she heard his breathing quicken. 'I could insist on my rights here and now.' As she tensed in protest he made a shushing noise into her ear, as if gentling a startled creature of the wild. 'But I won't, at least not yet, on condition you agree to consider my proposal.'

'And if I still say no?'

He turned her around and pressed a kiss to her forehead, chaste, gentle, but so filled with promise that she trembled from head to toe. 'I'll have to find a way to change your mind. Your doctor insists I spend the next few days right here so I'll have plenty of opportunity.'

The promise was as sweetly tempting as it was dangerous and she knew that, whether Jase lived at the palace or not, he would lay siege to her emotions as effectively as he was doing now. Men like Jase didn't take opportunities, they made their own.

His message was clear enough: she may as well surrender to him now and save herself the trouble. Except that she was a princess and as stubborn as he was. 'You may not win,' she said huskily, wondering if she believed it herself.

Evidently he didn't. 'This isn't about winning and losing. It's about trusting yourself to love again. It's time somebody helped you to do it.'

'Somebody being you, I suppose?'

His gaze caught and held hers. 'Call it fate, karma, whatever. You know the old saying, "when the pupil is ready, the teacher will appear"?'

What he might teach her hardly bore thinking about. 'It's a poor bargain unless I stand to gain something if I win,' she threw at him.

His eyes snapped with amusement and something she could swear was respect. 'Fair enough. If you still feel the same by the time the doctor gives me the all-clear I'll release you from any obligation to me under the challenge.'

The thought of him abandoning his claim to her was surprisingly bleak but she forced a smile. 'Will you stop building the resort at Crystal Bay?'

'No,' he said shortly. 'That's non-negotiable. The area needs the stimulus. Your freedom is the sole prize. Agreed?'

A lump the size of Sapphan threatened to choke her, but she nodded. 'Agreed.'

CHAPTER TEN

ON THE surface Talay's task was simple enough—to remain distant from Jase until the doctor pronounced him fully recovered. She hadn't counted on Jase making it so difficult. She couldn't put her finger on exactly what he *did* but he managed to make her feel more cherished than she had ever felt before.

Small gestures, such as a single flower left at her place at dinner or on her pillow at night, were part of his campaign. At first she couldn't work out how he got around the palace so easily in his injured condition. Then she caught her maid, delivering a rose to her bedside with a note attached in Jase's handwriting. It turned out that he had conscripted most of her staff to his cause. 'Why are you helping Mr Clendon, Ria?' she asked her maid.

The woman blushed. 'He's so charming, Your Highness. It's utterly romantic to see him so much in love with you.'

Charming, yes—romantic, certainly—but in love with her? It was inconceivable since he had proposed marriage while reserving the right to see other women. 'Mr Clendon doesn't love me,' Talay said, picking up the single perfect rose. Its languorous perfume enveloped her in a sensuous cloud. 'This is only a game with him.' With her as the prize, she thought but didn't add.

The woman frowned. 'I'm sorry if I did the wrong thing, Your Highness.'

Talay hastened to reassure her. 'You didn't. Mr

Clendon can be...persuasive.' How persuasive she knew only too well, 'He'll be leaving the palace soon. Until then, there's no harm in playing along.'

Left alone, Talay removed the note from the flower's delicate stem. The message said, 'Meet me in the rose garden for dinner tonight.'

Dinner in the garden? Was the chef in league with Jase too? To her annoyance Talay found her cheeks growing hot at the thought of dining with Jase in such a romantic atmosphere. At night, the perfume of the roses would provide a sensual aphrodisiac of which Jase must be well aware. She was rubbing the velvet petals of the rose against her face, breathing in the seductive fragrance, when she came to her senses. This couldn't go on.

Jase was in what had become his usual place, out on the terrace working on his computer. He looked up, his expression shifting from dark to light when he saw who it was. She placed the rose atop his keyboard. 'This has to stop.'

His look was innocence personified. 'You don't like roses? They're supposed to be your favourite flower.'

'That's the whole point. You've obviously researched my preferences as thoroughly as you researched everything else before coming to Sapphan. But I don't wish to be another of your...projects.

'What do you wish to be, Talay?'

'Left alone.' Even to her ears, the phrase lacked conviction. No wonder Jase looked doubtful. 'Why are you so determined to pursue me when we mean nothing to each other?' she asked. For lack of conviction, that took the cake. She might find his playboy ways offensive, his personal style forceful, and his blatant sensuality overwhelming, but she couldn't pretend that he had no effect on her.

He set the computer aside and twirled the rose between strong fingers. 'Marriages are made for many reasons, business among them.'

She made no attempt to disguise her hurt feelings. 'Is this business to you?'

His intense gaze bored into her. 'Whatever is between us goes way beyond business and we both know it. There's no other man in your life. You don't want any other man. The King won't rest until you're safely married so why not to me?'

She swallowed hard. 'You make it sound simple.'

He leaned forward. 'It *is* simple, so why not say yes?'

She shook her head, refusing to be caught out so easily. 'You're well aware that in Sapphan marriage requires no more than the informed consent between the two parties to become fact. If I say yes now I become your lawful wife.'

His eyes danced. 'The thought never crossed my mind.'

'I'm sure it didn't.' She stood up. 'I saw the doctor arriving after breakfast. What did he say about your ribs?'

'They're healing well enough for him to cut down on the painkillers I'm taking. He prescribes no strenuous exercise for a few more days.'

So it wasn't over yet. Talay tried to be sorry but a frisson of pleasure surged through her. So he would remain in the palace for a little longer. 'Then we can have that dinner in the rose garden,' she agreed, wondering if she was crazy. Meeting him there was playing with fire, but the flames were extraordinarily seductive.

He nodded as if her agreement had never been in

doubt. 'Wonderful. The chef has volunteered to prepare some Australian dishes for a change. I think you'll enjoy them.'

Bread and water would be wonderful in such a setting, with such a man, she found herself thinking. She chased the thought away. 'I came to ask if you'd like to play a game of chess?'

Jase stretched with some care, not entirely out of pain despite his bravado, she noticed. 'Thanks, but the doctor says I'm OK to go for a drive so I plan to visit the resort site. A few matters require my attention.'

Her response was involuntary. 'Jase, you can't. I forbid it.' Travelling the rough road could damage his half-healed ribs.

His eyes darkened and a frown etched his forehead. 'Trying to pull royal rank again, Princess? I thought we agreed the resort is non-negotiable.'

'You agreed. I didn't,' she tossed back at him. Admitting that her concern was for his welfare would be far too revealing. It was better to let him think she objected to the trip because of her opposition to the resort plans.

He shrugged. 'Suit yourself. I'm going out there, anyway.'

Her chin lifted. 'Then I'll come with you.'

'The hell you will. I don't need a watchdog tagging along.'

It was truer than he suspected. She was only going to protect him from himself. 'All the same, as your adviser to the project, I'm coming. Otherwise you may find it impossible to obtain a car and driver to take you there.'

He glared at her. 'You tempt me to drive myself just to prove you wrong. Unfortunately, the doctor's potions don't make it advisable, even at a reduced dosage.'

'Then you'll take me with you?'

'What choice do I have? Tell Sam to meet us at the east gate in fifteen minutes.'

She nodded, biting back the reminder that this was her palace and her staff he was ordering around. At some level she realised she found his masterful behaviour stimulating. All her adult life others had looked to her to take charge. Having someone else make decisions for her was a refreshing change. How she would regard a lifetime of it, she wasn't sure, but for now it was something of a novelty. 'I'll change and meet you there,' she told him.

She wouldn't have been surprised if Jase had convinced Sam to leave without her, but they were waiting at the east gate as agreed when she came down. She had exchanged her house-sarong for navy trousers and a silk blouse in her favourite cerise. A tiny frog was embroidered on the pocket.

Jase's appraisal was warmly appreciative. 'You should wear that colour more often. It flatters you.'

The compliment lifted her spirits to a degree she should have found alarming. But she smiled. 'Thank you.'

'What's the frog for?' he asked when they were under way. Sam had discreetly closed the partition between them, she noted. It reminded her that Sam was also under the ridiculous impression that Jase was in love with her.

'It's a family joke,' she said, reluctant to explain it to him. 'As a child, I told my uncle I'd have to kiss a lot of frogs before I found a prince. Now he won't let me forget it. He loves to give me things with frogs on them.'

'And have you?' Jase asked in a low tone.

'Have I what?'

'Kissed many frogs? Obviously, you haven't found your prince yet.'

She sighed involuntarily. 'I've decided I probably never will.' That it may already have happened she didn't want to consider. Jase was not, repeat *not*, her prince. So what was he to her? She had no easy answer.

She kept silent during the drive to Crystal Bay, but was disturbingly conscious of the man at her side. Their seat belts kept them apart and the roomy car with its excellent springs didn't throw her against him nearly as much as on the last journey. But she might as well have been touching him from the way his presence resonated through her so strongly.

By the time they reached Crystal Bay she wanted to feel his arms around her so much it hurt. It was just as well that Sam was with them or she might have done something incredibly stupid, like asking Jase to kiss her again.

If she agreed to marry him there would be no need to ask. She had no doubt he expected a real marriage, at least in the physical sense. Her senses ran riot at the very idea. His experience would make him a skilful lover, and she knew him well enough to be sure he would be a thoughtful one, caring as much about her pleasure as his own.

But was it enough? Could she endure the lonely times when he was out of the country, imagining him with someone else? With no children to console her, the prospect seemed bleak. Yet he was right when he said it was the ideal marriage for her. After losing so many people close to her, she didn't want to risk such anguish again. So what better solution than to ally herself with a man who wanted to marry for expediency rather than love? The tug of war inside her refused to be resolved.

She was glad the building activity provided a distraction from her confusing thoughts. But it was hardly a distraction to be welcomed. 'They've done so much,' she said on a startled note.

Where only a week ago there had been cleared land and frenzied activity, now the skeletons of buildings rose out of the rainforest, like dinosaur bones at an archaeological site. She recognised some of the people from the sea-nomad village and glanced at Jase in surprise. 'You're employing the villagers already?'

'I told you I would,' he confirmed. 'We're also looking at relocating the pearl trading centre to the resort. The woman who operates it loves the idea.'

Talay couldn't help noticing how much good-natured banter and laughter there was among the workers. She also noticed that many of the workers were in their late teens and early twenties, the age when many left the village to find work in the cities. 'Your employees look happy enough,' she concluded reluctantly.

'You sound surprised. What did you expect, overseers with whips?' he asked, but his mild tone didn't disguise the censure in his voice.

'Of course not,' she responded. 'I just didn't expect the villagers to be so happy about the resort.'

Her response seemed to satisfy him. 'They are happy because they have a future. The elders are happy because their children don't have to go away to work. You're the only person in the province who has a problem with this.'

She knew what the problem was. Her eyes told her he was right, but admitting it would take away the one barrier remaining between them and she was terrified of where it would lead.

'I don't—'

He didn't let her finish. 'Come with me.'

Grasping her hand, he towed her along a narrow path through the rainforest to a clearing. He halted in front of a building which was almost finished. She recognised it immediately. 'The honeymoon bure?' Why had he brought her here?

'I can hardly remain at the palace or the Martine villa indefinitely,' he said grimly. 'This bure will serve as my headquarters until the owner's suite is finished.'

'Why bring me here?'

'I had hoped…'

He didn't finish the thought. Had he hoped the sight of the building would convert her immediately to his cause? It was attractive enough, she saw, when he took her inside. The vaulted ceiling was modelled on a chief's house from ages past, with exposed beams and tallowwood floors polished to a soft patina. The effect was spacious and light-filled. The thoughtful location of windows and doors admitted soft breezes from the ocean, making air conditioning unnecessary.

At one end of the suite a simple mattress was raised on a wide wooden platform with a gauzy mosquito net draped over the head end, creating an island of intimacy. The angle of the platform ensured that the occupants would awaken to a glorious view of the sun rising over the ocean in the morning. Other furnishings were shrouded in dust covers, and a coffee-table was piled with files and plans.

She had been brought up to admit when she was wrong. 'It's a beautiful suite,' she said on a sigh because her rampant imagination insisted on imagining the suite being used for its intended purpose. 'All right, I admit the resort may be better for the region than I first thought.'

He went to a kitchen nook where a kettle and coffee-mugs testified to his recent occupancy. 'Would you like some tea?'

A lesser man would have capitalised on her admission, she thought. She could be equally gracious. 'Yes, thank you.'

His stiff movements reminded her that he was far from fully recovered. She moved towards him. 'Let me do that. You should be resting.'

'I'm fine,' he insisted gruffly.

'I haven't led such a sheltered life that I can't make tea,' she pointed out.

Ignoring her, he finished the task and brought two cups to the living area. She pushed aside some of the files to make room for them as he eased himself onto a chair with some care. 'I guess we both have a few preconceptions we need to change.'

'I'm sorry I jumped to conclusions about the resort,' she said.

He nodded. 'I understand your motives but you also need to know that I care about this area, too. When I first set eyes on it I thought I wouldn't mind spending the rest of my life right here.'

'And now?'

His head came up and his eyes blazed. 'Now the prospect is even more compelling.'

The bure had been designed for intimacy. Her senses started to spin and the building activity seemed to retreat into the distance. It was easy to imagine them as newlyweds, alone for the first time in this island of seclusion as Jase's look roved over her, heating her blood to an alarming degree. She began to regret asking Sam to wait at the car.

'Jase,' she murmured, hardly aware of having used his name.

'Talay.'

Without ceremony, he stood up and gathered her into his arms, his mouth seeking hers as if he had been hungry to do so for a long time. Her lips parted instinctively and his tongue began a teasing dance with hers which sent shock waves rippling through her.

His hands felt hot against the small of her back, pressing her against his hardness until every corded muscle was outlined against her. The intimacy of the embrace took her breath away, literally.

He transferred his mouth to her eyes, running his lips over her closed lids until her breathing became so fast and shallow she thought she might faint. It was as well his strong arms buoyed her up because her legs felt like jelly. She knew an instant of panic. 'Jase, I don't think—'

His mouth found hers, silencing her with another kiss which robbed her of all reason. 'Then don't think. Just feel. Even a princess knows how to feel, doesn't she?'

She did, but what she felt frightened her half to death. She was used to being royal and in control, not to feeling vulnerable and so out of control she was like a Catherine Wheel, spinning dizzily amid a myriad pinpoints of light.

Jase was her anchor and she clung to him, her arms linking around his neck as if she would fall if she let go. He didn't seem to mind as his hands slid lower, cupping her hips to press her ever more intimately against him.

When he lifted her legs and guided them around his hips, she knew a moment of confusion then instinct took over and she crossed her ankles behind him, letting him lift her off the ground, his fingers locking beneath her seat to pull them closer together.

Freeing one hand, he reached behind her and she heard the sound of a door locking. Her eyes flew open. 'What are you...?'

His mouth claimed hers again, silencing her. 'Shh. You're only supposed to feel, remember?'

She could do little else as he carried her in this fashion to the bed, where he eased them both down onto the soft mattress. Somehow her shirt buttons were unfastened and her trousers slid past her hips until the sea breeze whispered against her heated body. Jase's hands traced the pattern of the breeze, caressing her gently but irresistibly until her whole being felt ablaze with desire.

Was this love? she wondered dazedly around the confusing sensations gripping her. Her friends had tried to describe to her, without success, the all-consuming need to possess and be possessed, to know another person as intimately as you knew yourself, but Talay had never really understood until now. It was like a fever in the blood, torturous, bittersweet, terrifying.

She felt as if she was losing herself in the needs Jase aroused in her. Where was the ice princess now? Lost in a maelstrom of sensations she was powerless to resist. Holding onto Jase provided the only safe haven, and she linked her arms around his broad shoulders, pulling him down to her to embrace his sanity-saving solidity.

'Talay, are you sure about this?' he asked hoarsely as he continued to caress her in ways which imposed mind-shattering degrees of pleasure-pain on her already overloaded senses. 'If you want me to stop you'd better say so now, while I still can.'

The very thought made her arch her back in protest.

Didn't he share her need to strip away the last barriers between them? She shook her head. 'Please, Jase.'

He took her plea for the acquiescence it was, and slid her trousers all the way to her ankles where he helped her to kick them aside. Her bra came next, the wispy white lace floating unheeded to the floor to join the chaos of his clothing. The touch of his hair-roughened skin against her smoothness further inflamed her until it seemed impossible for the fires inside her to bank any higher, without consuming her utterly.

She drew a strangled breath as he sought and pleasured the most intimate parts of her, and she found it was indeed possible for the fires to go higher than she could ever have imagined.

When he lifted himself over her at last she felt as if she were about to explode, but her untutored body reacted automatically. Unable to stop herself, she tensed and cried out. Instantly, he swore under his breath and levered himself away from her. The change was so abrupt that she felt shocked until she remembered his recent injury. Was he in too much pain to continue? Guilt immediately assailed her. This was hardly the rest the doctor had prescribed. 'What is it, Jase? Are your ribs hurting badly?'

'Some, but it doesn't matter. This is your first time, isn't it?'

Understanding slowly seeped into her tangled thoughts. His injuries weren't the problem. She was. 'Everyone has a first time,' she said in a low voice.

He swung his legs over the side of the platform bed and rested his forearms on his knees. 'Not with me. At least, not like this.'

There was only one possible conclusion. Her inexperience had disappointed him. 'I thought you wanted to marry me.'

'I did. I still do. But it never occurred to me...' His voice tailed away but not before she had heard the censure in it.

'That there was any such thing as a twenty-six-year-old virgin,' she finished, unable to keep the bitterness out of her tone. 'In your world I must seem like a dinosaur.'

'Which is precisely my point,' he growled. 'We're worlds apart, but until now I hadn't realised how much difference it made. Of course you couldn't have experience with a man. It would be all over the front pages of your newspapers.'

It didn't change the facts. She didn't measure up to the other women he'd known. The thought left a bitter taste in her mouth. 'I'm sorry if I disappointed you,' she said stiffly.

He regarded her oddly. 'You haven't. I disappoint myself. This wasn't supposed to happen, Talay.'

'Then we can both be thankful it didn't,' she snapped, as a great wave of weariness overtook her. With as much dignity as she could muster, she reached for her clothes. She could hold herself together as long as he didn't apologise. She knew the flaw was hers. Somehow she had thought he wouldn't let it matter. It hurt to discover how wrong she was.

She began to get dressed. 'It's time I got back to the palace.' To her ivory tower, she added inwardly. Jase was right, it was where she belonged. But it had never seemed more lonely and isolated than at that moment.

CHAPTER ELEVEN

WATCHING Talay drive away, Jase could have kicked himself. There must be a dozen ways he could have handled the situation, all of them better than letting her think that he regarded her virginity as some kind of shortcoming.

She had left in too much of a hurry to let him explain that he wanted her first time to be a tender, joyful experience, not rushed through in a half-finished building because testosterone had got the better of him.

If she'd given him the chance, he could have told her he wanted her so much it hurt. The pain in his ribs was minor, compared with the urgency to possess her which throbbed through every bone in his body. He hadn't taken her, not because he didn't want to but because it wasn't right. She was a princess. She had saved herself for her husband this long. How could he betray her trust for his own selfish pleasure? Since she wasn't interested in explanations, he had no option but to escort her back to her car in silence.

As he walked back to the honeymoon bure alone, it occurred to him to wonder when he had turned into such a paragon of virtue. Probably right after meeting the grown-up Talay Rasada. Before then he would never have held back for anything but his own reasons. Now she even had him proposing marriage, and the hell of it was, meaning it. Not because he wanted the freedom to pursue other women, as he'd told her. But because he wanted her.

Wanted her or loved her?

He shied away from answering the question even in his own mind. After today it was probably academic anyway. She wouldn't want anything more to do with a man who had rejected the greatest gift any woman had to offer.

He should have insisted on returning to Alohan with her, he thought, furious with himself. Instead he had handed her into the car with Sam at the wheel, assuring her that the foreman would drive him back when his business was concluded. Lord only knew what thoughts were going through her head as she sat in that vast back seat alone. She was probably berating herself for her inadequacy when he should have been there to set the record straight.

The buck stops right here, he thought, facing the truth head on. He had known she was relatively innocent but had never expected to be her first lover. The discovery had caught him so off balance he hadn't said or done a single thing right. Which left only one option. As soon as they were alone tonight he would tell her the truth.

Satisfied with his decision, he sat down at the coffee-table and reached for the latest quantity surveyor's report. The figures should have absorbed all his attention but his gaze kept straying to the platform bed. One of the pillows retained the imprint of Talay's head and the coverlet was gathered in imitation of her slender shape. The faint jasmine scent of her perfume hung in the air. He inhaled deeply.

Looking down at his hands, he was amazed to find the papers quivering in his unsteady grasp. The painkillers must be wearing off, he told himself. The dull ache in his side reminded him vividly of how many hours had passed since his last dose. Funny, he hadn't even noticed the pain while Talay was here. He had only started hurting again after she left.

Pain wasn't the word Talay would apply to her state of mind. Mortification was more like it. How could she have been so stupid as to think Jase would want to make love to someone so untutored in the art?

If she became his wife he would teach her, she didn't doubt, but the memory of this afternoon would stand between them as a reminder that he hadn't proposed out of love but only to protect himself from the world's matchmakers. He would be doing her the same favour. It should have been enough but somehow she sensed it wasn't.

Distraught, she barely noticed that they were nearing the Martine villa until she caught sight of a familiar car in the circular driveway. 'Stop, please,' she ordered Sam. 'Allie and Michael must have come home early.'

This time Sam took no chances and checked to ensure that the Martines were at home before he parked the car around the back of the villa, saying that he would have a cup of tea with the Martines' chef, Elita, while Talay visited her friends.

Allie hugged her in delight. 'It's fantastic to see you, Tal. I was going to call you tonight.'

Talay explained about sighting the car as she drove past. 'I didn't expect you back until next week.'

Allie made a face. 'Paris was everything I hoped for and we had a wonderful time, but I didn't count on my pregnancy making me too ill to enjoy half of every day. Michael thought it was wise to come home.'

'Is everything all right?'

Allie patted the small mount of her stomach. 'I'm assured it's perfectly normal for a first baby. But what about your news? Elita told me about Jase being challenged by Luc over you, and winning. Apparently, it's the talk of the province. How on earth did it happen?'

Haltingly, Talay explained everything, including how she had taken Allie's place to meet Jase and been photographed kissing him. Allie's eyes went round. 'He didn't think I would kiss him, I hope?'

Talay had the grace to blush, recalling Jase's initial suspicions of his friend's wife. 'I'm sorry I did such a thing,' she said contritely. 'It will never happen again.'

Allie grinned. 'Hardly, since it means you now belong to Jase. Does he know about that part of the deal?'

Talay looked away. 'He knows. Oh, Allie, he wants to marry me.'

Allie patted her arm. 'You could do worse, you know. From what Michael tells me, Jase is a good man.'

'For a playboy,' Talay stated.

Allie shook her head. 'Don't believe everything you hear. Some men like to create a smokescreen around their true natures.'

'And some men really prefer to play the field rather than commit to one woman,' Talay observed soberly.

'You think Jase is like that?' When Talay nodded Allie looked bemused. 'He's one of my husband's oldest friends and you know more about him than I do.'

Talay chewed her lower lip thoughtfully. 'I doubt it. I don't think Jase lets anyone get close to him, except perhaps physically.'

Allie's jaw dropped. 'You mean you let him...that you and Jase...'

'We aren't lovers,' Talay shot in, uncomfortably aware that they would have been if not for her inexperience.

'I don't know if I'm relieved or disappointed. I can't

leave the country for a minute, without you getting into all kinds of trouble. It's just as well I'm home in time to save you from yourself.'

After today, Talay was unlikely to need saving. But it was wonderful to have her confidante back. As a princess she could confide in few people, making Allie doubly special. She hugged her friend. 'I'm glad you're home, anyway, even if you're too late to save me from myself.'

Allie read between the lines. 'Are you in love with Jase?'

Talay shook her head, feeling her skin scorch. 'Never. He wouldn't welcome it.'

'But he wants to marry you. He must feel something for you.'

She nodded. 'I'm sure he does, but never love. Marrying me would stop people aiming their daughters at him all over the globe, without restricting his lifestyle too much.'

'And what would you gain from marrying him?'

Talay had asked herself the same question. 'Uncle Philippe would stop matchmaking on my account.' She ticked off the main benefit on her fingers.

'And?'

She would have Jase as her consort, even if he didn't love her. He may not have royal blood but he was born to rule. There was ample evidence in the way he had adapted to palace life, handling the surroundings and staff as if he had been doing it all his life. 'I would be Jase's wife,' she said simply.

'It sounds as if he's getting the best of the bargain. In fact, I know he is,' Allie said loyally. 'Are you sure you can handle it? What if you fall in love with him?'

Talay straightened regally. 'I'll just have to make sure I don't.'

She steered the conversation off the sensitive subject of Jase and onto the Martines' recent travels. Admiring Allie's photographs of Paris and her purchases from the famous fashion houses occupied the rest of the visit.

When it was time to leave, Talay refused Allie's invitation to dinner, suggesting instead that she and Michael dine at the palace soon. On the way back to the palace, Talay remembered that she and Jase were supposed to have dinner together in the rose garden tonight. If he still wanted her company, that was. After today, she wouldn't be surprised if he had changed his mind about everything.

But he hadn't, she discovered when Sam returned her to her apartment. She recognised a four-wheel-drive vehicle from the resort site, parked beyond the portico, testifying that Jase's foreman had returned him safely to the palace.

As she passed under the portico she spotted Jase and his foreman on the terrace, poring over unfurled plans. She debated whether to join them then considered the impact her arrival would have on Jase's foreman. He was probably finding it hard enough to concentrate on his work in the opulent surroundings as it was, without the presence of royalty complicating things further.

She was also reluctant to face Jase just yet, she admitted to herself. She drew an impatient breath. Here she was considering her effect on the foreman. No one ever considered that *she* might be shy among strangers or find it hard to talk to them. As a royal she was supposed to be comfortable in any situation when, in truth, it was more often a superb job of acting. The Americans had a useful saying—'bluff your way through'. Pretend everything was fine. That's what she would do when she dined with Jase in the rose garden this evening.

To boost her confidence, she dressed in a formal sarong of shimmering white silk figured in gold. It showed off her olive skin to perfection and the artful wrapping flattered her slender figure. Her maid dressed her hair in a waterfall design, sleek and high at the back, curling under at the ends, with a spray of fine curls around her face. Like the sarong, it was deceptively simple, held in place by a few carefully placed pins. Completing her light make-up, she stood back and studied the effect in the mirror.

'Not bad,' she murmured.

Her maid looked affronted. 'Your Highness, you look beautiful.' She lowered her eyes. 'I trust he is worth it.'

Talay pretended innocence. 'He? I'm merely dining with my house guest tonight, Ria.'

The woman's eyes danced. 'Of course, Your Highness. In that case, as his hostess, you will take his breath away.'

Maids were paid to flatter, Talay told herself, but Ria's comments raised her spirits as she made her way to the rose garden. As a mother of four children, Ria was usually stoic and down-to-earth, managing to find ways to let Talay know when she didn't look her best. To have her unreserved approval was worth a lot.

She saw it again in Jase's expression when she met him in the gazebo in the centre of the rose garden and felt warmth steal over her. She didn't know what she had expected. Censure, perhaps. The coldness of rejection. Certainly not appreciation. It had taken courage to appear before him as if nothing had happened between them. She needed every bit of it now to keep her feet moving.

It was a perfect night, balmy with a soft breeze barely stirring the foliage. In the velvet darkness the roses spun tendrils of seductive perfume around her, drugging her with their scent and making it hard to think straight. In contrast, the gazebo was an island of flickering light, candles from the table casting a pool around the intricately carved structure.

At its centre a table had been laid for dinner with a white linen cloth, translucent china, bearing the Rasada crest in gold, and silverware which gleamed in the candlelight. Two cushioned chairs were set opposite each other at the table.

Jase was seated on one and her heart gathered speed as he stood up. His impeccably cut dinner jacket sat easily across his wide shoulders, the dark colour melding with the shadows beneath the soft glow of satin lapels. A dress shirt as white as the purest gardenias framed his rugged features, the candlelight making him look craggier than usual and a little forbidding, she couldn't help thinking.

Jase reminded her of a statue of an ancient hero which guarded the city square. Jase possessed the same air of warrior-artist, as if he could fight or make love with equal assurance. She gulped air at this last thought, unwillingly remembering the heaven she had so nearly found in his arms. Found and lost, she reminded herself as he came around the table to her.

She kept her voice steady with an effort. 'Good evening, Jase.'

He held a chair out for her. 'Your Highness.' This time his use of her title held none of his usual mockery. 'Your appearance gives the roses serious competition,' he murmured close to her ear as he seated her.

Considering that the roses were prize-winning examples of the grower's art and the pride of the palace, this was a compliment indeed. She felt her skin glow and looked down at her place. 'I doubt it, but thank you anyway.'

'It isn't a compliment, it's a fact,' he growled as he slid into the chair opposite her. 'As I should have had the sense to realise this afternoon. You're a beautiful, delicate hothouse flower and need to be treated with the same care.'

She lifted furious eyes to him. Part of Jase's appeal was that he *didn't* treat her as a hothouse flower but as a beautiful, desirable woman. She hated the idea of him changing his behaviour towards her simply because she wasn't a woman of the world after all. 'There's no need to apologise,' she said coldly.

He looked mildly amused. 'It isn't an apology. More like a...promise,' he said after a significant pause.

Heat coiled through her and her fingers trembled. She curled them around the stem of a wine glass, more for something to occupy them than because she wanted the wine. In fact, she felt as if she had drunk several glasses already, such was his unsettling effect on her. It took every ounce of royal training to say coolly, 'A promise? This sounds intriguing.'

He removed the wine glass from her hand and replaced it with his strong fingers, his blunt nails teasing her sensitive palm. She tried to pull free but his fingers tightened. 'If you keep adopting that royal air of polite indifference, I may be tempted to demonstrate exactly how intriguing. Would your staff be shocked to find their princess making love to a man in the rose garden?'

She was on fire now, the heat from his hand radiating along her arm and flowing into every part of her body.

'They can't be shocked by the impossible,' she said in a low voice.

In the flickering light his eyes held hers. 'Are you sure it's impossible?'

'You must know it is. This afternoon—'

He cut in before she could finish, reminding him of how she had disappointed him. 'This afternoon you bewitched me so completely I forgot that you aren't an ordinary woman, Talay.

It was out before she could stop herself. 'I wanted you to forget.'

'You succeeded better than you know. Only afterwards I realised you can't have led an ordinary life. My own experience with the paparazzi should have forewarned me but I wasn't thinking too clearly today.' Opening her hand on the table, he traced a pattern across her palm with his index finger. 'I guess I need more practice at wooing royalty.'

He needed more practice? 'How...how much more do you think you need?'

He lifted his hand, running the same finger across the crease of her lower lip in tantalising imitation of a kiss. 'A great deal.'

She swallowed hard, the words clogging in her throat, but they needed to be said. 'Then it isn't...over... between us?'

He gave a lazy smile which promised the sun, the moon and the stars. Her heart turned over. 'Princess, it has barely begun.'

The tight knot of tension which had built up in her all afternoon dissolved in a rush, but was quickly replaced by suspicion. 'Why, Jase? You can have any woman in the world. Why would you want someone who hasn't the experience or skill to satisfy you?'

'You're judging on limited evidence,' he ground out. 'I nearly made the same mistake.'

So he had judged her and found her wanting. She had to know. 'What changed your mind?'

'You did. After you left I tried to work but you haunted my thoughts.'

It was some satisfaction to find she wasn't alone in this, but she kept the observation to herself.

'I realised I hate it when you come on all royal on me because it points up the gulf between us. I'm from the back blocks. Brought up in boys' homes with no real education but what I've given myself.'

'It's hardly a crime to be self-made,' she pointed out, more distressed than she cared to admit. She had never dreamed he would imagine himself inadequate in any way.

'Don't get me wrong. I'm stating facts, not apologising,' he grated out. 'But your background is entirely different from mine, as this afternoon made me aware. So I need to change my strategy, which is what I plan on doing.'

'I'm not sure I understand.'

He leaned towards her, taking both her hands in his and meeting her eyes with an assurance that threatened her breathing. 'I intend to court you as a princess deserves to be courted. Forget my proposal of marriage. Pretend I haven't asked you yet. Before I ask again I want you to know exactly what you're getting into.'

She knew a disappointment so fierce it was like a gash across her heart. What if he decided not to ask again? 'Don't I get a say in this?' she asked around the pain.

'You made your point in my bed this afternoon, even if it wasn't the one you meant to make. You're an innocent in the art of love, Talay, and if I take advantage of your innocence I'll never forgive myself. As it is, I have a hard time living with what nearly happened between us.'

So did she but for different reasons. 'It wasn't all one-sided,' she reminded him bleakly.

His expression softened slightly but it was a shift from stone to wood. 'Don't you think I don't know it? If I didn't I wouldn't be here now.'

'You thought of leaving?' She couldn't keep the shock out of her voice.

He nodded tautly. 'I considered it. Maybe it's still the right thing to do.'

Was this his way of letting her down lightly? 'Next you'll be telling me I deserve better.'

'You do,' he said shortly. 'At least Luc Armand belongs in your world.'

That he could think such a thing was unbelievable. 'Perhaps, but not in my life. I could never love Luc.' Not after meeting Jase, she thought but didn't add.

'There are other men of your own type.'

She stood up, almost sending her chair crashing. 'If you want out, say so now, Jase. By custom, I'm the one bound to you, remember? Not the other way around.'

Moving so swiftly he was a blur, like a hunting tiger, he came to her side of the table and swept her into strong arms. His mouth crashed down on hers so hard it blotted out all conscious thought. There was no time to brace herself for the onslaught of sensations, only to let herself be carried along on the raging current.

Her lips parted in shock, allowing his tongue to invade her mouth with such devastating urgency that desire speared to the centre of her being. She clung to him, too stunned to return the kiss, so it became a one-sided taking from his giving. He didn't seem to care as he ravished her mouth, thrusting his fingers through her hair and down the sides of her face as if too much of her would never be enough.

By the time he ended the kiss she was a morass of needs and desires, her legs weak and her body trembling. He kept his arms tightly around her, as if sensing that she would buckle if he let her go. 'Does that feel like someone who wants out?' he demanded roughly.

It wasn't an answer and yet it was. She was probably reading into the kiss what she wanted to believe, she accepted, but she was powerless to stop herself. 'No,' she admitted. It felt like a man who would like to do much more than kiss her, but was restraining himself by a superhuman effort.

'Then don't say stupid things.'

'No, Jase.' The submissive response wasn't entirely faked as he guided her into her chair and returned to his side of the table. She felt as exhausted as a marathon runner at the end of a course. The kiss had drained her resistance, as perhaps Jase had intended.

He wasn't totally unaffected, she saw as he picked up his wine glass. His fingers were white around the delicate stem, as if he needed to break something to release his own inner demons.

He took a deep breath and raised the glass to her in a toast. 'Today never happened,' he said huskily. She opened her mouth to protest, desperately wanting to cling to the intimacy they had known, but he silenced her with a look. 'It's for the best, Talay. Trust me.'

It was a measure of how much she did that she was able to raise her own glass and join him in drinking to the day that never was.

CHAPTER TWELVE

TALAY awoke to the feeling that something was wrong. Then it came back to her. Last night she had agreed to pretend that Jase had never proposed marriage to her. Unfortunately she couldn't erase it from her memory so easily.

She should welcome his willingness to overlook her inexperience as a lover, and court her properly, but it bothered her for some reason. Maybe she didn't want to forget those moments in his arms as readily as he apparently did. Her very inexperience was what made them so special. The touch of his hand, his kisses scorching her mouth—she wanted to cling to them all as first times which would never happen again in exactly the same way.

She sat up in bed. That was the key. However raw her responses may have been, she didn't want to forget and Jase did. What did it mean? She wouldn't get any peace until she found out.

Affairs of state conspired to keep her away from Jase until mid-morning. He had breakfasted alone in his suite and she had glimpsed him working on the terrace, but was drawn away to her own office by a problem in the province. The whole time she was dealing with it Jase waited at the back of her mind. She was in a fever of anticipation by the time she slammed down the telephone in her office. 'Unless it's the King, I'm unavailable for the next hour,' she told her secretary. Her legs felt curiously stiff as she headed for the guest wing.

She was unprepared to find Jase packing.

'What's the matter?' she asked, catching sight of the open suitcase as soon as he admitted her to the suite.

He raked long fingers through his hair. 'I got a call an hour ago. The Thailand resort had run into trouble. They need me to intervene in some pretty big industrial relations problems which could derail the whole project. If that happens, hundreds of people could lose their jobs.'

Talay moved forward, a vast weariness overtaking her. This crisis struck her as oddly convenient for some reason. 'How long will you be away?'

'Only overnight. I wouldn't go at all if the local staff hadn't assured me that my intervention could make a difference.'

It did to her. Why not to his international army of employees? She snapped herself out of her apathy. 'Before you go, I have something to tell you,' she began. 'About last night—'

He was beside her in two strides, taking her by the shoulders. 'No, Talay.'

The pressure of his fingers bit into her tender skin, but she welcomed his touch as solid. Little else about this scene seemed real. She lifted her head. 'Do you intend to come back?'

The back of his hand grazed her mouth and his fingers trailed across her cheek. 'Whatever makes you think I won't?'

It wasn't an answer, yet she feared that it was. 'Then why won't you let me tell you...?'

The fingers returned to her mouth, this time pressing her lips to silence. 'It's the wrong time. This troubleshooting mission can't wait.'

'And I can?' Her tone was uncontrollably bitter.

'If I had a choice you wouldn't have to,' he assured her. 'But I don't so I'm going to deal with the problem then get back here as fast as I can.' He dropped a kiss onto her forehead. 'Maybe it's for the best, giving you some time to think.'

The message was clear. He wanted time to think. She had already reached a decision. 'I don't need time.'

'Take it, anyway. You might change your mind.'

She tried to inject a teasing note into her voice and suspected she failed. 'What's the saying? You never know your luck.'

His look hardened. She had surprised him and he looked as if he didn't relish having the tables turned. The thought gave her surprisingly little satisfaction. 'Are your ribs healed enough for you to travel?' she asked.

He nodded tautly. 'The doctor called in this morning and, provided I'm careful, I can resume normal activities. No cliff dives, though.'

She grimaced at the reminder of how much he had suffered because of her. No wonder he was anxious to put it all behind him. 'I've never thanked you properly for accepting the challenge,' she said diffidently.

He dismissed her attempt with a gesture. 'No need. I got you into it, after all.'

'It was hardly your fault.'

'You weren't photographed alone,' he reminded her. 'I think we were doing something like this.'

Without warning, he crushed her to him and found her lips with unerring precision, kissing her with a passion that left her reeling. But only for a moment. She recovered enough to answer his passion with her own, putting into the kiss everything he wouldn't allow her to say—that she was his, not only because of the challenge, but because she wanted to be. Here in his arms she was finally home.

When he released her his eyes were dark with questions. 'Talay, I—'

This time she was the one to silence him with a touch. 'You're right. We need time. Go and deal with your crisis. We'll talk when you return.'

She almost said *if* you return but stopped herself in time. He said he would come back and Jase was a man of his word. The assurance was all that kept her from breaking down as she watched him finish packing and prepare to leave.

He had ordered a car to take him to the airport so there was no chance of a private farewell. In any case, she had put into the kiss everything she wanted him to know. It would have to serve until he returned. She drew a regal air around her like a cloak, managing to say goodbye as if to any ordinary palace guest.

'I'll make this as fast as I can,' he vowed, for her ears alone. 'Take care of yourself.'

'You, too,' she murmured softly, and stepped back.

The driver closed the passenger door and the tinted windows screened Jase from her sight, which was just as well because she was afraid that her composure wouldn't last much longer.

'What do you mean, Jase has gone?' Allie asked when she arrived at the palace a short time later. Her visit was unexpected but Talay had never felt so pleased to see anyone.

'He was called away to a crisis at his resort in Thailand,' Talay explained.

The hollowness in her voice wasn't lost on her friend. 'You make it sound as if he isn't coming back.'

Something broke inside Talay. 'He says he will, but how can I be sure?'

'But the two of you...'

'There's no two of us,' Talay interrupted. 'Last night he asked me to forget he ever proposed marriage.' Among other things, she thought but didn't add. There was no need for her friend to know the full extent of her humiliation.

Allie's arms came around her. 'Princess or not, you need a hug right now. You may think you know Jase Clendon but this doesn't sound like the Jase my husband has talked about for all these years.'

Talay returned the hug briefly then freed herself and pulled the cloak of reserve around herself again. 'You wouldn't think so if you'd seen the alacrity with which he left this morning.'

'Michael's friend isn't a liar,' Allie pointed out. 'I'm sure there really is a crisis.'

So much was already spoiled that Talay didn't want to add the longstanding friendship between Allie's husband and Jase to the casualty list. 'You're probably right,' she agreed. 'In any event, I'm glad you're here.'

Allie smiled ruefully and held out a leather folder. 'You left these behind at the villa. I thought you might need them.'

Talay recognised them without interest. 'My latest jewellery designs.'

'I peeked. They're some of your best work.'

They were the designs Jase had admired when he still believed she was Allie Martine. Remembering the long, lonely evening when she had created them, Talay wondered how she could face a lifetime of such evenings, wondering where Jase was and with which woman. Maybe Jase knew what he was doing after all. In her heightened emotional state his departure was impossible to accept, but maybe with time...

'Let's put the television on. The King is speaking at noon,' Allie suggested.

Talay was chagrined at having forgotten her uncle's regular address to the nation. Normally she watched every broadcast. She followed Allie into the sitting room adjoining her office, and picked up the remote control to activate the set.

Afterwards she had little memory of anything the King had said. Her mind was too distracted by thoughts of Jase, winging his way first to the capital, Andaman, then on to Thailand. She tried to tell herself that Allie was right, that he would return when he had dealt with the crisis there. It was hard to make herself believe it.

As the last strains of the national anthem died away she reached for the control again but her hand froze as a news flash symbol appeared on the screen, followed by a grim-faced announcer. 'Just to hand is a report of a plane missing with twenty-four passengers on board. The plane was on a scheduled flight between Andaman and Thailand this morning.'

Talay's gasp was involuntary. 'Oh, please, no. The only flight between Sapphan and Thailand at this hour is Jase's.'

'Details are still sketchy but the pilot reported engine trouble when the flight was a few minutes away from the capital,' the announcer continued. 'Contact was lost before the extent of the problem could be established. A search is now under way for the missing plane.'

Allie's hand closed around Talay's. 'Maybe they're having radio trouble. They could be all right, just unable to contact anyone.'

Talay's eyes brimmed but she clung to control. Going

to pieces wouldn't help Jase now. He was still alive. Her mind refused to deal with any other possibility. 'There must be something we can do,' she whispered.

Allie nodded. 'At least we can establish if he was on the flight.'

Hope soared in Talay's breast. Why hadn't she thought of that? 'I'll have my secretary call,' she said, turning towards the intercom.

Allie shook her head. 'I'll do it. If you can be me, I can be your secretary.'

Before Talay could protest, Allie had obtained the number from the palace switchboard and was frowning into the phone as she waited to be connected. It took an age. The airline must be swamped with calls.

'I'm calling on behalf of Her Royal Highness, Princess Talay Rasada,' Allie said importantly, when she was finally put through. 'The palace needs to establish the whereabouts of a passenger on your missing airliner.'

Talay held her breath as Allie gave Jase's name. Allie waited, nodded a couple of times then seemed to go paler, before hanging up the phone. 'Normally they don't release the names of passengers over the phone,' she said, sounding shaken.

Tension made a tight fist around Talay's heart. 'But this time?'

'They didn't have his name listed. But they did have a listing for two people under Jase's company name.'

'Go on,' Talay urged impatiently. Whatever Allie was trying to tell her couldn't be any worse than the news she had just received.

'They said Jase was booked on the flight in the company of a young American woman,' Allie finished. 'Tal, there has to be some mistake.'

There was, but she had made it, Talay understood. No

wonder he hadn't wanted to discuss his proposal of marriage this morning. It was no consolation that he had warned Talay that he wanted the freedom to see other women as he chose. She hadn't wanted to believe him, she realised belatedly. She had managed to convince herself that, with her, he would be different. Now he was flying off with another woman, quite possibly on genuine business. But Talay knew only too well where close contact with Jase could lead, and the knowledge filled her with anguish.

None of it mattered now, she told herself, distress threatening to swamp her. Jase's plane was lost and he might never know she loved him.

The thought almost made her choke. She did love Jase but she had refused to face it until now, when it was probably too late. Losing someone she loved in a disaster involving a plane was her worst nightmare, the very reason she resisted giving her heart to anyone. Once in a lifetime was enough to have to endure. Now it was happening again. She felt herself sway.

Allie was beside her in an instant, her arm strong around her friend's shoulders. 'Would you like me to call a doctor? I don't want you going into shock.'

Talay shook her head. 'I'm all right, really. I have to be strong—for him.'

Allie watched her closely. 'You're in love with Jase, aren't you?'

How well Allie knew her. 'Yes,' she said. 'How did you...?'

'Your face when you heard his plane was missing. You looked as if your world had come to an end.'

'Maybe it has,' Talay whispered. 'He has no idea how I feel.' He had proposed marriage for mutual convenience. Her love would be the last thing he wanted.

'He must care for you, too,' Allie pointed out, 'otherwise he wouldn't have asked you to marry him.'

Talay laughed bitterly. 'He was already regretting it last night.'

'He actually said so?'

'In every way but in words. I'm certain it's why he left in such a hurry. Now I'll never know for sure.'

'There's no point thinking the worst until we have more to go on,' Allie stated emphatically.

She was right. The plane could have set down on one of Sapphan's offshore islands. There were hundreds, some with remote airstrips left over from the Second World War. If Jase's plane had landed on one of them it could be some time before contact was re-established.

Talay resisted the urge to pace. She loved Jase. She wanted him to live, even if it meant that Talay would never see him again. The thought was almost beyond bearing, but she would find a way to endure it if only he survived.

Allie knitted her fingers together. 'Should I call the airline again?'

It was all Talay could do not to jump into a car and drive downtown to camp at the airline's office, but she shook her head. 'Let them do their jobs. They'll call the palace as soon as they have any news.'

'You're so brave. If my Michael was on that plane...'

'You'd do the same thing,' Talay assured her. She wasn't brave at all, simply schooled by a lifetime of royal training not to go to pieces in a crisis, outwardly anyway. If she was falling apart inside it was for her alone to know. She touched Allie's arm. 'I'm glad you're with me.'

'It seems I'm fated not to meet Jase,' Allie observed,

then saw Talay's face. 'Oh, Tal, I didn't mean... He will be all right.'

'Was it like this for you when you met Michael?'

Allie nodded. 'I was practically love at first sight. If you remember, I was promised to a distant cousin but Michael challenged him for my hand.'

'You never did tell me what Michael had to do to win you,' Talay said.

Allie made a face. 'I hate talking about it. They had a car race and my cousin's car went into the sea. He was recovered safely but he couldn't accept losing and went overseas to work. I haven't seen him since.'

'What would you have done if Michael had lost the challenge?'

Allie shrugged. 'Probably run away with him to Australia. Love makes you do crazy things.'

For the first time in her life Talay understood what her friend meant. Love for Jase could make her forget country, duty, even morality. A tide of heat travelled up her neck into her face as she remembered the power of his kisses, which had swamped all rational thought. What she wouldn't give to be held in his arms again.

'I didn't want this,' she said in a voice barely above a whisper.

'Oh, Tal, love isn't something you want. It sweeps you away on a tide you can't possibly resist. Just because you're a princess doesn't mean you're immune to human frailties.'

'Funny, Jase said the same thing.'

Allie moved restively. 'He's right. I know you got hurt badly, emotionally as well as physically, when your parents were killed, but they wouldn't want you to live without love for the rest of your life.'

'I thought it would protect me from enduring such loss

again,' Talay explained. 'It hasn't worked too well, has it?'

'Nor should it,' Allie replied. 'You have so much to give. If Jase can't see it, he isn't the man my husband thinks he is.'

Allie's determined use of the present tense wasn't lost on Talay, and she appreciated it more than words could say. It wouldn't bring Jase back to her but she knew she could live with losing him, if only he was alive and well.

She was unprepared for the shrill summons of the telephone and jumped convulsively. With a wary glance at her, Allie picked it up. 'Princess Talay's office.'

Talay resisted the urge to snatch the receiver from her friend. Her hands shook so badly she would probably drop it, anyway. 'Do they have news?' she asked, when Allie had hung up.

Allie's eyes shone. 'It seems the plane made an emergency landing on an island airstrip north of Andaman. Their radio is out but a passenger was able to call on a mobile phone. They're all right. But it will be some time before a relief flight can get to them.'

Talay barely heard the rest of the message. Jase was all right. The words burned themselves into her mind until she wanted to scream with relief. She forced herself to ask, 'And Jase's...travelling companion?'

'They don't have details yet and the passenger's phone battery was running low. A few bumps and bruises from the forced landing but no serious injuries were reported. So I guess they're both all right.' She took a deep, steadying breath. 'All we can do now is wait until the relief plane reaches them.'

Talay felt light-headed suddenly and forced her head down between her knees, fighting the nausea which churned in her stomach. After a few minutes she was able to lift her head. 'We know everything we need to know,' she said shakily. 'The details hardly matter.'

'Don't you want to know why Jase was travelling with this woman?' Allie asked.

Talay was afraid she already knew the answer. 'It's supposed to be business. Whether it is or not doesn't change reality. Jase doesn't love me.'

Allie made a frustrated sound. 'You don't know for sure. You didn't admit your own feelings until you thought you'd lost him.'

'It doesn't mean Jase feels the same way. Face it, Allie, happy endings are for fairy tales, not real life.' She stood up.

'What are you going to do?'

'If I wait here any longer I'll go crazy. I'll have Sam drive me to Crystal Bay. I need to choose the pearls for my new jewellery collection.'

Allie looked shocked. 'Now? What if the airline calls with more news?'

'They can reach me on the car phone.' It wouldn't be good news for her, she was well aware. Now she knew that Jase was all right, she didn't need to hear any more. What she needed was the solace of activity. She sensed that Allie understood.

'I'll come with you,' her friend announced.

Talay's glance went to Allie's gently curving stomach. 'The bumpy road won't be good for the baby. I'll be fine.'

'If you're sure?'

'I'm sure. You've been wonderful, but Michael will never forgive me if I allow you to overtax yourself in your condition.'

Allie nodded and hugged Talay. 'Call me if you hear anything, or need anything.'

Promising she would, Talay saw her friend out, then summoned Sam to fetch the car. Her heart wasn't in choosing pearls for her jewellery collection but she had to do something active or she would go mad. No matter how much she yearned to, a princess didn't break down and cry her eyes out because she'd lost the only man she'd ever loved.

CHAPTER THIRTEEN

THE journey was more comfortable than Talay expected. Some of the fist-sized stones had already been cleared away and the road to Crystal Bay widened and graded, she noted distantly. Gritty dust still rained against the car windows, forcing her to keep them closed.

She welcomed the cocoon of the air-conditioned car, not wanting the tropical sounds and scents to remind her of her last journey along this road when she had come from Jase's bed. Now, as then, the hunger for him threatened to consume her, however much she tried to force it away. On a sudden impulse, she slid aside the partition between her and Sam. 'The turn-off to the new resort is up ahead. I'd like to call there on the way to the village.'

Sam was too well trained to ask for an explanation, which was just as well as Talay had none to offer. She only knew she needed to feel close to Jase, probably for the last time, and the only way open to her was to visit his creation.

She could hardly believe she had been so opposed to it at the start, she reflected as Sam parked the car under a full-sized tamarind tree which hadn't been there yesterday. It was one of a grove of mature trees which had been transplanted to the site.

Already the rainforest was reclaiming the area, she noticed. Lush stands of palms and flame trees hugged the buildings and a thick understorey of tropical plants softened and greened the surroundings. Jase was right.

Soon you would need to be practically on top of the resort to see it.

Recognising her car, the foreman hurried towards her and brought his palms together at chest height. 'Your Highness, welcome. I was not forewarned of your visit or I would have prepared for your arrival.'

She waved away his apology. 'I'm on my way to the pearl trading post and decided to check on progress here first.'

The man's expression lightened. 'Then you have saved yourself a wasted journey. The trading post was moved to this site this morning. See for yourself.'

Jase didn't waste time, Talay reflected as she allowed herself to be shown to the new centre. Although the traditional thatched-roofed style had been retained, the new building looked far more comfortable and inviting than its predecessor, both for the family who ran the business and for their customers. When Talay entered, the owner smiled broadly. 'Welcome, Your Highness. Forgive the chaos but we are not yet open, although I will help you in any way I can, of course.'

'I had hoped to choose the pearls for my latest jewellery collection,' Talay informed her, 'but there's no hurry. A few days from now will do very well, when you've had time to organise your wares.'

The woman looked anxious. 'If Your Highness is sure it's no inconvenience?'

Talay softened the moment with a smile. 'Quite sure. You must be looking forward to the resort opening.'

'The whole village plans to celebrate,' the woman agreed, beaming.

Another victory for Jase, Talay thought soberly. A twinge of the old opposition surfaced. 'Won't you miss your peaceful village life?'

The woman's shoulders lifted expressively. 'The grave offers peace enough for anyone. Until then, it is better to live a little.'

'Have you no objections to the changes?'

'Perhaps one. I am told Mr Clendon plans to go away after the resort opens. He is a good man. I wish he would stay here.'

Talay took a step backwards in shock. It was so exactly what she wished herself that she wondered if the woman had somehow read her mind. 'It's impossible. Mr Clendon has many...other interests,' she said in a choked voice.

Luckily the woman didn't press her for details, plainly imagining she referred to Jase's chain of international resorts. Loyal to her husband of several decades, the woman would find it hard to understand Jase's desire for freedom. Well, that made two of them.

The centre felt suffocating suddenly, although the whisper of air conditioning should have made it impossible, and Talay was glad to escape to the open air. 'I want to consult some papers in Mr Clendon's temporary office. Wait for me while I take care of it,' she told Sam.

The bodyguard offered to retrieve the papers for her but accepted her assurance that she could locate them more easily. 'I may be a while so why don't you drive to the village and visit your sister?' Talay said, on a sudden afterthought.

Sam's grin broadened. 'She'd be happy to see me. She's expecting her third child,' he explained. 'However, if you should need me, Your Highness...'

She waved him away with a gesture. 'Take as long as you like. I'll be perfectly all right here. The resort staff will look after my needs.'

Sam inclined his head respectfully. 'As you wish.'

As he drove away she sighed with relief. She wasn't sure why she wanted to spend some time alone at Jase's bure but it was suddenly imperative that she did. The foreman showed her the way and unlocked the building for her, then acceded to her request to be left by herself, assuring her he could be reached by the house phone if she needed anything. For now, the only outside line was at the site office, he informed her regretfully.

'All I require is here for the moment,' she told him as he bowed his way out. The one thing the place lacked, no one but Jase could supply.

When she was finally, blessedly, alone she rested her back against the closed door of the bure, letting her breath escape in a ragged sigh. She was alone so seldom that she treasured the moments. But there was more going on here. In the bure she felt closer to Jase, and knew it was the reason why she had felt compelled to come.

Jase had said he intended to use the bure as his headquarters, and already she imagined that the indelible mark of his personality was stamped in the design of the building. She felt an almost overwhelming urge to snatch up a sweater she remembered him leaving here on their last brief visit and bury her face in its folds. Her fingers trembled as she reached for it but she pulled them back with a determined effort.

She had told Sam she wanted to consult some papers but it had only been an excuse to visit this room where she had first felt close to Jase. Instead, she felt like an intruder. Since Sam wouldn't be back to collect her for some time, there was nothing she could do except stay. Another horrible possibility occurred to her. What would she do if Jase turned up with another woman and found her here?

Only the thought that she would be gone before he

could possibly return gave her the strength to remain in the bure. It was bad enough imagining him in someone else's arms, without being faced with irrefutable evidence.

This thought prompted another. She picked up the house phone and was rewarded by the foreman's voice. At her request he agreed to telephone the airline for a progress report on the downed plane and its passengers and let her know the answer immediately. She thanked him and hung up. In her haste to be alone she had forgotten that Sam would take the car phone with him, leaving the airline with no way to apprise her of developments.

Moments later the foreman called back, assuring her that the plane had been reached and all passengers were safely accounted for with only minor injuries and, not surprisingly, a few cases of shock. The passengers were returning to Andaman aboard the relief aircraft.

'Did they mention Mr Clendon and his companion?' she asked, proud of barely hesitating over the question.

'They simply said all passengers were accounted for, Your Highness,' the foreman repeated. 'If you wish, I'll make further enquiries.'

'There's no need,' she hastened to assure him. 'As long as everyone is safe, there's no point hampering the rescue operation with needless calls.'

'As Your Highness wishes.'

She thanked him and hung up the phone, feeling spent. Her prayers had been answered. Jase was safe and on his way back to Andaman where he would, no doubt, resume his interrupted journey. Hearing that he was travelling with a woman shouldn't have come as a surprise. Even before he had withdrawn his proposal of marriage his terms had been clear. Talay had agreed to them with her eyes open.

It was only knowing how much she loved him that had changed everything.

Now she could no more live with the thought of him seeing other women than she could fly. Which left only one solution. If she still wanted it, Jase's deal permitted her to be released from the obligations of the marriage challenge after he was cleared as fit by the royal physician. This morning the doctor had cleared him to travel so the decision was hers.

If she convinced Jase she didn't want to marry him, he would have to release her as promised. They would both be free, although she hated to think how long it would be before she was truly free of his devastating impact on her.

All that mattered was his safety, she told herself as a great wave of weariness washed over her. The strain of the day was fast catching up on her. She fought the tendency of her knees to buckle, and finally gave in. The platform bed beckoned.

She went to it and stretched out, trying not to remember that Jase intended to sleep here once he established the bure as his headquarters, although she persisted in picturing his head touching the same pillow. It was a losing battle and her last thoughts, before she slid down into darkness, were of sharing this bed with him.

In her dreams, his lips found hers in a gossamer kiss that triggered flutters of desire deep inside her. She resisted them, knowing even in sleep that he was not for her. But the pressure of his mouth increased and she forced her eyes open, belatedly noticing that this was an uncommonly vivid dream.

Shock coiled through her as her eyes met Jase's. 'What are you doing here?'

'Awakening Sleeping Beauty in the customary fashion,' he replied, straightening.

Her sleep-drugged thoughts whirled. 'But how...? When...?'

He eased himself onto the side of the bed and took her hands, pressing his lips to each palm in turn. His touch turned her insides to fire but she pulled free and hugged her arms around her trembling body. 'I must be dreaming.'

He shook his head. 'If this was a dream, I'm sure you'd do a better job of conjuring up a prince.'

He was real. She sat up. 'How did you get here so fast? The plane...'

'I wasn't on it. When I heard what had happened to it I knew you'd be concerned so I took the first domestic flight back to Alohan.'

Dizziness threatened to claim her until she forced it away. 'Alohan? But I checked with the airline. They said you and a woman from your company were together on the plane that made the emergency landing.'

The emotion charging her voice as she mentioned another woman made his eyebrows rise. 'We were supposed to travel together, but not in the sense you evidently think. She's my CEO in Thailand and was in the United States when the industrial crisis blew up. When she heard I'd been called in she stopped off in Andaman on her way back to Bangkok to see if I wanted her resignation. To prove she still has my complete confidence, I decided to let her handle the crisis and consult with me by phone if she feels the need.' He gave an unsteady grin. 'It wasn't a hard decision to make. The further away from here I got, the more I wanted to turn around and come back.'

Talay's brow furrowed. 'But the airline said you were on the plane as well.'

He nodded. 'The flight was full so I gave my ticket to a young Thai man who was desperate to get home to his wife who'd just given birth.'

Talay was sure she looked as bemused as she felt. 'You could have telephoned to let me know you were all right.'

'I tried but you'd left the palace so I tried your car phone. Sam told me where to find you.'

'Good old Sam,' she said shakily. She would need to discuss with him later his idea of protecting her. Leading Jase to her didn't qualify.

He seemed to think it did. 'Why did you come here?'

To be closer to him, she thought, but couldn't bring herself to admit it. 'I had some loose ends to tie up,' she substituted.

His look warmed her. 'Me, too.'

Suddenly she saw herself through his eyes—stretched out like a cat on his bed, her eyes sleep-filled and her hair tousled from the pillow. She began to rise but he gestured her back. 'Stay there. You look beautiful in my bed.'

'Which is precisely why I shouldn't stay in it,' she struggled to say. She didn't belong here on any permanent basis.

His eyes argued with her. 'It's the perfect place to begin a marriage, Princess.'

Which was precisely why she should leave as quickly as possible. She glanced away. 'Last night we agreed your marriage proposal never happened.'

'Doesn't mean I can't say it again. Talay Rasada, will you—'

'Don't please.' Her cry silenced him as intended. Ever since she had believed him lost in the downed plane, she had struggled to come to terms with her own needs. Yes, she wanted him. Yes, she...loved him. There was no denying it. But they were also the reasons she couldn't endure the kind of marriage he wanted.

Jase swallowed. 'Talay, what is it? I could have sworn—'

'I've had time to think,' she interrupted swiftly. 'It won't work.'

His whole body seemed to sag and bitterness etched the dark eyes. 'Is it because you're a princess and I'm a commoner from Australia?'

His pain vibrated through her body as if it were her own. 'Jase, my position has nothing to do with this. Under our law, no royal can marry another member of the royal family so there's ample precedent for us.'

'Then it can only be because you don't love me.'

This was going to be much harder than she'd anticipated, but he deserved nothing less than the truth. 'It's precisely because I do love you that I can't marry you.'

He moved closer. 'What did you say?'

She lifted her head, meeting his eyes with all the regal defiance she could muster. It very nearly wasn't enough. What she saw in his eyes almost melted the last of her resistance. 'I said I love you,' she affirmed huskily. 'I realised it when the plane went missing and I thought you were...' She couldn't go on. Waiting for news of Jase's flight had brought back the horror of losing her parents and almost being killed herself by a terrorist bomb as they had boarded a plane. It was an effort not to touch Jase to reassure herself that he was really all right.

'Then, in the name of all that's holy, why won't you marry me?'

'Because of the kind of marriage you want,' she managed to say, far more matter-of-factly than she felt. 'When I thought you had gone away with another woman I knew I couldn't tolerate the thought of you with anyone else. When you told me it was what you wanted I thought I could handle it. Today I found out I can't.'

There, he knew it all now. She waited for his confirmation that it was the right decision, the only possible one to make under the circumstances. Instead, he stood upfrom the bed. Amazingly he was smiling.

'So that's the only reason you think our marriage won't work? What if I tell you I reserved the right to see other women purely as a defence mechanism?'

'Against what?' she asked in a voice of bemusement.

'Against loving you so much it scares the hell out of me,' he grated out. 'Waiting for my flight at Andaman Airport, it occurred to me that this was the sort of freedom I'd told you I wanted—except I no longer wanted it. I only wanted you, as I've done since I first set eyes on you.'

Her mind flashed back to the hospital in Australia and to a scarred sixteen-year-old. 'You couldn't have wanted me then,' she whispered. 'You found my injuries so repulsive you couldn't get away fast enough.'

He shook his head. 'You couldn't be more wrong. I was so smitten by you I could hardly speak.' He took her shoulders, turning her gently to meet his eyes which burned with the power of his memories.

'I first saw you reading to a tiny little kid who sat on

your lap, staring at you as if you were Christmas. Unlike the other patients, you took no notice of all the hullabaloo of athletes and photography. All your attention was on pleasing that child and making her forget her pain for a few precious minutes. I've never been so moved by anything in my life.'

Her eyes widened. 'I didn't know you were watching me.'

He nodded. 'Precisely my point. You weren't doing it for show but because you wanted to help. Considering the extent of your own injuries, I was staggered that you could put someone else's pain before your own. No amount of outer damage could dim the inner beauty I saw radiating from you at that moment.'

The humiliation of his perceived rejection warred with what she was hearing. 'Yet you wouldn't be photographed with me because I looked so dreadful.'

He looked appalled. 'I refused for your sake, not mine. I could only think how I'd feel if the roles were reversed. The last thing I'd want would be to have my suffering captured on film for all the world to see.'

It was an explanation she had never considered. 'You did it to protect me?' she said in a voice barely above a whisper. 'All these years, I believed...'

'What, Princess?'

He gave her title the sound of a caress, enabling her to go on although she could barely make herself meet his eyes. 'I thought my scars were so horrible you couldn't bear to look at me.'

He stiffened. 'You couldn't help what happened to you, any more than I could help my start in life.' He massaged his chin ruefully. 'After I left the hospital I kept thinking about you and wondering how you were getting on. When I found out you lived near Crystal Bay

I hoped we'd meet again. It never occurred to me that we'd find ourselves on opposite sides over the resort.'

'Not any more,' she assured him, feeling herself colour. 'I can see the good you're doing there. You've removed any doubts I had.'

He tilted her chin with one hand. 'Except one. About me.'

Fire radiated from his hand along her neck and shoulders. 'I never doubted you, Jase. Myself, perhaps, but never you.'

'Then end the doubts now. Say you'll marry me.'

It took every ounce of resolve she possessed to shake her head. 'I can't.'

His hand dropped away. 'Why not? What possible obstacles remain?'

'Only one. You know I love you. It won't work if it's one-sided.'

He drew a breath that sounded strangled. 'You don't know, do you? I started falling in love with you when I saw you in the hospital, but never dreamed we could have a future together until now. I love you and I will love you till the sun turns to ice and there are no stars left in the sky. As long as I live there will never be another woman for me but you.'

Her spirits soared. What woman had ever received such a declaration of love? Feeling suddenly shy, she lowered her eyes, only to raise them again to meet his, drawing strength from the adoration she saw there. 'Then my answer is, yes, I will marry you, Jase.'

He took her hands and drew her off the bed to him, holding her a heartbeat away. 'You realise what this means?'

She nodded, her heart starting to race. 'It means, under Sapphan law, marriage has been proposed and lawfully accepted. From this moment we are legally husband and wife.'

His gaze turned her blood to fire. 'The law applies to foreigners, too. I checked.'

That he had thought to do so brought a lump to her throat. Raising herself on tiptoe, she kissed him. 'As of this moment you are no longer a foreigner, my husband.'

He fumbled in his pocket and withdrew a black velvet box which he held in front of her. 'I don't know if this is your tradition, but it's definitely mine.'

He opened the box to reveal the heart-shaped baroque pearl he had purchased in the village. Now it glowed against the white-gold band of a ring. He lifted it out, letting the box drop as he slid the ring onto the third finger of her left hand. 'With this ring, I thee wed,' he vowed.

His hand closed around hers and, with a growl, he pulled her hard against him, crushing her mouth with a kiss that set the blood soaring in her ears. She had barely absorbed the impact when he scooped her up and set her back upon the bed, his lips never leaving hers for an instant.

Her arms linked around his neck and she returned the kiss with a passion she hadn't known herself capable of experiencing. Jase was her husband. They were legally bound to each other for as long as they lived. It would be barely long enough to contain the joy she sensed they would share.

She knew an instant of panic as he undressed her. 'Jase, I still don't...'

His finger slid across her lips. 'Shh, my darling. It will be all right, I promise.'

She felt her eyes start to brim. 'I disappointed you before. I couldn't bear it if it happened again.'

He frowned. 'I didn't stop because I was disappointed but because I realised I couldn't take what rightfully belonged to your husband.'

A thrill coursed through her. He had understood. 'And now?'

'Now we have gifts to give each other.' Hers was the precious gift of her innocence. His was the skill to ensure she remembered this moment with joy for an eternity.

He knew a moment of uncertainty. She deserved the sun, the moon and the stars. Was he the man to give them to her? She had made her choice freely, and she had chosen him. Her faith in him sent the blood roaring through his veins, banishing all doubts. He reached for her.

Her breathing was fast and shallow, but no more than his as he covered her body with his own. How slight and delicate she was, like a porcelain figurine, although the warmth radiating from her denied the comparison. He slid a hand lightly down the length of her, delighting in the way she trembled.

She moaned softly. 'Jase, please.'

He hovered over her mouth. 'Please what?'

Her glorious eyes went round as saucers. 'You'd have me beg?'

'I'd have you any way that pleases you. Like this...and this...'

She arched her back with delight as he found and pleasured her most secret places. He took a long, long time, making sure she was not only ready but dizzy with wanting him. By that time he felt close to exploding himself.

This time there was no going back. When he came to her she surrendered to him utterly, trusting him to take her beyond the first hurting to a realm of wonder she had never dreamed existed. As the waves crashed over her she knew she had never felt so utterly fulfilled.

Afterwards, she lay dreamily in the crook of his arm, delighting in the warm hardness of his body against her softness. 'Is it always like that?' she asked in wonder.

He chuckled. 'It may even get better.'

It seemed impossible, but she was willing to trust his greater wisdom in such matters. Then another thought clouded her mind and she lifted herself onto an elbow beside him. 'Jase, we didn't take any precautions.'

His eyebrows lifted. 'So?'

She might be new to love-making but biological theory wasn't beyond her experience. 'I thought you didn't want children.'

He grazed a finger along her cheek. 'I don't want them being born into a loveless environment like the one I endured. I don't think there's any risk with us, do you?'

Loveless was hardly the word to describe what they shared. 'Not a chance,' she agreed.

'Then there's no problem, is there, Princess?'

The thought of bearing his child...his children...was exhilarating beyond words. 'I suppose not.'

He rolled towards her, taking her into his arms. 'You suppose? If there's the slightest doubt I'll have to try harder to convince you.'

And he did. She really had no doubts left but she wasn't about to stop him now, just when she was getting the hang of this. Love-making was obviously a skill that improved with practice, and Jase clearly meant them to have lots of practice. A lifetime of it, in fact. It seemed barely enough.

EPILOGUE

'HE'S perfect,' Talay breathed, counting fingers and toes. 'Imagine, he's only two hours old.'

Jase leaned closer to the glass wall that partitioned the nursery from the hospital corridor. 'Hello, Jason Michael Martine. Welcome to the world.'

'I'm Talay and this is Jase,' Talay told the baby. 'We're going to be your godparents.'

Smiling indulgently, as the baby was fast asleep, Jase took her arm. 'Michael and Allie have had enough time alone now. We should offer our congratulations.'

With a lingering look at the baby, she linked her arm in her husband's. 'Just think, in four more months it will be our child in the crib.'

He patted her gently rounded stomach approvingly. 'Philippe seems like a good name for our first son.'

She laughed. 'Or Philippa. We could have a daughter.' She had refused the doctors' offer to tell her, enjoying the anticipation and wanting only for their child to be healthy.

'Either way, the King will be delighted,' Jase agreed, remembering the monarch's joy when they'd broken the news of Talay's pregnancy to him.

Talay frowned. 'I'm still not sure I should be speaking to my uncle after what he did.'

Jase grinned and his arm tightened around her waist. 'He did bring us together.'

'He manipulated me. He wanted you for my husband

all along but he knew perfectly well that if he'd said so I'd have run a mile in the opposite direction.'

'Instead, he forbade you to have anything to do with me, and look what happened. Wise man.'

She nodded. 'That's why we let the family elders choose our marriage partners in Sapphan.' She snuggled close to Jase. 'I'm only thankful Uncle Philippe chose you. But he might have let me know. I almost let you slip through my fingers.'

'He didn't share the secret with me, either,' Jase pointed out with a laugh. 'But it didn't matter. I wasn't about to lose the best thing that ever happened to me.'

Remembering how close they had come, she shivered. At the same moment she felt a movement deep inside her, so slight it almost escaped her. Then she laughed and pressed Jase's hand to her womb. 'I just felt our baby kick.'

'That settles it. It's a boy, making his presence felt.'

She sighed happily. 'Just like his father.' As they continued down the corridor she knew she wouldn't mind in the least.

HARLEQUIN  PRESENTS®

HARLEQUIN PRESENTS
men you won't be able to resist
falling in love with...

HARLEQUIN PRESENTS
women who have feelings
just like your own...

HARLEQUIN PRESENTS
powerful passion in
exotic international settings...

HARLEQUIN PRESENTS
intense, dramatic stories that will keep you
turning to the very last page...

HARLEQUIN PRESENTS
The world's bestselling romance series!

From rugged lawmen and valiant knights to defiant heiresses and spirited frontierswomen, Harlequin Historicals will capture your imagination with their dramatic scope, passion and adventure.

Harlequin Historicals...
they're too good to miss!

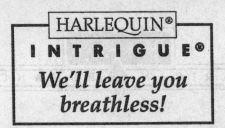

If you've been looking for thrilling tales of
contemporary passion and sensuous love stories
with taut, edge-of-the-seat suspense—
then you'll *love* **Harlequin Intrigue!**

Every month, you'll meet four new heroes
who are guaranteed to make your spine tingle
and your pulse pound. With them you'll enter
into the exciting world of Harlequin Intrigue—
where your life is on the line
and so is your heart!

THAT'S INTRIGUE—DYNAMIC ROMANCE AT ITS BEST!

HARLEQUIN®

I N T R I G U E®

LOOK FOR OUR FOUR FABULOUS MEN!

Each month some of today's bestselling authors bring four new fabulous men to Harlequin American Romance. Whether they're rebel ranchers, millionaire power brokers or sexy single dads, they're all gallant princes—and they're all ready to sweep you into lighthearted fantasies and contemporary fairy tales where anything is possible and where all your dreams come true!

You don't even have to make a wish...
Harlequin American Romance will grant your every desire!

Look for Harlequin American Romance
wherever Harlequin books are sold!

HAR-GEN

HARLEQUIN SUPERROMANCE®

...there's more to the story!

Superromance. A *big* satisfying read about unforgettable characters. Each month we offer *four* very different stories that range from family drama to adventure and mystery, from highly emotional stories to romantic comedies—and much more! Stories about people you'll believe in and care about. Stories too compelling to put down....

Our authors are among today's *best* romance writers. You'll find familiar names and talented newcomers. Many of them are award winners—and you'll see why!

If you want the biggest and best in romance fiction, you'll get it from Superromance!

Available wherever Harlequin books are sold.

Look us up on-line at: http://www.romance.net

HS-GEN

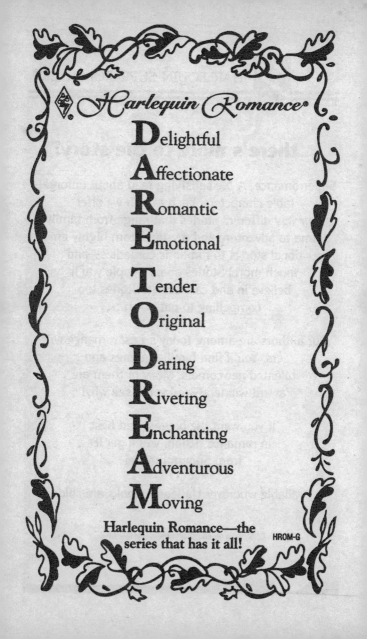

Humorous love stories by favorite authors and brand-new stars!

These books are delightful romantic comedies featuring the lighter side of love. Whenever you want to escape and enjoy a few chuckles and a wonderful love story, choose Love & Laughter.

Love & Laughter—always romantic...always entertaining.

Available at your favorite retail outlet.

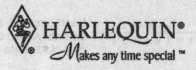

Look us up on-line at: http://www.romance.net

HLLGEN

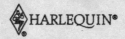

Not The Same Old Story!

Exciting, glamorous romance stories that take readers around the world.

Sparkling, fresh and tender love stories that bring you pure romance.

Bold and adventurous—Temptation is strong women, bad boys, great sex!

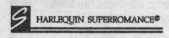

Provocative and realistic stories that celebrate life and love.

Contemporary fairy tales—where anything is possible and where dreams come true.

Heart-stopping, suspenseful adventures that combine the best of romance and mystery.

Humorous and romantic stories that capture the lighter side of love.

Look us up on-line at: http://www.romance.net   HGENERIC